Algebra

By Christopher Sexton

ISBN-13: 978-1516981069
ISBN-10: 1516981065

Table of Contents

Extras page 1- 52; 377- 464

We all have an innate curiosity about how the world works. An interesting science experiment begins with a question and gives you the tools to explore that question. This book is designed for you to experiment with numbers and algebra in a shared journey of discovery. These extras will help you along the way.

Number Sense page 53-124

Numbers give you the power to understand the power of earthquakes, the speed of catamarans and the endlessness of infinity. And if you can do arithmetic with numbers, you can do algebra.

Expressions and Equations page 125-170

Once we start using variables as symbols for numbers we can represent genies, road trips and skateboarding dogs with expressions. Then we can write an equation to answer questions about them.

Linear Functions page 171-280

The world needs change to function. A car's distance increases as a time passes, a glider drops as it flies towards the horizon and a gym membership costs more over time. Functions are equations which help us visualize these changes.

Data and Statistics page 281-330

There are a billion jelly beans in the world, but you don't need to look at them all to know that 11% are blue. The practice of statistics is using ample samples and yummy summaries to describe and predict the world.

Exponentials and Quadratics page 331-376

Exponential growth can help your bank account and your science experiment blow up. Quadratic functions are another important non-linear function. Exponential and quadratic functions form the basis for arithmetic and geometric sequences.

Guidelines

Courage To Core is designed for high school and middle school students to work together in groups. It can be used in pairs or with the whole class working in tandem, guided by a teacher. Students can be encouraged to discuss and come to consensus before seeking help from a teacher. A short opening lecture at the beginning can review important skills needed for the mission and another at the end can help cement the experience. Mini lectures in the middle of class to small groups or the whole class can address questions and curiosities. As the mission is designed to be group paced, work may be completed over several days, or students can be asked to complete parts of the mission at home.

Additional Student Resources

Selected mission answers and additional Q and A help for students at:

www.couragetocore.com

(Answers are password protected with password "CTC")

The Philosophy of Courage To Core

We all have an innate curiosity about how the world works. We are wired to experiment at the edges of our knowledge, to look for patterns and to draw conclusions. Mistakes are the welcome surprises which help us refine the experiment. In a student-centered classroom, students collaborate to ask questions, gather data, interpret results and articulate understanding. Success at the edge of knowledge demands persistence and creativity. Courage To Core provides a context for students to work together to become the agents of their success and the owners of their cognition.

About Courage To Core

Courage To Core was designed and refined over 15 years of teaching at domestic and international high schools. If you like Courage To Core, please share your feedback and spread the word! If you've got an idea for improvement let me know as missions are continually adapted to meet the evolving needs of students and teachers. Contact me at www.couragetocore.com. Thanks again!

Take-Home Messages for Missions[1]

NS1: Adding Fractions — 7.NS.1

Comparing and adding or subtracting fractions is made easier by expanding both fractions so there is a common denominator. Reducing fractions requires identifying and canceling a common factor. Terminating or repeating decimals can be converted to fractions and vice versa.

NS2: Multiplying Fractions — 7.NS.2

Multiplying fractions requires multiplying numerators and denominators. Cross-canceling is a useful simplification tactic.

NS3: Exploring Fractions Through Probability — 7.SP.5; 7.SP.6; 7.SP.7

The probability of an event can be expressed as a fraction. The probability of two independent events both occurring is the product of the probability of each one. The probability of either of two mutually exclusive events is the sum of the probability of each.

NS4: Rational Numbers — 8.NS.1

Rational numbers are ratios of two integers, and as decimals they can be terminating or repeating. Irrational numbers like $\sqrt{2}$ and π have infinite non-repeating decimals.

NS5: Dividing Fractions — 7.NS.2

Dividing fractions is really just a fraction of fractions, expanded by the reciprocal of the denominator to convert the denominator to 1.

NS6: Exponential and Scientific Notation — 8.EE.1; 8.EE.3; 8.EE.4; N.Q.3

Expressing relative differences can be accomplished with exponential notation, or specifically with scientific notation, which is particularly helpful in unit conversion. Significant figures express the level of precision of a measurement.

NS7: Roots and Rationalizing Denominators — N.RN.2

Square rooting numbers has different impacts on numbers depending on if they are bigger or less than 1. Rooting of numbers that aren't perfect squares produces irrational numbers. Rationalizing denominators is accomplished by expanding the fraction.

NS8: Order of Operations — 6.EE.2

Simplifying numerical expressions using the order of operations correctly requires a correct interpretation of the expression and careful application of all the arithmetic procedures you've learned so far.

---

[1] Common Core standards at right. Although Courage To Core is designed for high school students, missions sometimes focus on material from prior grades in order to reinforce understanding and prepare students for future missions.

NS9: Distribution and Factoring 7.EE.1

The fact that multiplication is distributive over addition makes simplifications a lot easier. Factoring is distribution in reverse, and is as useful as distribution.

NS10: Rational Exponents N.RN.1; N.RN.2

Decimal or fractional exponents can be approximated by comparison with nearby integer exponents. Fractional exponents mean a root and integer exponent combined. The rules for fractional exponents are like those for integer exponents.

EE1: Evaluate, Simplify Polynomials and Exponents 6.EE.1; 6.EE.2; 7.EE.1; 8.EE.1

Evaluating a variable expression means substituting a number for a variable then completing the arithmetic prescribed by the expression. Simplifying variable expressions means using same arithmetic you use with numerical expressions.

EE2: Equations 6.EE.5; 6.EE.6; A.REI.1

Algebra is built around expressions that take values as input and return values as output. Given a desired output, we can reverse the operations of the expression in order to find the input value.

EE3: Polynomial Expressions A.APR.1

Polynomials can be identified and classified by performing simplifications in accordance with the properties of arithmetic and by understanding a few essential terms. Once polynomials expressions and equations are correctly identified and classified, algorithms can be applied to make our work with them easier.

EE4: Linear Equations 8.EE.7; A.REI.1

Simplifying expressions is important when solving equations, and performing arithmetic operations on both sides of an equation produces an equivalent equation. Linear equations can have no solution, one solution or infinite solutions.

EE5: Eliminating Fractions, Eliminating Negatives 8.EE.7; A.REI.1

Eliminating fractions or negatives is based on the idea that arithmetic operations applied to both sides of an equation commonly result in equivalent equations.

EE6: Equations with Multiple Variables A.CED.4; N.Q.1

Real world formulas are equations for which we can substitute values to find quantities of interest. We can solve these equations in advance for quantities of interest so that the formulas are easier to use.

LF1: Linear Functions A.CED.1; A.CED.2; F.BF.1a

Functions are built from expressions which take values as input and return values as output. They can be represented using equations, tables and graphs. We can write and solve special equations to answer essential questions about functions.

LF2: Rate of Change 8,F.2; 8.F.4; A.CED.2; F.BF.1a

The rate of change of a linear function is its slope. it is the ratio of $\frac{\Delta y}{\Delta x}$ for two points on the line, or the vertical change over the horizontal change in the graph. It shows up as the coefficient of the linear term in the equation, while the y-intercept shows up as the constant term.

LF3: Finding Functions 8.F.3; 8.F.4; A.CED.2; F.BF.1a

The equation for a line includes a slope (m) and a y-intercept (b). Slope-intercept form for this equation is $y = mx + b$. The equation can be found from any two points on a line. These points can be given in a table or determined from a graph.

LF4: Linear and Non-Linear Models 8.F.5; N.Q.2; A.CED.1

Functions are models of quantities like density and force which exist in the real world. Sometimes these functions are linear and sometimes not. Recognizing linear and other types of functions makes it easier to work with them.

LF5: Linear Models and Problem Solving 8.F.4; A.REI.6

Functions can be used to model relationships in the real world. Often, we create functions with one-way causality in mind, expressing this idea by saying that one variable (y) is a function of another variable (x).

LF6: Horizontal and Vertical Lines 8.F.4

The graph of a constant function with a slope of 0 is a horizontal line. The graph of a line with no slope is a vertical line.

LF7: Domain and Range 8.F.1; F.IF.1; F.IF.2

The domain of a function is the set of all x values that the function accepts as input. The range is the set of all y values that the function returns as output. You can determine the domain and range by substituting values for x, by noting the function type, and by examining the graph.

LF8: Linear Models and Systems A.REI.6; A.REI.11

A system of linear equations is composed of two equations in two variables. To solve the system, we solve for an x-value that returns the same y-value for both functions, then find the corresponding y-value. The solution is an ordered pair which makes both equations true. It is the point of intersection of the two lines.

LF9: Forms for Lines and Methods for Systems 8.EE.8; A.REI.6

Any linear equation can be manipulated algebraically to create an equivalent equation. At the point of intersection of two lines, the x and y-values are the same, which allows us to use the substitution method to solve the system. Furthermore, any two linear equations are true at their point of intersection, which means we can add them to create another true equation. This gives us the elimination method for solving systems.

LF10: Parallel and Perpendicular A.CED2; F.BF.1a

Parallel lines have the same slope while intersecting lines have different slopes. By rotating a given slope triangle 90° we can see that perpendicular lines have slopes that are negative reciprocals of each other. Lines which are coincident have the same slope because they are the same line.

LF11: Graphing Linear Inequalities A.REI.12

The graph of a linear inequality is a shaded region representing all the points which make an inequality true. We graph the linear equation first then shade above or below using methods appropriate to slope-intercept or standard form.

DS1: Introduction to Statistics S.IC.1; S.ID.1

The practice of statistics is the process of collecting and processing data, and using that data to make educated guesses about how the world is working and how it might work in the future. We take a sample from a population and then produce numbers, also called statistics, like the mean, median, mode and range from the sample.

DS2: Shape, Center and Spread S.ID.1; S.ID.2; S.ID.3

Numerical statistics standing alone or represented graphically tell us about the shape, center and spread of the data. The mean, median, mode, Q1, Q3, IQR and outliers as pictured in box and dot plots help us understand these.

DS3: The Normal Distribution S.ID.3

Frequency plots and probability distributions show us how data is distributed and the likelihood of different outcomes based on the given distribution. The shape, center and spread of the data can be described by referring to the normal distribution, the mean and the standard deviation. Distributions can approach normal as the number of experiments increases.

DS4: Sampling S.ID.2

A greater sample size produces a tighter distribution with a smaller standard deviation. This allows us to estimate population parameters with greater confidence. We are always about 68% confident that the actual parameter is within ± 1 SD of the mean.

DS5: Linear Regression S.ID.6; S.ID.7; S.ID.8; S.ID.9

A correlation between two variables mean that changes in one correspond to changes in the other. This can be seen in the graph. Causality means that the change in one causes the change in the other—this can't be known just from the graph. We can use linear regression to estimate the line of best fit and find the correlation coefficient, thus summarizing the relationship between the two variables.

EQ1: Exponential Functions HSF.LE.A.1

Linear functions have the form $y = mx + b$ while exponential functions have the form $y = a(b)^x$. For linear functions, the input is multiplied by a number (the slope), and the y-intercept b is the initial value returned for $x = 0$. For exponential functions, the input counts the number of times an initial value a is multiplied by a constant b.

EQ2: Comparing Linear, Quadratic, Exponential HSF.LE.A.1, HSF.LE.A.2

Linear functions have graphs that are straight, and increase at a constant rate called the slope from an initial value called the y-intercept. Increasing exponential functions have graphs which curve steeper as x increases, because x counts the number of times an initial value is multiplied by a constant. Quadratic functions with a positive leading coefficient have a required quadratic term and increase symmetrically to the left and right of a lowest value.

EQ3: Vertical and Horizontal Shifts HSF.BF.B.3

Vertical and horizontal shifts can be performed on any function given its equation, table or graph. Vertical shifts are applied to the y-values while horizontal shifts are applied to x-values.

EQ4: Arithmetic and Geometric Sequences HSF.LE.A.2

Arithmetic sequences are like linear functions, and geometric sequences are like exponential functions. Arithmetic sequences feature a common difference (slope) while geometric sequences feature a common ratio (multiplier). Sequences have an initial value at $n = 1$ instead of $x = 0$, which means the formulas show a horizontal shift of 1. The domain is restricted to the natural numbers in both, and the formulas for each are written using subscripts to identify terms. There are additional formulas to find the sum of the first n terms.

Step-By-Step Mission Guides[2]

NS1: Adding Fractions

Students will explore the application of fractions in real world contexts. They will learn to expand or reduce fractions, how to compare them, and how to add or subtract them, through fraction bars. Students will also convert between decimals and fractions.

Lecture	Some review of any of the following: fraction as part of a whole, fraction bars, lowest common denominators, converting between improper fractions and mixed numbers, converting terminating decimals to fractions or converting percentages to fractions
1	Facilitate a smooth process of collecting the class birthdays, either on the board, each person verbally sharing, or a clipboard passed around while you complete the opening lecture.
2-7	As you are circulating these are good problems for you to use to check to see that everyone recorded the data correctly.
Lecture	Students can be given the chance to share interesting observations about the set birthdays in the class.
7-10	These problems use fraction bars for comparing fractions. Expanding and reducing fractions reviewed.
11	This problem should generate some debate in groups. Some students may answer this question by visual estimates of the fraction bars above, others may find a common denominator and compare the numbers directly, or compare the numbers in other ways, for example noting that $\frac{3}{5} > \frac{3}{6}$
12-13	These problems use the same fractions as in #11, but students must expand them and compare them, then add them.
14-17	More reducing and expanding, with new fractions.
18	A concise pair of examples demonstrating how to compare fractions and add fractions.
Lecture	Reinforce #18. Review percentages to fractions and decimals to fractions if needed for #19.
19	Converting percentages and decimals to fractions. Two repeating (non-terminating) decimals are included: $0.\overline{3}$ and $0.\overline{6}$.
20-22	Extensive practice converting, comparing, adding and subtracting fractions. The mixed number problems in #22 can be challenging.
23-24	Encourage students to write a thorough set of steps for each skill, comparing and adding.
Lecture	Students can share their sets of steps.

NS2: Multiplying Fractions

Students will explore the meaning of multiplying fractions through doubling and halving, and beyond, with fraction bars. They will learn procedures to multiply fractions and cross-canceling to simplify when multiplying.

Lecture	Some review of any of the following: familiarity with fraction bars, reducing and expanding fractions, converting between mixed #'s and improper fractions, converting terminating decimals to fractions or commutative and associative properties of multiplication.
1-3	The first three problems use fraction bars to double or halve fractions and invite students to explain why the arithmetic works as it does. Encourage students to arrive at clear, concise explanations and come to agreement in their group.
4-6	These problems challenges students to visualize expanding a fraction in order to halve it. This process offers a justification for multiplying denominators when multiplying fractions.
Lecture	Summarize results of 4-6.

[2] Selected problems, suggested approaches. Unless otherwise indicated, only pencil and eraser are required and calculators can be used sparingly or not at all. Bolded words in missions represent new vocabulary for the course.

7	A summary of results so far, followed by practice problems. Note that mixed numbers and decimals may be challenging for some groups.
8-10	These problems provides a demonstration of how the commutative and associative properties of arithmetic justify cross-canceling, and then an opportunity to practice.
11	Encourage students to work carefully together to come up with some rules and tips for multiplying fractions.
Lecture	Students can share results from #11, then preview #12.
12	Converting verbal expressions to numerical expressions.
13-14	Practice, with some potential challenges cross-canceling or converting mixed numbers to improper fractions or decimals to fractions.

NS3: Exploring Fractions Through Probability

Students will explore the meaning of fractions and of multiplication and addition of fractions in the context of probability.

Materials	Two coins per group, two dice per group. One pack of colored candies or beads per group optional.
Lecture	Review of any of the following: converting fractions to decimals to percentages, reducing fractions, multiplying fractions, adding fractions or probability as a fraction and percentage. This mission is designed to explore probability and give context for students' work with fractions.
1-4	The first four problems are a context for introducing the idea of probability, first through a ball toss game at a fair, then through experimentation with flipping two coins to determine probability of two heads. Focus on materials management and efficiency of experimentation.
5-8	These problems explore the difference between experimentally determined probability and probabilities that are determined mathematically based on known probabilities of random events. The also touch on the idea of randomness and the meaning of probability. Energetic debate encouraged.
Lecture	Summarize ideas explored thus far. Introduce tree diagram in #9.
9-12	Here we use a tree diagram to calculate probabilities of two independent events (results of two coin tosses). #12 implicitly introduces the addition rule for mutually exclusive events. Both of these are covered extensively in subsequent problems—these are designed for students to think about the ideas before being provided the rules.
13	Defines independent events and gives the rule for calculating probability of two independent events. Also introduces the notation for probability.
14	Again applying the addition rule.
Lecture	Summarize results without completely explaining the addition rule. Hand out dice and explain the next experiment.
15-17	Experiment with dice rolling and then gives the addition rule for mutually exclusive events.
18-21	Probability of rolling a 7. Then practice problems.
22	M&M's optional. You can assign this one for homework—students can Google the answer.
23-28	Practice problems and they summarize rules for probability.

NS4: Rational Numbers

Students will explore different types of real numbers. They will focus on rational numbers and contrast them with irrational numbers, as well as converting repeating decimals to fractions.

Materials	Optional: limited calculator or use recommended online calculator for #10.
Lecture	Review of number types (rational, irrational, real, integer, natural). Review of any of the following: long division to convert fractions to decimals, multiplying or dividing on both sides of an equation, notation for repeating decimals or adding and multiplying fractions.
1	Number types. Students write definitions, give examples and place the terms in a Venn Diagram. I take natural numbers to be positive integers.

2	Long division to convert decimals to fractions.
Lecture	Work the examples of the procedure for converting repeating decimals to fractions in #3.
3-4	Converting repeating decimals to fractions.
5	Comparing numbers $(>, <, =)$ using the previous tools as needed.
6	Placing numbers on a number line using the previous tools as needed.
7	Adding and multiplying fractions using the previous tools as needed.
8-9	Adding and multiplying mixed numbers (write and review steps).
10	Encountering the repeating decimal in fractions by using a calculator.
11-16	Summarizing number types with true/false statements, composing definitions, listing characteristics, and exploring beyond the current categories.

NS5: Dividing Fractions Using Reciprocals

Students will characterize dividing fractions as a fraction of fractions. They will expand the overall fraction by the reciprocal of the denominator in order to convert the denominator to 1.

Lecture	Review any of the following: adding fractions, reducing, expanding and multiplying fractions, definition reciprocal, multiplying and dividing with negatives. The first problems involve creating equivalent fractions.
1	Reviewing adding fractions.
2-3	Step by step process of reducing or expanding fractions, leading to simplifying divided fractions by selectively multiplying by reciprocals. Students should be encouraged to write out this multiplication as expansion in order to avoid the pitfall of memorizing the "multiply by reciprocal" rule without understanding the arithmetic logic of it. Simplifying variable expressions later will be easier for students as a result.
4	Define reciprocal and give example.
Lecture	Summarize the tactics used. Introduce a nested fraction. Review negatives in fractions and nested negatives.
5-6	Practice, with some nested fractions requiring students to break the problem down into multiple steps.
Lecture	Possible walk through of a nested fraction in #7.
7	Nested fractions.

NS6: Exponential and Scientific Notation

Students will explore relative differences between quantities and represent those using exponential notation. They will explore scientific notation as another way to denote relative differences, including using negative exponents. Metric unit conversions and significant figures will also be explored.

Materials	Metric rulers for #21 (measuring hand spans in groups)
Lecture	Any of the following: simplifying exponential expressions, scientific notation, negative exponents or converting metric units. The first problem explores computing absolute and relative differences.
1-2	Absolute and relative differences. Students can be encouraged to debate when each is a more useful measure of difference, with examples.
3	Students can be encouraged to debate the correct answers and come to agreement on a guess for their group. Answers are available at www.couragetocore.com.
4-6	Absolute and relative difference again. Converting meters to mm is part of the process.
7	Simplifications of exponential expressions. A good place to check background knowledge.
Lecture	Prepare students to note that 10^5 is 10^2 times as big as 10^3 or 10^3 is $\frac{1}{10^2}$ times as big as 10^5. Also, that 6^3 times 6 is 6^4. This is probably enough for them to embark successfully on the experiments which follow.

8-14	This section posits a fish which grows by a factor of 6 when fed a candy, and grows by a factor of $\frac{1}{6}$ when fed a peanut. Students should be encouraged to grapple with the stories and relationships detailed here at their own pace. Check work carefully.
15-18	0 exponent, negative exponents introduced.
19	How many times as big (again).
Lecture	Converting metric units in scientific notation.
20	Converting metric units in scientific notation.
21-25	Significant figures experiments and practice (measuring hand span).
26	How many times as big (again).
27	Multiplying numbers in scientific notation, with significant figures.
28-29	Student summary of comparing, converting and significant figures.
Lecture	Check summaries.

NS7: Roots and Rationalizing Denominators

Students will characterize rooting as the inverse of raising to a power and explore the impact of rooting on numbers in the context of a real world example. Roots of perfect powers will be used as references for estimating irrational roots, and roots will be simplified. Students will learn to rationalize the denominator.

Lecture	Review of any of the following: squaring natural numbers, squaring fractions, rooting perfect powers or properties of exponents and roots, specifically as in $2^3 5^3 = (2 \cdot 5)^3$ and $\sqrt{2} \cdot \sqrt{5} = \sqrt{10}$.
1	This exercise requires students to square different numerical expressions, then root the result to observe the inverse relationship between squaring and rooting.
2-7	These problems develop a simple real world formula involving roots. It is designed to contextualize roots and help students see the impact of rooting, namely that as the radicand (above 1) increases the result increases, but more and more slowly. Students develop a simple formula but are not yet challenged to graph the function. Deeper exploration of functions comes later in the course.
Lecture	Summarize the real world model and impact of rooting.
8	Simplifying roots of perfect powers, and roughly estimating roots by using perfect powers as references.
9-10	Rooting fractions between 0 and 1, and observing that the results are larger than the radicands. Assist students in understanding how to root fractions and explain the results.
11-12	Simplifying roots and expanding fractions in preparation for learning to rationalize the denominator.
13	Rationalizing the denominator.
14	More practice simplifying.
15	More practice rationalizing the denominator.
16-18	Students summarize results: Impact of rooting on numbers greater than 1 or between 0 and 1. Approximating roots using perfect powers as a reference. Simplifying roots and rationalizing denominators.

NS8: Order of Operations

Students will explore simplifying numerical expressions requiring correct application of the order of operations.

Lecture	Review any of the following: adding fractions, rationalizing denominators, adding like terms or PEMDAS.
1	Simplifications of various numerical expressions. This is a good chance to observe areas where students may need particular review.
Lecture	Potential areas of review include: expanding fractions to add them, adding like terms, fractions to a power, powers of 1.

2	Students are challenged to create their own numerical expressions which employ the indicated operations in the indicated order.
Lecture	Have students share out the expressions they created.
3-4	Additional practice problems.

NS9: Distribution and Factoring

Students will explore distribution and factoring as essential truths about multiplication and addition. They will practice the procedures of distribution and factoring with a wide variety of numerical expressions, drawing on the variety of arithmetic techniques they've learned so far.

Lecture	Review of any of the following: adding fractions, multiplying roots as in $\sqrt{2}\cdot\sqrt{3}=\sqrt{6}$, factoring roots as in $\sqrt{6}=\sqrt{2}\cdot\sqrt{3}$ or multiplying exponential expressions as in $x^2x^3=x^5$.
1-4	Students experiment with distribution and factoring to understand the process and verify equivalence of results with or without distribution or factoring.
5	Distribution and factoring applied to irrational numbers, variables and numbers in scientific notation. Here we establish the power of distribution and factoring to help us simplify certain expressions.
6	Establishes factoring as a way to understand combining like terms.
7	Practicing factoring by identifying the greatest common factor. A good problem to check progress and review skills.
Lecture	Review selected problems from #7. Perhaps preview selected problems from #8.
8-10	Practicing factoring and distribution.
11	Continued practice with a final simplification step added to demonstrate the power of factoring as a tool to assist simplifications.
12	Student summary.
Lecture	Students share summary comments.

NS10: Rational Exponents

Students will explore rational exponents in the real world context of earthquake magnitude. This will lead students to an understanding of both decimal and fractional exponents. They will learn how to approximate values resulting from such exponents as well as simplifying expressions with such exponents.

Lecture	Review any of the following: product of powers rule: $2^2 2^3=2^5$, power of powers rule: $(2^2)^3=2^6$ meaning of negative exponents: $2^{-3}=\frac{1}{2^3}$ or meaning of 0 exponent: $2^0=1$.
1	Review the rules related to integer exponents.
2	Gives a context for understanding the application of non-integer rational exponents: Earthquakes and the Richter Scale.
3-4	Explores the application and meaning of non-integer rational exponents.
5	Without a calculator students approximate values for $10^{1.3}$ and similar, using perfect powers as references. The problems which require squaring then approximate rooting, like $10^{\frac{2}{3}}$ will inspire discussion and debate.
Lecture	Address any questions in #5.
6	Students summarize results so far.
7-8	Returning to the application of non-integer rational exponents to earthquakes and the Richter Scale.
9-11	Practice problems, many reviewing simplifications with integer exponents as well as practice with non-integer rational exponents.
12	Students summarize application of exponents with Richter Scale
13	Students summarize rules of exponents.
14-15	Challenge problems.

EE1: Evaluating and Simplifying Polynomials

Students will experiment with evaluating, generating and simplifying algebraic expressions.

Lecture	Some review of any of the following: arithmetic with fractions, negatives, exponentiation, properties of arithmetic or order of operations. The first problem contains important vocabulary (in bold) which can also be discussed.
1	Students may select the same term to define in their group, or different terms and share. The last three problems involve fractions and exponents.
2-3	Encourage students to come to agreement on the wording of their explanation.
4	Requires extensive multiplication and division with fractions, and conversion of decimals to fractions or cross-cancelling to ease the process.
5	Encourage students to come to agreement on the wording of their steps.
Lecture	Review dividing fractions, cross-cancelling and conversion of decimals to fractions.
6-11	This set of problems explores combining like terms, distribution and factoring.
12	Practicing simplification of numerical and variable expressions.
Lecture	Review the distinction between a numerical and variable expression.
13-14	Properties of arithmetic. Careful that students do not assume addition is distributive over addition, or that multiplication is distributive over multiplication.
15-18	Mixed practice. Simplification with exponents invites mini lectures.
Lecture	The properties of arithmetic apply equally to variable expressions as to numerical expressions.

EE2: Equations

Students explore expressions that take values as input and return values as output. This leads equations designed to answer the question: "What input produces the desired output?" Solving simple linear equations can be thought of as follows: Given a desired output, we perform inverse operations in reverse order to arrive at the needed input.

Lecture	Review of any of the following: creating, evaluating and simplifying polynomial expressions, order of operations, adding, multiplying and dividing fractions or the distributive property.
1-3	This set of problems imagines a genie (in reality, a function) who takes a number of candies as input and returns candies as output. Students know the output desired but must give the genie the correct input.
4-6	This section provides a brief history of algebra and gives students the fundamental idea of solving equations to answer the question: "What value of x makes an equation true?"
7-8	Students reverse the order of operations in order to solve equations, and explain their steps.
9	Students build expressions then create equations.
10-11	Revisiting the idea of reversing the order of operations in order to solve equations.
Lecture	Summarize results so far. Prep for reversing operations involving division or fractions, as in $2(3x+4)=5$ or $\frac{2x+7}{3}=8$. remind students that multiplying by a reciprocal is the equivalent of division. Demonstrate to students that multiplying by $\frac{a}{b}$ is the inverse of multiplying by $\frac{b}{a}$. I strongly encourage students to focus on reversing operations at this stage, but this can be supported by the idea of balancing equations by performing equivalent operations on either side of the $=$ sign.
12-13	Practicing solving equations.
14	Practice building expressions and equations.
15-16	More practice solving equations.
17	More practice solving equations. Note that with many of these, simplifications of numerical or variable expressions first will ease the solving.
Lecture	Summarize results.

EE3: Polynomial Expressions

Students learn to identify, distinguish and classify different polynomial expressions. They simplify polynomial expressions and ensure they are in standard form.

Lecture	Review of any of the following: simplifying polynomial and exponential expressions or solving linear equations. Early problems focus on distinguishing, defining and categorizing polynomials. The first problem connects identification of expression and equation types to the solution algorithm used.
1	Fill in the blanks with some vocabulary in a paragraph which describes how identifying types of expressions can lead to an algorithm to solving related equations.
2	Students distinguish between a list of polynomials and non-polynomials and describe distinguishing characteristics of polynomials. Their first attempt at this should not necessarily be perfect.
3-4	Fill in the blanks which compare animal identification and typing to polynomial identification and typing.
5	Students classify polynomials by number of terms and degree. Note that those far all polynomials are given in standard form so that leading term is implicit at this stage.
6	Fill in the blank to summarize results through examples.
7	Students summarize results.
Lecture	Review vocabulary and summary of results.
8-9	Simplifying polynomials and writing them in standard form.
10-15	Review where students create examples to meet written descriptions.
Lecture	Have students share examples from #10-15. Introduce degree of leading term as degree of polynomial.
16-17	Practice simplifying, writing in standard form and classifying polynomials. Lots of fractions here.
18-20	Extending the lesson to writing equations to fit descriptions.
Lecture	Have students share equations and solutions from #18-20.

EE4: Linear Equations

Students will expand on the basic strategy of reversing operations to solve linear equations. They will add the tactics of simplifying on either side of the equation, and of eliminating a linear term by addition or subtraction when there are linear terms on both sides. Linear equations can have no solution, one solution or infinite solutions.

Lecture	Review any of the following: solving linear equations of the form $ax + b = c$ or simplifying polynomial expressions.
1	Solving linear equations of the form $ax + b = c$. Note that fractions or scientific notation may prove challenging.
2	Simplifying polynomial or numerical expressions first in order to ease solving. Note that there is one equation that has no solution. This will be addressed later but is thus previewed here.
3	Fill in the blanks to summarize strategies and tactics of solving equations. So far students have the general strategy of reversing order of operations. They also have the tactic of simplifying first, and, when there are linear terms on both sides of the equation, the tactic of adding or subtracting to eliminate a linear term on one side of the equation.
4-5	Practice eliminating a linear term. Note that there is one equation that has no solution. This will be addressed later but is thus previewed here.
Lecture	Summarize results, perhaps having students share the example they created in #4. Introduce examples of no solution and infinite solutions as desired.
6	Equations with no solution, and equations with infinite solutions, with examples.
7-9	Practice. Help students distinguish between equations that have infinite solutions and those that have a solution of 0.
Lecture	Summarize results.

EE5: Eliminating Fractions, Eliminating Negatives

Students will expand their tactics for solving linear equations.

Lecture	Review of any of the following: solving linear equations of the form $ax + b = cx + d$, adding and subtracting fractions, arithmetic with exponential notation. The first three problems involve playing games with partners and establishing algorithms which guarantee a win or draw, as preparation for identifying algorithms for solving linear equations.
1-3	Game playing in order to establish win or draw algorithms.
4-5	Introduction of tactic to eliminate fractions in equations. Practice.
6-7	Introduction of tactic to eliminate negatives in equations. Practice.
Lecture	Summarize results.
8-10	Practice problems.
Lecture	Summarize results.

EE6: Equations with Multiple Variables

Students will explore a variety of real world formulas from physics and geometry to gain insights into the power and application of algebraic techniques for finding values of interest. They will solve for variables of interest after substituting as well as re-arranging equations prior to substitution.

Materials	Optionally, groups can attempt to measure Circumference and diameter of a circle to calculate pi in #7. Coffee lids rotated on a page, strings around cans, and of course a ruler. Alternately, give them values with your own demo.
Lecture	Review any of the following: evaluating expressions, units for length, velocity, acceleration, energy, etc., scientific notation, significant figures or solving linear equations. The first two problems challenge students to write a couple of familiar equations and to identify some famous given equations.
1-2	Students write a couple of equations with which they are familiar, and try to describe the meaning of several given equations which are hopefully familiar to them.
3	A matching assignment of physics and geometry equations with their descriptions and units. Students can be encouraged to work together on this and then you can provide answers at the end.
Lecture	Answers to previous problem and discuss as desired.
4-5	Simple application of one formula, $F = ma$, to find F in #4. In #5 students must find a using the Force found in #4.
6	Substituting and solving using the various page 2 formulas, with examples.
7-8	Continued practice with page 2 formulas. Note that for one of these they need to measure the Circumference and diameter of a circle and calculate pi. Coffee lids rotated on a page, strings around cans, and of course a ruler. Alternately, give them values with your own demo.
9-10	Students explore why solving for a variable before substituting makes sense. (To avoid having to repeat the same substitution and solve process every time...)
11-12	Practice solving the page 2 formulas for selected variables.
13	Practice substituting and evaluating using the page 2 formulas.
14-16	More practice, more open-ended.
17	Google formulas. You can demo this for them or they can do it on their own at home.

LF1: Linear Functions

Students will explore how expressions can be used to express a relationship between input values and output values, using the context of a genie who performs transformations on the numbers of candies brought to him. Students will experiment to understand the relationship between input and output for linear functions using graphs, tables and equations, and begin to learn the language and notation for describing functions.

Lecture	Review of any of the following: familiarity with linear expressions, evaluating linear expressions, graphing points or solving linear equations.
1-3	Students are introduced to a genie who accepts a number of candies as input and returns a number of candies as output. They review terminology related to linear expressions and describe the impact of the expression on the input.
4-6	Students organize input and output in a table with examples. They label axes and graph the resulting points and connect the points with a line extended to the edges of the graphing window. They look for patterns manifest in the table and graph.
7-10	Students imagine and compare different genies, then specifically compare two different genies using the table and graphs as above.
Lecture	Review results so far. Preview the essential question: "What single input produces the same output in both functions?"
11	Students try to answer the essential question above creatively, with the graph or with trial and error.
12	Closing out the two genies recently explored.
13-17	Students create different linear equations for which a specific ordered pair is a solution. In our scenario, these are different genies who produce the specified output for the specified input. They compare these genies.
18-22	Comparing two more genies.
23-26	Function summary so far, then practice working from a single point, a table or a graph to create a function. Tables and graphs are relatively simple, so complete algorithms are not introduced. Students are encouraged to experiment.
Lecture	Functions so far. Preview four essential questions.
27	Four essential questions about functions: Given y, what is x? Given x, what is y? Given two functions, what single x produces the same y in both? What y is produced?
28-34	Using tables, graphs, or equations to answer those questions. No algorithms yet.
35-44	Mixed practice.

LF2: Rate of Change

Students will explore how rate of change is revealed in the equations, tables and graphs of linear functions. Students will learn about the rate of change in the contexts of real-world functions modeling the distance a car has traveled over time. They will represent slope conceptually as velocity, algebraically as $\frac{\Delta y}{\Delta x}$ and graphically using a slope triangle.

Lecture	Review of any of the following: evaluate linear expressions, graph linear functions using a table, notion of input and output for functions, graphing points or velocity.
1	A basic review of the structure of a linear function in the context of a genie who accepts candies as input and returns candies as output.
2-8	Students are challenged to experiment to determine the rate of increase in output as input is increased by various amounts. Students review how the coefficient for the linear term defines the magnitude of this increase in output. (Note use of distributive property.) More details on this appear later.
Lecture	Summarize idea of rate of change as a measure of the increase in output as input increases. Preview calculating this from tables, but formula comes later.

9-14	Rate of change from table. Points from graph to table to calculate rate of change. Association of constant rate of change with linear function.
Lecture	Preview velocity in km/hr using odometer readings. Preview graphing points.
15-33	Modeling a car or bike with a constant rate of change using odometer readings, a given function, a graph with a slope triangle, points in a table or written as ordered pairs. Lots of graphing here. Slope named in 22. Y-intercept named in #29.
34	Summary of the ways slope is revealed in descriptions, equations, tables and graphs.
35-52	Continued practice and summary.
Lecture	Students share out summaries from #52.

LF3: Finding Functions

Students will use their knowledge of linear functions and slope to build functions from given information.

Lecture	Review of any of the following: graphing linear functions using a table or finding slope of a line from table, graph or equation.
1-4	Students are challenged to use their prior knowledge to experiment with creating functions to fit data from a table. Subsequent problems will help students refine an algorithm.
5-9	Students are challenged to use their prior knowledge to experiment with creating functions to fit data from graphs. Subsequent problems will help students refine an algorithm.
10	An important summary of how to find slope. This material is covered extensively in LF2: Rate of Change.
11	A review graphing lines using a table and identifying slope and y-intercept.
12	Slope-intercept form for a line.
13	Finding function from a table again, without specifying an algorithm yet.
14-15	Finding the function according to steps in algorithm.
16-20	Practice finding the function from the table. Students may confuse $(0,3)$ and $(3,0)$. Also, students may not notice when Δx in the table becomes something other than 1.
21	Finding function from a graph. Students are encouraged to use approximate values for two points, in part to understand graphs as approximations of the function (points on the grid notwithstanding).
22-23	Summarizing results.

LF4: Linear and Non-Linear Models

Students will explore real world functions drawn from physics in order to compare and contrast linear and non-linear relationships.

Lecture	Review of any of the following: equations from Physics as in #1, units for length, velocity, acceleration, energy, etc., scientific notation, significant figures, identifying linear functions, graphing linear functions, polynomial expressions or finding slope from a graph or table.
1	A matching assignment of physics and geometry equations with their descriptions and units. Students can be encouraged to work together on this and then you can provide answers at the end.
2	Students graph a simple quadratic using a table, in the context of a genie who accepts candies as input and returns a different number of candies as output.
3-6	Students find the slope between points on the quadratic and observe the non-linearity of the function.
Lecture	Review observed non-linearity of the function just explored. Introduce physics concepts explored in #7. Introduce terminology: "as a function of."
7	Specific details of the real-world functions students will explore in the subsequent problems.
8-11	Exploration of density concepts and quantities through a 3-D drawing and visualization of air density.
12-13	Graphing air density as a function of elevation. An explicit function is not provided.
14	Concepts of density as students guess densities of different materials and rank them.

15-19	Force as a function of mass, on earth. Given mass, compute force. Given force, write an equation and solve for mass.
20-24	Force as a function of mass, on the moon. From a given mass, compute the force required to lift it on earth, and determine the mass a person can lift on the moon.
Lecture	Summarize results. Quadratic genie, non-linear. Density as a function of elevation, non-linear. Force as a function of mass, linear. Introduce concepts of power as discussed in #25-26. Demonstrate solving a simple rational equation like $4 = \frac{5}{x}$.
25-28	Power as a function of time. Students need to solve a simple rational equation in #28.
29-32	Graphing a line and quadratic on same axes. Comparing, contrasting, estimating intersection point.
Lecture	Summarize again. Introduce concepts of momentum as in #33-36.
33-35	Two different equations for momentum: $p = mv$ and $p = Ft$.
36	Momentum as a function of velocity from tables. Students find the constant mass for each function.
37-38	Students summary.
Lecture	Summarize.

LF5: Linear Models and Problem Solving

Students will create linear models to fit data and use these models to answer questions.

Materials	Protractors, at least one per group. The alternative is demonstrating the measurement and giving values as needed.
Lecture	Review any of the following: graphing functions, finding functions from graphs or tables, using equations to answer questions about functions or measuring angles.
1-4	Modeling and comparing gym memberships. Equations to find month of equal cumulative cost, and the cumulative cost.
5-9	Modeling base jumpers flying along a line, given slope. #6 introduces angle of depression without measurement required. #8 introduces x-intercepts. Finding x-intercepts and meaning of x-intercepts in context.
Lecture	Summarize results so far. Show how to measure angle of depression.
10-14	Continued base jumper modeling. #14 introduces angle of depression with measurement required.
15-20	Continued base jumper modeling with angles of depression.
21-24	Base jumper modeling from points.
Lecture	Summarize results. Discuss causality and terminology: "as a function of."
25-28	Exploration of causality.
29-31	Simple linear model.
32	Students summary problem.
Lecture	Summary.

LF6: Horizontal and Vertical Lines

Students will explore horizontal and vertical lines as compared to linear functions with non-zero slopes and defined x-intercepts and y-intercepts.

Lecture	Review any of the following: structure of a linear expression, graphing linear functions, finding linear functions from graphs or tables or using equations to answer questions about linear functions.
1-4	Constant functions explored from verbal descriptions translated to tables and graphs.
5	Introduces the term "constant function."
6-9	Exploring slope and y-intercept for a constant function, comparing with linear functions having non-zero slopes.

Lecture	Summarize results. Possibly introduce lines with small slopes to show how to distinguish their graphs from graphs of constant functions.
10-12	Review of precision in graphing so that lines with small non-zero slopes do not appear horizontal.
13	Comparing slopes of lines. Lines presented as equations or as two points, so this section also reviews finding slope.
Lecture	Review finding slope. Review finding x-intercepts and y-intercepts if needed in advance of #14.
14	Finding x-intercepts and y-intercepts. Lines presented as two points or as equations, including horizontal lines.
15-16	Students explore vertical lines from verbal description translated to table and graph.
17	Students are given a summary of vertical and horizontal lines.
18	Students practice vertical lines.
19-23	Practice determining slope, x-intercepts, y-intercepts, and writing equations to fit descriptions.
24	Students summarize results.
Lecture	See #17.

LF7: Domain and Range

Students will explore the domain and range of various functions through equations, tables and graphs.

Lecture	Review any of the following: identifying types of polynomial functions, graphing and linear and non-linear functions using a table, real numbers or identifying points from the graph of a function
1-6	Identifying functions by type from their equations and beginning to explore the limits of input and output for different functions.
Lecture	Discuss results so far, preview graphing of linear, quadratic and rational function in #7 a bit.
7-11	Graphing functions of three types and exploring the limits of the input and output.
12	Students informally write rules for input for several functions.
13	Students identify the lowest value that a function returns as output.
Lecture	Preview or partial walk through of #14-15.
14-15	Domain and range defined. Tactics for finding domain and range discussed.
16	Domain and range from graph.
17	Domain and range for many functions, from equations.
Lecture	Introduce non-functions.
18-20	Domain and range from graphs, including non-functions.
21	Function notation.
22-24	Review and student summary.
Lecture	#14-15 revisit.

LF8: Linear Models and Systems

Students will create linear functions, tables and graphs to model real-world scenarios. They will solve systems to answer questions about these scenarios and explore the meaning of solving a system along with the substitution method for solving systems when lines are in slope-intercept form.

Lecture	Review any of the following: creating linear functions from a verbal description, graphing linear functions using a table or solving linear equations.
1-8	Students create linear models to fit a real-world scenario, and represent them with equations, tables and graphs. They solve the linear system and interpret and confirm the solution.
9	Explanation of systems of linear equations including rational for substitution and meaning of solution.
Lecture	Summarize #9 and introduce #10.

10	Students walk through given steps for solving a system using simple substitution when equations are in slope-intercept form.
11-13	Practicing solving systems.
14-16	Practice modeling real world scenarios and solving the resulting systems.
17	Students create a system and solve.
18	Students explain difference between solving an equation in one variable and solving a system.
Lecture	Summarize results.

LF9: Forms for Lines and Methods for Systems

Students will explore the algebraic rational for substitution and elimination methods for substitution, then they will practice these methods.

Lecture	Review of any of the following: graphing linear functions, slope-intercept form for a line, basic concept of finding intersection point using substitution when lines are in slope-intercept form (LF8), solving linear equations requiring distribution or solving linear equations in two variables for either variable
1-6	Graphing review comparing lines with different y-intercepts, different slopes.
7	Equivalent equations.
8	Standard form for a line.
9-11	Converting between standard and slope-intercept form.
Lecture	Review results so far. Preview graphing standard form by finding x and y-intercepts.
12-13	Graphing standard form using x and y-intercepts.
14	Identifying standard or slope-intercept form for a line.
15	Graphing, both forms.
16	Solving the system, substitution, informally introduced.
17-19	Substitution, formally introduced, then practice.
Lecture	Summarize results so far, introduce elimination concepts as in #20.
20-24	Elimination formally introduced, multiplying equations prior to elimination introduced, practice.
25-26	Selecting a method and explaining selection. Practice.
27-28	Student summary.
Lecture	Summarize results.

LF10: Parallel and Perpendicular

Students will explore the relationship between parallel and perpendicular lines through comparing and rotating slope triangles. Students will practice finding equations for lines which are parallel or perpendicular to existing lines.

Lecture	Review any of the following: graphing lines, finding slope of lines from graphs and equations or finding equations for lines from graphs or points
1-7	Graphing lines comparing slopes. Parallel lines have same slope.
8-9	Students try to determine if given lines are perpendicular.
Lecture	How to rotate a slope triangle $90°$ in preparation for #10. Rotation of slope triangle synonymous with rotation of line connected to it.
10	Students rotate a slope triangle and line to see if it is parallel to a given line after rotation, which would imply it was originally perpendicular to the given line.
11-13	The slope triangle for a given line is rotated $90°$ and a perpendicular line is created.
Lecture	Summarize results. Introduce terminology "negative reciprocal" and "coincident" in anticipation of #14.
14	Summary of possible relationships between lines in coordinate plane.

15	Students compare lines and identify relationships from equations.
16-24	Practice given equations, points or graphs and various instructions.
25	Student summary.

LF11: Graphing Linear Inequalities

Students will explore the relationship between points on and above or below a line. Students will learn to represent these relationships using inequalities.

Lecture	Review any of the following: graphing points, graphing lines in slope-intercept form or graphing lines in standard form.
1-3	Students graph a line then graph points above or below the line. They compare the y-values to those of the points on the line with the same x-value.
4-5	Students are challenged to explain how the location of points with respect to a line can be expressed using equations or inequalities.
6-8	Formalizing the idea that points on a line make an equation true, points above or below make a corresponding inequality true.
9	Students make a first attempt to explain how you would graph an inequality.
10	Example of graphing an inequality.
Lecture	Summarize results.
11	Practice graphing an inequality.
12-13	Inequalities with $\geq$ and $\leq$.
14	Practice.
15-17	Equations and inequalities in standard form.
18	Systems of inequalities.
20-24	Practice and student summary.
Lecture	Summarize results.

DS1: Introduction to Statistics

Students will informally explore data through statistics with which they are already familiar. They will collect and process data using the mean, median and mode.

Materials	White board, chalk board, or other surface students can use to record data the whole class can see.
Lecture	Review and of the following: percentages, feet to centimeter conversions or mean, median and mode.
1-3	Students brainstorm examples of measurements which we use to quantify things in the real world.
4-5	Students consider that the purpose of statistics is to answer questions about the world, and brainstorm questions which statistics can answer. They see examples of statistics and pose follow-up questions.
Lecture	Summarize results so far. Introduce table on the board for recording student heights and explain their responsibilities for transferring and organizing the data in #6-9. This includes organizing the data from smallest to largest as well as using different colors (or pen or pencil) to distinguish gender. This also includes a frequency plot of the data, and mean, median and mode.
6-9	Students record their heights on the board, then transfer the data to a table a frequency dot plot with gender distinguished by color. Students then find the mean, median and mode for the entire group, then for boys and girls.
Lecture	Summarize results. Introduce outlier informally.
10-13	Range of values. Informally introduce outliers. Students are challenged to identify the middle 50% informally.
14	Students find mean for their group, then mean for their group and another combined. Then they check to see if the means of their sample was closer to the class mean after increasing sample size. Informal discussion.

15	Students discuss how this data would be different for a class of 12th graders. (If this is a class of 12th graders, posit the question for a class of 9th graders.)
Lecture	Summarize results. Explain homework assignment in #16-17.
16-17	Students gather data from their homes. For example, number of pets. This can be completed in class but is designed as homework.
18	Students select and process data from the previous problem. Students are still working informally with middle 50%, and outliers.
Lecture	Summarize results.

DS2: Shape, Center and Spread

Students will gather data, compute statistics, represent the data graphically and interpret patterns in the data.

Lecture	Review any of the following: mean, median and mode; the practice and purpose of statistics or dot plots (DS1). Preview data gathering (heart rates) in #3.
1-2	Students will measure their heart rate and speculate at the causes of variation, both for an individuals and between individuals.
3-5	Students will add all the heart rate data to the board then record the data from smallest to largest, coded by gender. Some statistics are computed and the data is graphed as a BAR GRAPH.
6-8	Students compare data by gender. Students compare statistics for a smaller sample and larger sample to the overall class population. Students discuss how the data would be different for 12th graders (or any grade level different from theirs).
Lecture	Anticipate table formalizing terms in #9.
9	Table formalizing statistical terms.
10-12	Students given fictional heart rate data and box plot.
13-16	Daily mid-day temperatures box plots for each month. Reading data from the graph.
Lecture	Summarize results in anticipation of student-created box plots.
17-18	Student given temperature data for one month and have to find statistics and create box plot.
19-20	Students given winning marathon time data, find statistics, create box plot.
21	Students invent data, find statistics and create box plot.
22	Students summarize results.
Lecture	Summarize results.

DS3: The Normal Distribution

Students will explore the normal distribution through real-world examples and the binomial distribution as number of experiments increases. Students will connect frequency plots to their corresponding probability distributions, and explore the shape, center and spread with the mean and provided standard deviation.

Materials	At least one coin per group, for flipping.
Lecture	Review any of the following: graphing frequency plots (bar graphs); finding mean or finding simple probabilities from frequencies.
1	Students graph a frequency plot from given data.
2	Students attempt visually approximate the mean, then they calculate it. Hopefully they discover they can multiply the values times their frequencies in order to expedite summing, but if not they will see it demonstrated later. Note graphing of mean as a labelled vertical line.
3	Informal discussion of distribution of the data.
4	Formula for mean used in an example.
5	First probability question.

6	Introducing normal distribution. Symmetrical decreasing tails around mean, which is also mode.
Lecture	Summarize results. Introduce the terms "trials" and "experiments." Introduce the coin flipping in #7.
7-8	Coin flipping experiment, frequency plot, finding probabilities then graphing probability distribution.
9-10	Same as above with data provided. Introducing term "relative frequency."
11-14	Same as above, adding discussion of whether the data looks normally distributed.
Lecture	Summarize results. Introduce increasing number of experiments to make distribution more normal.
15-18	Increasing the number of experiments to see the binomial distribution emerge as normal.
19	Student summary of the relationship between frequency plots and probability distributions in anticipation of the binomial distribution (infinite number of experiments).
Lecture	Summarize results, preview #20 (binomial distribution).
20-22	Binomial distribution and informal standard deviation.
23	Range of values for ±1 SD.
24	Graphing vertical lines at ±1 SD. Approximating probability of outcomes between these two values.
25	Important summary table, shape, center and spread using comparison to normal, mean and SD.
Lecture	Summarize results.
26-35	Continued practice as above.
Lecture	Summarize results.

DS4: Sampling

Students will explore samples and statistics using real world data. They will take multiple samples from a roughly normally distributed population and observe that the samples statistics are normally distributed with a mean close to the population mean.

Materials	TI-84+ graphing calculator or other tool for calculating standard deviation (one per group). M and M's, skittles, colored beads or similar which allows samples of 10 or 20 to be taken by each person in group.
Lecture	Review samples, statistics, populations and parameters. Introduce the experiment with begins with #4.
1-3	Introducing basic idea of a ratio to represent the relative frequency of colors of candies. Reviewing terminology.
4-9	Experimenting with sample size 10. Discourage eating of candies at this stage since there is more work to be done ahead.
Lecture	Review mean and standard deviation if needed. Demo standard deviation with calculator to preview the procedure described in #10.
10-13	Finding, graphing and interpreting the standard deviation.
14-21	Repeating the process with samples of size 20.
Lecture	Summarize results.
22-23	Increasing number of samples.
24	Review normal distribution, mean and standard deviation.
25-26	Increasing sample size decreases the standard deviation.
27	Always the same percentage of values (roughly) within ±1 SD.
Lecture	Summarize results. Preview large number of samples and informally introduce continuous function.
28-30	Increasing sample size decreases standard deviation. Increasing number of samples informally brings us towards a continuous probability function.
31-33	A normal distribution modeling basketball player heights. ±1 SD encloses about 68% of the data.
34	Student summary: definitions.
35-36	Estimating mean mass of sodas from one sample. (Central limit theorem implied but not formally explored or identified.)
37	Student created data set.
Lecture	Summarize results.

DS5: Linear Regression

Students will visualize the relationship between two variables, look for correlation and speculate as to causality. They will explore the trend line as a set of statistics which summarize the data and allow for estimation of additional data points. They will use the calculator to find the line of best fit and the correlation coefficient and interpret these in the context of the real world data.

Materials	TI-84+ graphing calculator, one per group or demo to class.
Lecture	Correlation and causality preview for opening problems. One variable vs. two variable statistics.
1-4	Informal exploration of correlation: basketball player scoring vs. heights.
5	The graph of scoring vs. heights.
6	The graph of rebounds vs. heights.
7	Estimating rate of change for rebounds vs. heights.
Lecture	Summarize results. Revisit causality in context of rebounds vs. heights.
8	Causality.
9-10	Corn production vs. coffee production
11-13	Fuel efficiency vs. mass of car, with trend line and use of trend line for prediction.
Lecture	Summarize results. Preview calculator for linear regression and correlation coefficient.
14-17	Revisiting fuel efficiency vs. mass of car using calculator for linear regression.
18-19	Students create or invent correlated data and analyze.
20-21	Happiness vs. hours surfing.
22-23	Vocabulary vs. books read.
24-25	Students create or invent correlated data and analyze.
26	Student summary: definitions and examples.
Lecture	Summarize results.

EQ1: Exponential Functions

Students will explore exponential functions by modeling growth and decay scenarios. They will compare linear and non-linear models, and develop and standard form for an exponential function.

Lecture	Review of any of the following: writing linear equations as models, solving equations in multiple variables or function concepts and terminology.
1	Students will create linear models to fit a given scenario.
2	Students solve physics equations for a given variable and verbalize the resulting function.
Lecture	Summarize results. Preview exponential model with genie scenario in #3.
3-5	A genie multiplies the number of candies you have by a constant multiple times, and gives the resulting number of candies to you. The number of multiplications is determined by the number of magic pebbles (input) that you provide the genie. Students create expressions for these calculations and compute results. Then they generalize to create standard form for an exponential function.
6-9	Exponential models of growth and decay with bacteria and compound interest scenarios.
Lecture	Summarize results.
10	Creating exponential or linear functions to fit scenarios.
11-18	Graphing linear and exponential functions and comparing and contrasting them.
Lecture	Summarize results.

EQ2: Comparing Linear, Exponential, Quadratic

Students will compare and contrast exponential functions with linear and quadratic functions by using these function types to model different scenarios. They will compare the values returned for each function in real-world contexts, the tables and graphs, as well as the general forms and the meaning of the constants in each form.

Lecture	Review any of the following: forms and graphs for exponential functions, forms and graphs for linear functions, graphing a function by making a table, reading points from a graph or arithmetic with exponents.
1-4	Students compare two different savings plans: linear and exponential. A table organizes the comparison. Students may need guidance to calculate each successive value in the exponential plan, but they may be able to figure it out. Further details of the exponential model will be explored later in the mission.
5	A quadratic plan.
6-8	Comparing and contrasting the plans based on the tables.
Lecture	Challenge students to come up with equations for each plan. These are provided in #9.
9	Summarizing the equations and meaning of constants in each function type.
10-11	Comparing results of the typical arithmetic involved in linear, quadratic and exponential functions, to see the relative strength of addition, multiplication, squaring and higher powers.
12	Comparing linear, quadratic and exponential models based on short term, medium term and long term output.
Lecture	Summarize results. Challenge questions: Is it true that any increasing exponential model will at some point be greater than a quadratic one? And any increasing quadratic will always exceed an increasing linear one? Preview forms for the functions for next problem.
13-16	Forms, writing equations for functions, interpreting constants.
17-18	Determining equations for functions from a table.
19-24	Determining equations for functions from a graph. Graphing functions.
25	Identifying functions from table, graph, equation, form or scenario.
26-29	Students summarize results.
Lecture	Summarize results.

EQ3: Vertical and Horizontal Shifts

Students will experiment with applying vertical and horizontal shifts to functions represented using tables, equations and graphs.

Lecture	Review any of the following: forms and graphs for exponential functions, graphing a function by making a table, graphing simple quadratic functions, function notation. Prepare students to create and graph an exponential function modeling compound interest.
1-2	Students model a bank account which grows exponentially and complete a table and graph. Students then create a model which supplements the account with a fixed amount, completing a table and graph.
3-5	Students speculate on the differences and similarities between $y = x^2$ and $y = x^2 + 7$, then graph both using a table and observe and remark on the vertical shift.
6	Students perform vertical shifts given equations, tables or graphs for the pre-image.
7-9	Students are introduced to the correct notation and verbalization of vertical shifts.
10	Students describe vertical shifts as seen in equations, tables and graphs.
Lecture	Confirm understanding of notation and verbalization. Prepare students to experiment with horizontal shifts, without giving too much away so that students experiment in #11. Remind students of the meaning of the vertex of a parabola in preparation for #14.
11	Students perform a horizontal shift of a function.
12-14	Horizontal shifts are further explored and clarified through experimentation with parabolas. The vertex is identified. You can encourage students to revisit #11 after #14.

Lecture	Emphasize the observation that vertical shifts as applied to the equation for the function work as expected—to vertically shift by 4 units you add 4. However, horizontal shifts work somewhat counter-intuitively—to horizontally shift by 3 units you write the function with x-3 in place of x.
15-20	Continued practice including combining horizontal and vertical shifts, and identifying the vertex of a parabola from a transformed equation.
21-23	Representing shifts using function notation when no specific equation is provided.
Lecture	Review the previous problems as needed. Set up for shifts of more complex functions.
24	Shifts of more complex functions.
25	Student created function and shifts.

EQ4: Arithmetic and Geometric Sequences

Students will revisit exponential and linear functions by modeling real-world scenarios with equations, tables and graphs. They will use horizontal shifts to create sequences with initial values at $n = 1$. They will practice writing functions using sequence notation. Given selected values for a sequence, students will find the initial value, common difference or common ratio then write the formula for the sequence. They will use the formulae for finite series.

Lecture	Review any of the following: modeling with exponential and linear functions; equations and graphs of exponential and linear functions; horizontal shifts of functions; understanding domain; finding slope; solving equations involving exponents and requiring rooting.
1-3	Student model bacteria growth using an exponential then a linear model, comparing the equations, graphs and tables.
4-5	Students write equations for linear and exponential functions to model scenarios. They discuss where the y-intercept shows up in each type of equation.
6-8	Students model basketball scoring per game using linear and exponential models. They use tables to compare the arithmetic which increases each players score from game to game in each model.
Lecture	Summarize results. Prepare students for horizontal shifting as needed.
9-10	Students review horizontal shifts by graphing and comparing a given function and the given shifted image.
11-12	Students practice shifting functions, then discuss why it works that way.
Lecture	Summarize results.
13-16	Students model basketball scoring from game to game for two players in one season. For the second season, one players scores stay the same while those of the second are shifted.
17-19	Students model basketball scoring and the functions for both players are horizontally shifted one unit. Domain restriction previewed.
20-21	The horizontal shift and domain restriction is discussed, and the resulting functions defined as sequences. In a table with some blanks for students to fill in, sequences are further defined and notation provided and explained.
Lecture	Summarize results, ensuring the table in #21 is well understood.
22	Student write formulae for given sequences and find the 9th term for each.
23-25	Finding formulae for sequences given various terms—sometimes the first two and sometimes not. This is carefully guided with prompts and some equations and steps provided.
26-28	Practice problems.
29	Finding geometric sequence given non-consecutive terms and not given initial value. This is guided.
30-32	Practice.
Lecture	Summarize results. Introduce idea of finite series.
33-36	Derivation of finite arithmetic series, and practice.
37-43	Given formula for finite geometric series, mixed practice and challenge problems.

Cheat Sheets[3]

Number Sense

Comparing Fractions	Adding Fractions
$\frac{2}{3} \quad \frac{3}{5}$ LCD $= 15$ $\frac{2}{3}\left(\frac{5}{5}\right) \quad \frac{3}{5}\left(\frac{3}{3}\right)$ $\frac{10}{15} \quad \frac{9}{15}$ $\frac{10}{15} > \frac{9}{15}$	$\frac{3}{4} + \frac{5}{6}$ LCD $= 12$ (not 24!) $\frac{3}{4}\left(\frac{3}{3}\right) + \frac{5}{6}\left(\frac{4}{4}\right) =$ $\frac{9}{12} + \frac{20}{12} =$ $\frac{29}{12} = 2\frac{5}{12}$

Multiplying Fractions

Arithmetic	Rule	Example
Multiplying a fraction by a whole number	*Multiply the numerator by the whole number.*	$3 \cdot \frac{2}{7} = \frac{6}{7}$
Multiplying a fraction by a fraction whose numerator is 1	*Multiply the denominators.*	$\frac{1}{3} \cdot \frac{2}{7} = \frac{2}{21}$

$\frac{2}{3} \cdot \frac{5}{7} = \frac{10}{21}$ *Multiply numerators, multiply denominators.*

Probability Rules

Event A: first coin heads Event B: second coin heads	*The probability of A and B both occurring is the probability of A times the probability of B.*

$$P(A \text{ and } B) = P(A) \cdot P(B) = \frac{1}{2} \cdot \frac{1}{2} = \frac{1}{4}$$

Event A: rolling 5 and 6 Event B: rolling 6 and 5	*The probability of either A or B occurring is the probability of A plus the probability of B.*

$$P(A \text{ or } B) = P(A) + P(B)$$
$$P(11) = P(5 \text{ and } 6) + P(6 \text{ and } 5) = \_\_\_\_\_\_$$

[3] Some of the tables here are not completely filled in because they are part of assignments completed in the missions.

Converting Repeating Decimals to Fractions

$0.\overline{3}$ $0.\overline{3} = x$ (set the number equal to x) $3.\overline{3} = 10x$ (multiplying both sides by 10) $0.\overline{3} = x$ (re-written from above) $3 = 9x$ (subtracting the two equations) $x = \frac{3}{9} = \frac{1}{3}$ (solving for x)	$7.\overline{45}$ $0.\overline{45}$ (ignore the 7 for now) $0.\overline{45} = x$ $45.\overline{45} = 100x$ $0.\overline{45} = x$ $45 = 99$ $x = \frac{45}{99}$ $7\frac{45}{99}$ (put the 7 back in the answer)

Number Types

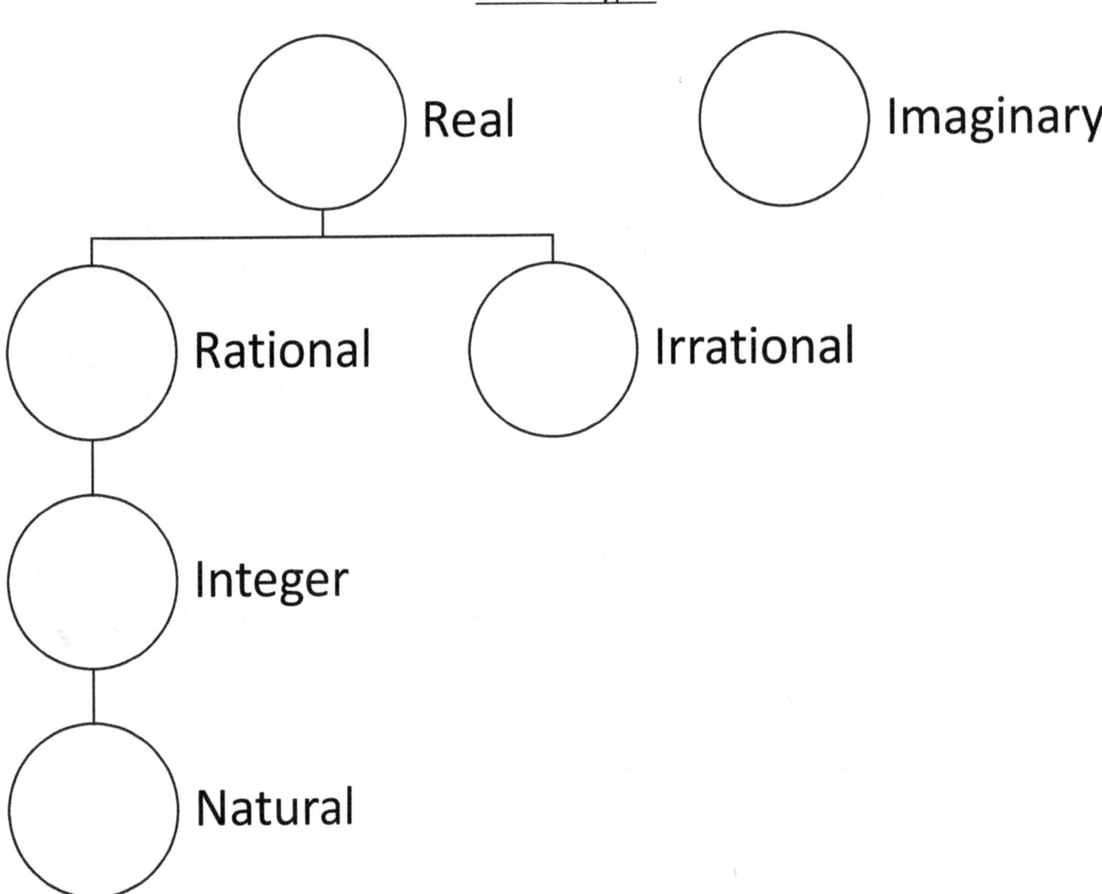

Exponential Notation

If you want to multiply 7 *by* 6 *forty times, you write* $7 \cdot 6^{40}$. *The exponent counts the number of times you are multiplying by* 6.
If you want to divide 7 *by* 6 *forty times, you write* $7 \cdot \frac{1}{6^{40}}$. *The exponent in this case counts the number of times you are dividing by* 6.

Distribution and Factoring

	distribution	factoring
with irrational numbers	$3(5+\sqrt{2}) = 15 + 3\sqrt{2}$	$12 + 18\sqrt{5} = 6(2 + 3\sqrt{5})$
with numbers in scientific notation	$10^3(5+6) =$ $5 \cdot 10^3 + 6 \cdot 10^3$	$22 \cdot 10^8 + 7 \cdot 10^8 =$ $10^8(22+7) =$ $29 \cdot 10^8$
with variables	$3(4+5x) = 12 + 15x$	$2x + 3x = x(2+3) = 5x$

Distributive Property of Addition

Forwards	Backwards
$3(4+5) =$ $3(4) + 3(5) =$ $12 + 15 =$ 27	$2(7) + 2(9) =$ $2(7+9) =$ $2(16) =$ 32

Commutative and Associative Properties

operation	**commutative**	**associative**
addition	$2 + 3 = 3 + 2$ $x + y = y + x$	$(1+2) + 3 = 1 + (2+3)$
subtraction (1st line changes it to addition)	$3 - 2 = (-2) + 3$	$(3-2) - 1 = 3 + (-2-1)$
multiplication	$3 \cdot 4 = 4 \cdot 3$	$3(4 \cdot 5) = (3 \cdot 4)5$

Solving Equations

Equation	The original order of operations	Working backwards	Solving the equation
$3x - 7 = 11$	Starting with x, first we multiplied by 3, then we subtracted 7.	Working backwards would be adding 7 then dividing by 3.	$3x - 7 = 11$ $+7 \quad +7$ $3x = 18$ $\frac{3x}{3} = \frac{18}{3}$ $x = 6$

Essential Tools for Solving Equations

Tactic	Tactic	Strategy
Simplifying each side as much as possible.	*When there are linear terms on both sides of the equation, eliminate one on one side by adding or subtracting.*	*Reversing the order of operations in order to isolate x.*

Essential Questions about Functions

Given one function:	Given two functions:
What input returns an output of ___?	*What single input will produce the same output for two functions?*
What output is returned for an input of ___?	*What output does the input above produce?*

Slope

Verbally	Algebraically	Algebraically Again	From a table	From a graph
Slope Rate of change Velocity	$m = \frac{\Delta y}{\Delta x}$	$m = \frac{y_2 - y_1}{x_2 - x_1}$	$\Delta x \leftarrow$ [table: x, y; x_1, y_1; x_2, y_2] $\rightarrow \Delta y$	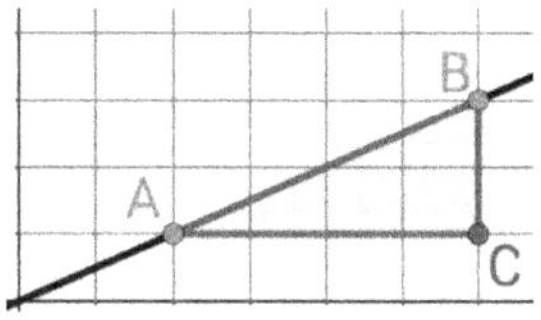 $m = \frac{vertical\ change}{horizontal\ change}$ $m = \frac{BC}{AC}$
Example	a car travels 120 km in 3 hrs $m = \frac{120\ km}{3\ hr}$ $m = 40 \frac{km}{hr}$	Point A: (20,130) Point B: (30,170) $m = \frac{40}{10} = 4$	$2 \leftarrow$ [table: x, y; 5, 12; 7, 18] $\rightarrow 6$ $m = \frac{6}{2} = 3$	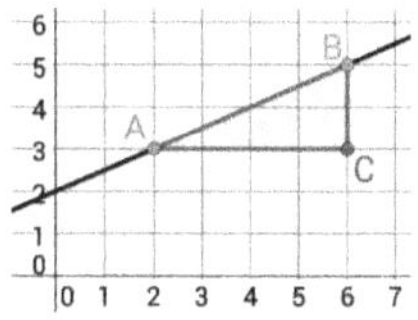 $m = \frac{2}{4} = \frac{1}{2}$

How to Find the Equation for A Line

	x	y
A	2	9
B	5	21
C	10	41

Example	Algorithm
$m = \frac{\Delta y}{\Delta x} = \frac{12}{3} = 4$	1) Find slope
$y = mx + b$	2) Write general slope-intercept form for a line.
$9 = 4(2) + b$	3) Substitute known information.
$b = 1$	4) Solve for b.
$y = 4x + 1$	5) Write specific function.
$y = 4(10) + 1 = 41$ ☺	6) Confirm function is correct using other point.

Horizontal and Vertical Lines

type	description	example	slope	x-intercept	y-intercept
horizontal	constant function	$y = 7$	0	none, unless it's the line $y = 0$	same as the value of y in the equation
vertical	not a function	$x = 5$	none	same as value of x in equation	none, unless the line is $x = 0$

Domain and Range

function: $y = x^2 + 7$

domain	range
all Real numbers	all Real numbers ≥ 7
The domain is the input. I can put all Real numbers in for x and get a result. For example, I can square 4, -5, and even 0.	*The range is the output. When I input 0 I get 7, and that's the lowest output I can get. That's because if, for example, I input 5 I get 25 and if I input -5 I also get 25. So an input of 0 returns the lowest y-value, 7. After that, the output gets bigger.*

Finding Domain and Range

	tactic one	tactic two	function types to look out for	examples
domain	*Try to input various values and determine if there are any which the function doesn't accept as input.*	*Identify the function type*	square root functions	$y = 5\sqrt{x}$ all Real #'s ≥ 0
			rational functions	$y = \frac{8}{x}$ all Real #'s $\neq 0$
range	*Input various values and observe how the output changes as you change the input.*	*Identify the function type and visualize its graph to help you.*	quadratic functions	$y = x^2 + 11$ all Real #'s ≥ 11
			square root functions	$y = 2\sqrt{x}$ all Real #'s≥ 0
			rational functions	$y = \frac{7}{x}$ all Real #'s $\neq 0$

Function Notation

	how we write it	how we read it
function	$f(x) = 2x + 7$	"f of x is $2x + 7$"
substitution example	$f(3) = 2(3) + 7$	"f of 3 is $2(3) + 7$"
example continued	$f(3) = 13$	"f of 3 is 13"

Solving a System of Linear Equations

Example	Steps
$y = 2x - 5$ $y = 4x + 17$	*Make sure you've got two equations in two variables. Make sure you are being asked to solve the system, or find the intersection point, or determine when the functions have the same y-value.*
$2x - 5 = 4x + 17$	*Write an equation which sets the two functions equal to each other in order find the x which returns the same y for both functions. Solve the equation.*
	Substitute the x-value you found into one of the two functions to find the corresponding y-value.
	Optionally, you can confirm your y-value is correct by repeating the last step with the other function.
	Write your solution as an ordered pair (a point). Optionally, you can confirm your solution is correct by graphing the two functions and finding the intersection point visually.

Substitution Method to Solve a System

Example	Steps
$-2x + y = 5$ $x + 3y = 1$	*Make sure you've got two equations in two variables. Make sure you are being asked to solve the system, find the intersection point, or determine when the functions have the same y-value.*
$-2x + y = 5$	*Use algebra to re-arrange one of the equations so that either x or y isolated. (In this example, isolate y.)*
	Substitute the expression for y in the above line in place of y in the other equation.
	Solve the resulting equation for the single remaining variable. (In this example, solve for x.)
	Substitute your value for x into either function to find the corresponding y value. Write your solution as an ordered pair.
	Optionally, you can confirm your solution is correct by substituting the ordered pair into the other equation to see that it makes that equation true.
	Optionally, you can confirm your solution is correct by graphing the two functions and finding the intersection point visually.

Easy Elimination Method to Solve a System

Example	Steps
$x = 11 - 2y$ $3x - 2y = 1$	*Make sure you've got two equations in two variables. Make sure you are being asked to solve the system, find the intersection point, or determine when the functions have the same y-value.*
$x + 2y = 11$ $3x - 2y = 1$	*Re-arrange the equations so that x terms, y terms and constants are vertically aligned. (Standard form works.)*
$4x + 0y = 12$	*In the "easy elimination" method, you'll quickly notice that adding (or subtracting) the two equations will eliminate x's or y's. Add or subtract straight down to accomplish this.*
$4x = 12$ $x = 3$	*Now you can solve for the single variable that remained after elimination.*
	Substitute this into either equation to find y.
	Write solution as an ordered pair.
	Optionally, confirm your solution by substitution in the other equation.
	Optionally, confirm your solution by graphing.

More Complex Elimination Method to Solve a System

Example	Steps
$3x + 5y = 7$ $2x - 7y = 15$	*Make sure you've got two equations in two variables. Make sure you are being asked to solve the system, find the intersection point, or determine when the functions have the same y-value. If needed, re-arrange for vertical alignment of terms.*
$2(3x + 5y = 7)$ $3(2x - 7y = 15)$	*Find the least common multiple between the coefficients of either the x or y terms. In this case it was* 6 *for the coefficients of the x terms. Multiply each equation by what is needed to change those coefficients into* 6.
$6x + 10y = 14$ $6x - 21y = 45$	*Re-write the equivalent equations resulting from the multiplication.*
$0x + 31y = -31$	*Add or subtract straight down in order to eliminate the x terms or y terms. In this case we will subtract to eliminate the x terms.*
$y = -1$	*Solve the resulting equation for the remaining variable.*
	Substitute this into either equation to find x.
	Write solution as an ordered pair.
	Optionally, confirm your solution by substitution in the other equation.
	Optionally, confirm your solution by graphing.

Relationships Between Two Lines in the Coordinate Plane

relationship	graphs	slopes	intersection point
intersecting	The two lines cross.	$\frac{1}{2}$ and 3, for example. Any two lines whose slopes are not the same intersect.	Solve the system and you'll get one solution.
parallel	The two lines do not cross.	$\frac{2}{5}$ and $\frac{4}{10}$ for example. Any two lines whose slopes are the same are parallel, meaning they will not intersect.	Try to solve the system and you'll get no solution.
perpendicular (thus also intersecting)	The two lines intersect at a $90°$ angle.	$\frac{4}{7}$ and $-\frac{7}{4}$ for example. Rotating the slope triangle $90°$ for one line gives you a slope triangle for the other. The slopes are **negative reciprocals** of each other.	Solve the system and you'll get one solution, since the two lines do intersect.
coincident	The two lines are the same line.	Two lines we initially think might be different are actually the same. For example: $y = 2x + 3$ and $2y = 4x + 6$. They have the same slope, and the same y-intercept.	Solve the system and you'll get infinite solutions, since two lines that are the same line intersect at all their points.

Understanding Graphs of Equations and Inequalities

$$y = \frac{1}{2}x + 5$$

location of points	example	relationship to equation or inequality	example
on the line	$(4,7)$	make the equation true	$7 = \frac{1}{2}(4) + 5$ $7 = 2 + 5$ $7 = 7$ ☺
above the line	$(4,8)$	make the inequality $y > \frac{1}{2}x + 5$ true	$8 > \frac{1}{2}(4) + 5$ $8 > 2 + 5$ $8 > 7$ ☺
below the line	$(4,6)$	make the inequality $y < \frac{1}{2}x + 5$ true	$6 < \frac{1}{2}(4) + 6$ $6 < 8$ ☺

Forms for Lines

	general form	specific examples	inequality examples
slope-intercept form	$y = mx + b$	$y = 2x + 3$ $y = \frac{1}{3}x - 7$	
standard form	$ax + by = c$	$2x + 3y = 6$ $\frac{1}{2x} + \frac{2}{3}y = -8$	

Graphing a Line in Standard Form

option 1	example	option 2	example
Convert it to slope intercept form.	$2x - 3y = 12$ $3y = -2x + 12$ $y = \frac{2}{3}x - 4$	Find the y-intercept by plugging in 0 for x.	$2(0) - 3y = 12$ $-3y = 12$ $y = -4$
Graph the line.		Find the x-intercept by plugging in 0 for y.	$2x - 3(0) = 12$ $2x = 12$ $x = 6$
		Graph the line using those two points.	

Shading an Inequality in Standard Form

option 1	example	option 2	example
Convert it to slope intercept form. Be careful to flip the sign when you multiply or divide by a negative!	$2x - 3y > 12$ $-3y > -2x + 12$ $y < \frac{2}{3}x - 4$	Substitute any point not on the line for x and y. (0,0) is often the easiest choice!	$2(0) - 3(0) > 12$ $0 > 12$ ☹
Shade appropriately.	In this case we shade below because the sign is $<$.	Decide if that point made the inequality true or false.	$0 > 12$ is false.
		Shade where the point is if true. Shade where it isn't if false.	Shade on the side of the line opposite from where the point (0,0) is.

Statistics Terms

statistic	Definition. Does this statistic describe the shape, center or spread?	example
mean	The average. Calculated by adding all the values and dividing by the number of values.	
median	The middle number. Calculated by counting to the center. If there are even number of values, average the two numbers nearest the middle to get the median.	
mode	The most frequently occurring number. There can be more than one of these.	
range	A statement of the smallest to largest values.	
Quartile 1 (Q1)	The median of the first half of the data. Found by finding the median of all the values below the median.	
Quartile 3 (Q3)	The median of the second half of the data. Found by finding the median of all the values above the median.	
Inter-Quartile Range (IQR)	The range of values between Q1 and Q3. Found by subtracting Q1 from Q3.	
outlier	Values far from the median. Found in 4 steps: 1) Calculate the IQR. 2) Calculate 1.5 times the IQR. 3) Calculate Q1 minus that and Q3 plus that. Anything outside those two values is an outlier.	

The Normal Distribution, Mean and Standard Deviation

	describes	what it means	how to visualize it
roughly normal distribution	shape	A frequency plot which shows us a mean which occurs more commonly than the other outcomes and symmetrical tails showing frequencies decreasing as we go further from the mean.	A bell curve:
mean	center	The average value. At or near the peak of a roughly normal distribution.	A vertical line in the middle of the curve above.
standard deviation	spread	The typical deviation from the mean. Some values will be more frequent or others less frequent, but this is the typical deviation.	Two vertical lines on the curve above showing the values that are 1 above and 1 below the mean.

Samples and Statistics, Populations and Parameters

term	definition	precision level	example
sample	A preferably random selection of data from the population.		Identifying whether candies are green or not in a random selection of 20 candies.
statistic	A summary number that describes the sample.	An estimate. Unlikely to be perfectly precise or perfectly accurate.	The ratio of green candies in the sample.
population	All the data that could possibly be gathered.		Identifying all the candies in the world as green or not.
parameter	A summary number that describes the population.	perfectly precise and accurate.	The ratio of green candies in the population.

How to Find Mean and Standard Deviation with a Graphing Calculator

- Press STAT then select "Edit…"
- Type values into L_1. Repeated values must be repeated in the list.
- Press STAT then arrow right to see the CALC menu. Select "1-Var Stats."
- Confirm that the list is L_1. Arrow down to select "Calculate."
- The mean is $\overline{x}$. The standard deviation is σx.

How to Find the Line of Best Fit with a Graphing Calculator

- Press "2nd, 0" select DiagnosticsOn. Press ENTER. (You can skip this step from now on.)
- Press STAT then select "Edit…" to create list L_1 of the x-coordinates.
- Create list L_2 of the y-coordinates.
- Press STAT then arrow-right to CALC.
- Select "LinReg(ax+b)"
- Write down the values for a, b and r rounded to three significant figures.
- Press WINDOW to create a graph window and scale appropriately, like the graph in #11.
- PRESS "2nd, Y="
- Select "On," and make sure L_1 and L_2 are Xlist and Ylist.
- Press "Y="
- Type in "ax+b" using your values for a and b.
- Press GRAPH. You should see all the data points as well as the line of best fit.

Comparing Linear, Exponential and Quadratic Functions

<table>
<tr><td>type</td><td>linear</td><td>exponential</td><td>quadratic</td></tr>
<tr><td>function</td><td>$f(x) = 2x + 10$</td><td>$g(x) = 5(2)^x$</td><td>$h(x) = x^2 - 20$</td></tr>
<tr><td>graph</td><td> 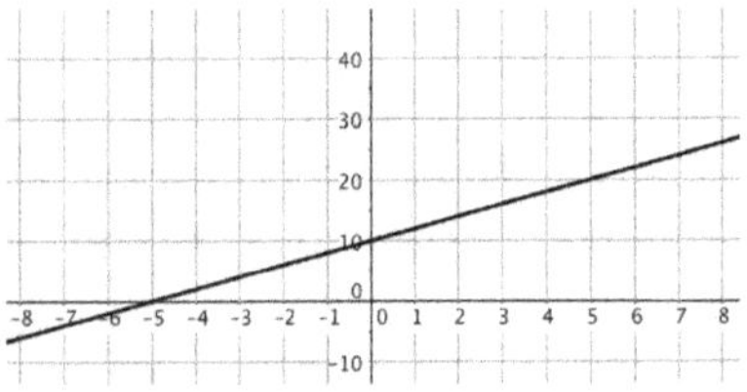</td><td> 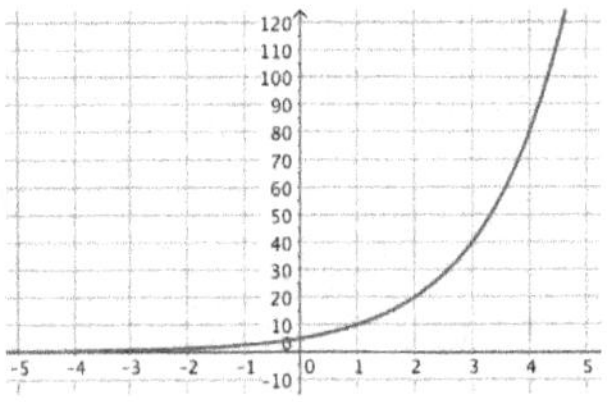</td><td> 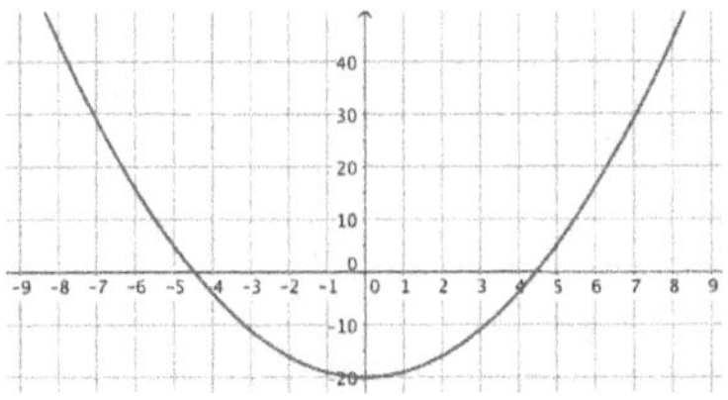</td></tr>
<tr><td>table</td>
<td><table><tr><th>x</th><th>y</th></tr><tr><td>-3</td><td>4</td></tr><tr><td>0</td><td>10</td></tr><tr><td>3</td><td>16</td></tr><tr><td>4</td><td>18</td></tr></table></td>
<td><table><tr><th>x</th><th>y</th></tr><tr><td>-1</td><td>2.5</td></tr><tr><td>0</td><td>5</td></tr><tr><td>1</td><td>10</td></tr><tr><td>2</td><td>20</td></tr></table></td>
<td><table><tr><th>x</th><th>y</th></tr><tr><td>-5</td><td>5</td></tr><tr><td>0</td><td>-20</td></tr><tr><td>5</td><td>5</td></tr><tr><td>10</td><td>80</td></tr></table></td></tr>
<tr><td>form</td><td>$y = mx + b$</td><td>$y = a(b)^x$</td><td>$y = ax^2 + bx + c$</td></tr>
<tr><td>in words</td><td>An initial value b has m added to it, x number of times.</td><td>An initial value a is multiplied by a constant b, x number of times.</td><td>Constants a, b, and c determine a polynomial consisting of a required quadratic term, and optional linear and constant terms.</td></tr>
<tr><td>example</td><td>Your brother puts $10 in a piggy bank and adds $2 every week.</td><td>The 5 starfish in a tank double every month.</td><td>A genie will square the number of candies you bring him, but keeps 20 of them before returning you the rest.</td></tr>
</table>

Arithmetic and Geometric Sequences

<table>
<tr><td>type</td><td>arithmetic sequence</td><td>geometric sequence</td></tr>
<tr><td>example scenario</td><td>Steph scores 3 points in his first game, and his points increase by 7 each game after that.</td><td>James scores 6 points in his first game, and his points double each game after that.</td></tr>
<tr><td>table for this example</td><td><table><tr><td>x</td><td>y</td></tr><tr><td>1</td><td></td></tr><tr><td>2</td><td></td></tr></table></td><td><table><tr><td>x</td><td>y</td></tr><tr><td>1</td><td></td></tr><tr><td>2</td><td></td></tr></table></td></tr>
<tr><td>function for this example</td><td>If he had scored 3 points in game 0:
$S(x) = mx + b$
$S(x) = 7x + 3$
Now horizontally shift that 1 unit, so that he scores 6 points in game 1, not game 0:
$S(x) =$ ____________</td><td>If he had scored 6 points in game 0:
$J(x) = a(b)^x$
$J(x) = 6(2)^x$
Now horizontally shift that 1 unit, so that he scores 6 points in game 1, not game 0:
$J(x) =$ ____________</td></tr>
<tr><td>sequence notation</td><td>$a_n = a_1 + (n-1)d$</td><td>$a_n = a_1(r)^{n-1}$</td></tr>
<tr><td>meaning of variables in sequence notation</td><td>Notice that the formula for a_n is the same as for $S(x)$ above, just slightly re-arranged with d in place of m and a_1 in place of b.

a_n is read “a sub n” and represents the nth term (or output) in the sequence.

a_1 is “a sub 1,” the first term in the sequence.

d is the common difference (or slope). It’s the amount that is added to a term to get the next one.

n is the term number (or input). In our example it’s the game number, starting with game 1.</td><td>Notice how this is the same as $J(x)$ above, with a_1 in place of a and r in place of b.

a_n is ____________________

$a_1$is ____________________

r is the common ratio (or multiplier). It’s the amount a term is __________ by to get the next one.

n is ____________________</td></tr>
<tr><td>example in sequence notation</td><td>$a_n =$ ____________________</td><td>$a_n =$ ____________________</td></tr>
</table>

Sums of Terms

Arithmetic	Geometric
$(a_1 + a_n) \cdot n = 2S_n$ or $S_n = \frac{(a_1+a_n)}{2} \cdot n$	$S_n = \dfrac{a_1(1-r^n)}{1-r}$

Courage To Core

Share the Discovery

NS1 Adding Fractions

Name(s)

Date Class/Period/Group

1) Below is a table you can use to record the birthdays of everyone in the class. Although you don't need to write in names for this exercise, there is space to do so if you wish.

January	February	March	April	May	June
July	August	September	October	November	December

2) What is the total number of birthdays recorded above? ____ How many of those birthdays are in April? ____ For your class, what **fraction** of birthdays are in April? ____ Use the space below to **reduce** this fraction. Then convert this number to a **decimal**. Then, write that decimal as a **percentage** rounded to tenths place.

3) What fraction of birthdays are in December? ____ Reduce, convert this to a decimal, then to a percentage rounded to tenths place.

4) What are the 3 months of summer? ___________, __________ and ___________. How many birthdays are in the summer? _______ What fraction of birthdays are in the summer? ____ Reduce, convert this to a decimal and a percentage rounded to tenths place.

5) What fraction of birthdays are in December **or** January? _________ The next question is very different from the previous one: What fraction of birthdays are in December **and** January? _____ Explain the meaning of "or" and "and" in the last two questions.

6) What fraction of the birthdays are:

on a day that is a prime number		on a day that is an even number	
on a day that is a whole number		on a day that is an odd number	
on a day that is a negative number		on the 5th day of the month	
within 30 days of today (looking forward)		within 30 days of today (looking forward and backward)	
the same day as your birthday (even if it's in a different month)		within 10 days of your birthday	
consecutive with another birthday		on the same day as another birthday	

7) Create another interesting fraction like the one's above. Describe the fraction and write its value below.

8) A chocolate bar needs to be divided up and shared between a group of 5 friends. What fraction does each friend get? _________

9) The division of a bar into 5 equal fractions can be represented visually as below. Write $\frac{1}{5}$ in each of the segments below.

10) Here is another **fraction bar**. Write the correct fraction in the segments.

11) Compare the two fraction bars above to answer the following question: Is $\frac{3}{5}$ greater or less than $\frac{1}{2}$?
______ How do you know?

12) Fraction bars can help you visualize a comparison between two fractions. The two bars below are divided into 10 segments. Shade $\frac{3}{5}$ of the first one, and $\frac{1}{2}$ of the second one. Which fraction is bigger and why? Use the word "tenths" in your explanation.

13) We can also use two fraction bars to add two fractions. Using fraction bars above, what is $\frac{3}{5}+\frac{1}{2}$? Write your answer as both an **improper fraction** and as a **mixed number**.

14) Fraction bars an also help you **reduce** fractions. For example, shade $\frac{4}{10}$ below. Then shade the equivalent amount in the second fraction bar. What fraction is represented by the second fraction bar?

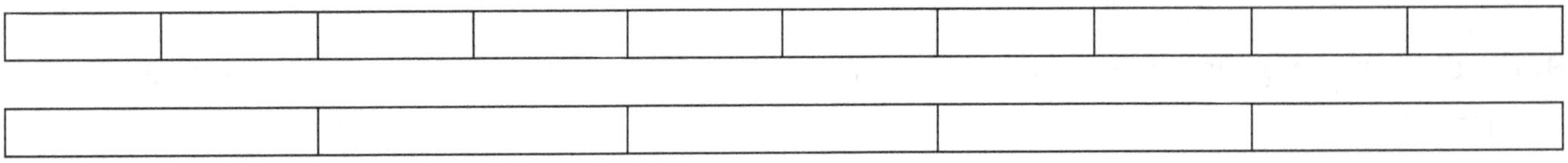

15) Now let's take a look at the parts of a fraction. Shade the fraction $\frac{1}{4}$ in each of the fraction bars below.

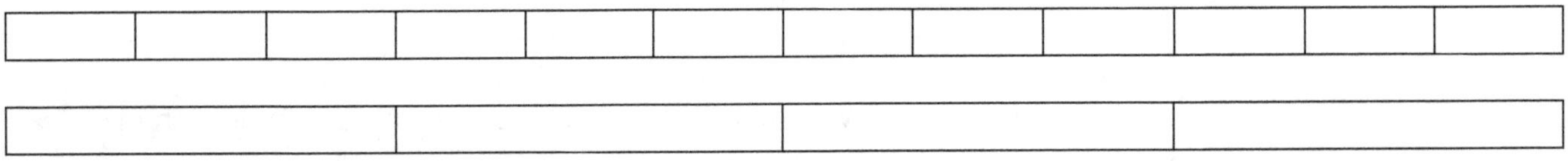

Use the above fraction bar to explain why $\frac{3}{12} = \frac{1}{4}$. Refer to the fraction bar as you describe the meaning of the **numerator** and **denominator** of the fractions $\frac{3}{12}$ and $\frac{1}{4}$.

16) Draw then shade two of your own fraction bars to compute $\frac{1}{2} + \frac{1}{3}$. Hint: How many segments in total should be in each fraction bar? Why?

17) In the last problem you found a **common denominator** between the two fractions and you **expanded** (the opposite of reducing) each fraction so you could easily add them. Compute $\frac{1}{2} + \frac{1}{3}$ again, this time by finding a common denominator without using fraction bars.

18) Below is a summary of the method of finding the **lowest common denominator** to compare or add fractions.

Comparing Fractions	Adding Fractions
$\frac{2}{3} \quad \frac{3}{5}$ LCD $= 15$ $\frac{2}{3}\left(\frac{5}{5}\right) \quad \frac{3}{5}\left(\frac{3}{3}\right)$ $\frac{10}{15} \quad \frac{9}{15}$ $\frac{10}{15} > \frac{9}{15}$	$\frac{3}{4} + \frac{5}{6}$ LCD $= 12$ (not $24!$) $\frac{3}{4}\left(\frac{3}{3}\right) + \frac{5}{6}\left(\frac{4}{4}\right) =$ $\frac{9}{12} + \frac{20}{12} =$ $\frac{29}{12} = 2\frac{5}{12}$

19) Often we will convert decimals or percentages to fractions in order allow us to compare them more effectively. Convert each of the following to fractions and reduce completely.

Decimal	Fraction	Percentage	Fraction
0.5		25%	
$0.\overline{3}$		60%	
1.5		10%	
$0.\overline{6}$		2%	
0.1		1%	
0.2		98%	

20) Compare the following pairs of numbers. Complete your comparison by using =, > or < . Show any work you did to change the way the fractions were represented.

$\frac{3}{8}$ $\frac{6}{16}$	$\frac{3}{8}$ $\frac{4}{8}$
$\frac{1}{2}$ 0.5	$\frac{21}{28}$ $\frac{3}{4}$
$\frac{150}{120}$ $\frac{45}{54}$	$\frac{56}{49}$ 1.3
14% $\frac{28}{200}$	$\frac{24}{27}$ $\frac{27}{24}$

21) Add or subtract the following fractions as indicated. Show your steps.

$\frac{5}{7}+\frac{3}{4}=$	$\frac{3}{14}+\frac{5}{7}=$
$\frac{5}{8}-\frac{3}{20}=$	$\frac{9}{28}-\frac{1}{21}=$
$0.5+\frac{3}{7}=$	$0.3+\frac{2}{11}=$
$2\frac{1}{3}+1\frac{2}{5}=$	$\frac{5}{10}-\frac{1000}{2000}=$

22) Add or subtract the following fractions as indicated. Show your steps.

$1\frac{6}{12}+3\frac{5}{15}=$	$\frac{21}{24}-3=$
$\frac{22}{11}-\frac{70}{49}=$	$\frac{0}{13}-\frac{24}{13}+1-\frac{1}{26}=$

23) Create an example of comparing two fractions with different denominators, and complete the comparison. Explain the steps in the process of comparing them.

24) Create an example of subtracting two mixed numbers with different denominators and complete the arithmetic. Explain the steps in the process of subtracting them.

The Take-Home Message: Comparing and adding or subtracting fractions is made easier by expanding both fractions so there is a common denominator. Reducing fractions requires identifying and canceling a common factor. Terminating or repeating decimals can be converted to fractions and vice versa.

Courage To Core

Share the Discovery

NS2 Multiplying Fractions

Name(s)

Date Class/Period/Group

1) Fraction bars are tools for visualization which can help us understand how to work with fractions. We can easily multiply fractions by whole numbers using fraction bars. The fraction bar below is divided into sevenths. Shade $\frac{3}{7}$.

2) If you double $\frac{3}{7}$, what is the result? _____ Doubling is of course the same as multiplying by 2. When you doubled $\frac{3}{7}$, what happened to the numerator? ____________ What happened to the denominator? __________________ Explain why.

3) Shade $\frac{4}{9}$ below. What is half of $\frac{4}{9}$? _____ Halving is of course the same as multiplying by $\frac{1}{2}$. When you halved $\frac{4}{9}$, what happened to the numerator? ______________ What happened to the denominator? _____________ Explain why.

4) In the previous example, we were able to compute $\frac{1}{2} \cdot \frac{4}{9} = \frac{2}{9}$ easily because the numerator was easily halved, but that's not always the case. For example, shade $\frac{3}{5}$ in the fraction bar below. Is it possible to shade exactly half of $\frac{3}{5}$ using this fraction bar? _____

Below is a fraction bar divided into tenths. Shade $\frac{3}{5}$ again. Is it true that $\frac{3}{5} = \frac{6}{10}$? ______

Now, to shade half of $\frac{3}{5}$, just shade half of $\frac{6}{10}$! Do so below. What is the result? _____

5) Shade $\frac{5}{6}$ below, then on the next fraction bar, shade half of $\frac{5}{6}$. What is the result? ________

6) Notice that when you double the number of segments in the fraction bar, the segments become half as big! So, when multiplying $\frac{1}{2} \cdot \frac{5}{6} = \frac{5}{12}$, the numerator (the number of segments shaded) stayed the same while the denominator (the number of segments in the entire fraction bar) was doubled. Use this idea to multiply $\frac{1}{3} \cdot \frac{4}{5}$. Fraction bars are provided below.

7) So far we've learned the following:

Arithmetic	Rule	Example
Multiplying a fraction by a whole number	*Multiply the numerator by the whole number.*	$3 \cdot \frac{2}{7} = \frac{6}{7}$
Multiplying a fraction by a fraction whose numerator is 1	*Multiply the denominators.*	$\frac{1}{3} \cdot \frac{2}{7} = \frac{2}{21}$

These two multiplications can be combined as in the following example.

$\frac{2}{3} \cdot \frac{5}{7} = \frac{10}{21}$ Rule for Multiplying Fractions: *Multiply numerators, multiply denominators.*

Use this rule to compute the following. Reduce your answer completely.

$\frac{3}{4} \cdot \frac{5}{11} =$	$5 \cdot \frac{2}{3} =$	$\frac{1}{5} \cdot \frac{2}{7} =$
$\frac{5}{2} \cdot \frac{2}{3} =$	$1\frac{2}{3} \cdot \frac{12}{7} =$	$0.5 \cdot \frac{9}{11} =$
$1.3 \cdot \frac{5}{7} =$	$\frac{6}{7} \cdot \frac{7}{6} =$	$\frac{2}{9} \cdot 9 =$

8) Unless you **cross-canceled** or reduced earlier in the process, some of your results in #7 involved big numbers which you had to reduce at the end. Complete each of the following computations and give a reduced answer. Notice that in the first one you can reduce each fraction right away!

$\frac{12}{3} \cdot \frac{14}{7} =$	$\frac{1}{3} \cdot \frac{12}{7} \cdot 14 =$	$\frac{12}{1} \cdot \frac{1}{7} \cdot \frac{14}{1} \cdot \frac{1}{3} =$

9) Did you get the same result for each of the above computations? ______ If not, try again! Because multiplication is commutative and associative we can re-arrange numerical expressions as shown above without changing the result. Try this for the computations below to make your arithmetic simpler. Note: It can be helpful to turn whole numbers into fractions here, for example $3 = \frac{3}{1}$.

$\frac{1}{3} \cdot \frac{5}{7} \cdot \frac{12}{10} \cdot 14 =$	$4 \cdot \frac{5}{6} \cdot \frac{1}{10} \cdot \frac{21}{8} =$	$\frac{12}{21} \cdot 7 \cdot \frac{1}{6} \cdot 3 =$

10) Cross-canceling is just a short-cut to this re-arranging. Instead of re-arranging then reducing fractions, we simply reduce straight-away as in the first example below. Complete the rest as by cross-canceling whenever possible.

$\frac{2}{3} \cdot \frac{12}{5} =$ $\frac{2}{1} \cdot \frac{4}{5} =$ $\frac{8}{5}$	$\frac{2}{3} \cdot \frac{3}{4} \cdot \frac{5}{6} =$	$\frac{12}{5} \cdot \frac{11}{4} \cdot \frac{10}{22} =$

11) Imagine you are advising your friend as to how to multiply fractions or fractions and whole numbers. What are 3 rules or tips you would provide him? Write them below with examples.

12) What is half of 6 ?____ What is half of $\frac{2}{3}$?____ What is a third of 12 ?____ All these questions can be thought of as numerical expressions involving multiplication. Re-write each of the following as numerical expressions then compute the result. Be sure to use the rules and tips you wrote above and give reduced answers.

What is half of $\frac{8}{9}$?	What is a third of $\frac{12}{13}$?	What is half of 7 ?	What is $\frac{3}{5}$ tripled?
What is $\frac{2}{3}$ of 42 ?	What is $\frac{3}{7}$ of $\frac{14}{27}$?	What is $\frac{5}{7}$ of $1\frac{1}{3}$?	What is half of $\frac{6}{7}$ tripled ?

13) Compute results for each of the following. Remember to use the rules and tips you established. Be sure to give reduced answers.

$\frac{2}{3} \cdot \frac{3}{4} =$	$\frac{3}{4} \cdot \frac{2}{3} =$
$\frac{12}{20} \cdot \frac{5}{3} =$	$\frac{32}{40} \cdot \frac{25}{20} =$
$1\frac{2}{13} \cdot 3\frac{5}{7} =$	$\frac{1000}{2}\left(2\frac{3}{1500}\right) =$
$\frac{9}{12} \cdot \frac{8}{7} =$	$\frac{56}{48} \cdot \frac{32}{12} =$
$\left(5\frac{4}{5}\right)\left(1\frac{2}{29}\right) =$	$\frac{81}{52} \cdot \frac{28}{27} =$

14) Compute results for each of the following. Remember to use the rules and tips you established. Be sure to give reduced answers.

$\frac{1}{2} \cdot \frac{2}{3} \cdot \frac{3}{4} \cdot \frac{5}{6} \cdot \frac{6}{7} =$	$\frac{100}{3} \cdot \frac{7}{17} \cdot \frac{3}{100} =$
What is $\frac{2}{3}$ of $\frac{5}{12}$?	What is $\frac{7}{8}$ of $\frac{16}{21}$?
What is $\frac{9}{24}$ of $\frac{20}{6}$?	What is $\frac{1}{2}$ of $\frac{1}{3}$ of 6 ?
Create two fractions which, when multiplied, return a result of 1.	Create two fractions which, when multiplied, return a result of 2.
Create and multiply two mixed numbers.	Create and multiply two improper fractions.

The Take-Home Message: Multiplying fractions requires multiplying numerators and denominators. Cross-canceling is a useful simplification tactic.

NS3 *Exploring Fractions Through Probability*

Name(s)

Date Class/Period/Group

1) A game at a fair involves tossing a ball towards a sea of boxes which are protected by bars and hoping it bounces off the bars luckily until it falls into the winning box. Draw a square below and divide it into 16 smaller squares. Shade one square as the winning square. If you toss a ball in the game, what do you think the chance is that you will win? Come to agreement in your group and explain your answer. Write your **probability** as a fraction, a decimal rounded to thousandths place, and as a percentage.

2) In the game above, imagine you tossed a ball 160 times. How many times would you expect to win? _________ Why? Are you guaranteed to win this number of times? Why or why not? What do you think the chance is that you wouldn't win at all? Explain your answers.

3) Take out 2 coins and give one to someone else in the group. Guess the probability that if you both flip your coins, they will both turn out to be heads. Come to agreement in your group and write your answer here: _________ Flip the coins 20 times and complete the table below.

Trial	1	2	3	4	5	6	7	8	9	10	11	12	13	14	15	16	17	18	19	20
coin 1																				
coin 2																				

4) Based on your experiment alone, what is the probability of getting two heads? Write your answer as a fraction and as a percentage. (Example: $\frac{7}{20} = \frac{35}{100} = 35\%$)

5) Rewrite your experimental probability for two heads here. ________ Do you think every group got the same result? ________ Why or why not?

6) Is the probability of getting two heads a fixed number or is it different for different people at different times? ________ Why or why not?

7) Is a coin toss an **experiment** whose outcome is truly random? _____ Why or why not?

8) If we assume a coin toss is an experiment whose outcome is random we can actually calculate the true probability of getting two heads. What do you think the true probability of getting two heads is? ______ Why? If we run the experiment 100 times, is this outcome guaranteed to happen that percent of the time? ______ Why or why not?

9) Sometimes it is useful to make a **tree diagram** to organize all the possible outcomes of a random experiment and determine probabilities. Each outcome is represented by a path from the beginning (2 un-flipped coins) to the end (both coins flipped). How many different outcomes are shown below? __________ List them here (first one is HH): ________________________

10) Follow each path to determine the probability of the desired end result. What fraction of times would you expect to get 1st coin heads? _____ And what fraction of those times would you expect to get 2nd coin heads? _____ Is the probability of each outcome, HH, HT, TH and TT, the same? _____ Why or why not?

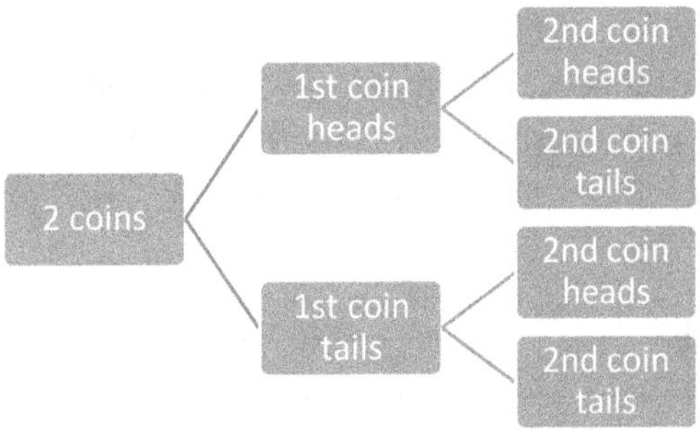

2 coins
1st coin heads
2nd coin heads
2nd coin tails
1st coin tails
2nd coin heads
2nd coin tails

11) You can calculate the probability of each final outcome by thinking of each of them occurring $\frac{1}{2}$ *of* $\frac{1}{2}$ of the time. That's $\frac{1}{2} \cdot \frac{1}{2} =$ _____ Give that answer in decimal and in percentage. ______ ______ Put the probabilities of each outcome next to each box at the end of the tree diagram.

12) What is the probability of two tails? _____ What is the probability of one or more tails? (Hint, the first or the second or the third outcome each have at least one tail!) _____ What is the probability of only one head? _____ What is the probability that the two coins are showing the same face? _____

13) On the previous page we saw that the probability of the first coin landing heads was $\frac{1}{2}$. The probability of the second coin landing heads was also $\frac{1}{2}$. Even if the first coin already came up heads it has no impact on the probability of heads for the second coin. This means the two **events** (first coin heads, second coin heads) are **independent**. The general rule for finding the probability of two independent events occurring is described as follows:

Event A: first coin heads
Event B: second coin heads

The probability of A and B both occurring is the probability of A times the probability of B.

$$P(A \text{ and } B) = P(A) \cdot P(B) = \frac{1}{2} \cdot \frac{1}{2} = \frac{1}{4}$$

Imagine an unfair coin which is weighted such that, when it lands on the table it is much more likely to come up tails rather than heads. You have two of these coins, and the probability that one of them when flipped will be heads is $\frac{3}{10}$. What is the probability that it will be tails? ____ Give the probabilities for each outcome below. Convert each to decimals rounded to hundredths place, then to percentages.

$P(H \text{ and } H) =$	$P(H \text{ and } T) =$	$P(T \text{ and } H) =$	$P(T \text{ and } T) =$

14) When you flip the two unfair coins above, what is the probability that they show different faces? _____ What is the probability that they show the same face? _____ Is this what you expected? _____ Why or why not?

15) Take out the two dice. What do you think the probability is of rolling a 12? _____ Experiment a bit and adjust your prediction. _____ What is the probability of just rolling a 6 with one die? ______ Use the rule for finding the probability of two independent events to calculate the chance of rolling two 6's (that is, the probability of rolling a 12).

16) Remember that when you were flipping coins you could get HT or TH and these were two different outcomes. (Think of them as a quarter and a penny, so clearly HT and TH will look different.) The same is true with dice. How many ways can you role an 11? _____ List them below. Then find the probability of rolling 11.

17) The above result give rise to the addition rule for probability. It applies to two events that are **mutually exclusive**, meaning either one or the other can happen but not both. (you can't get HT and TH for example, but you could get either one).

Event A: rolling 5 and 6
Event B: rolling 6 and 5

The probability of either A or B occurring is the probability of A plus the probability of B.

$$P(A \text{ or } B) = P(A) + P(B)$$

$$P(11) = P(5 \text{ and } 6) + P(6 \text{ and } 5) = \_\_\_\_\_\_$$

18) What is the probability of rolling a 7? (List each way to roll a 7, find the probability of each, then add them up...)

19) If you roll a six sided dice, what is the probability that you roll a 5? Write your answer as a fraction, a decimal rounded to thousands place, and as a percentage rounded to tenths place.

20) If you roll two six sided dice, what is the probability of rolling a 3?

21) If you flip three coins, what are all the possible outcomes? List them below—the first four are given to you. What is the probability of getting an outcome of THH? What is the probability of getting 2 Heads and 1 Tail in any order?

HHH THH
HTH
HHT

22) Your teacher will provide you with colored candies or beads. Based on your sample, what is the probability that if you select one candy at random, it is green? Give your answer as a fraction. Convert this to a decimal rounded to thousandths. Give your answer as a percentage rounded to tenths.

23) If a bag contains 15 green candies and 10 red candies, what is the probability that if you take out one candy it will be red? Convert to decimal rounded to thousands and percentage rounded to tenths.

24) If a bag contains 10 green candies, 10 blue candies and 15 red candies, what is the probability that if you take out one candy it will be red? Convert to decimal rounded to thousands and percentage rounded to tenths.

25) Suppose that for a particular brand of candy, the fraction of candies which are yellow in every bag is $\frac{1}{7}$. If I have two bags of candy, what is the probability that if I take one candy from each bag, they are both yellow?

26) Suppose I flip a coin and role a die. What is the probability that the coin comes up heads and the die shows 5 ?

27) Suppose I roll two dice. What are all the different ways I can get a sum of 10? What is the probability of getting a sum of 10?

28) Write and explain the rule for finding the probability of two independent events both occurring and give an example. Write and explain the addition rule for probability and give an example.

The Take-Home Message: The probability of an event can be expressed as a fraction. The probability of two independent events both occurring is the product of the probability of each one. The probability of either of two mutually exclusive events occurring is the sum of the probability of each.

NS4 *Rational Numbers*

Name(s)

Date Class/Period/Group

1) Write a definition and example for each of the following terms. Then complete the **Venn Diagram** to organize the terms correctly.

integers

rational numbers

natural numbers

irrational numbers

real numbers

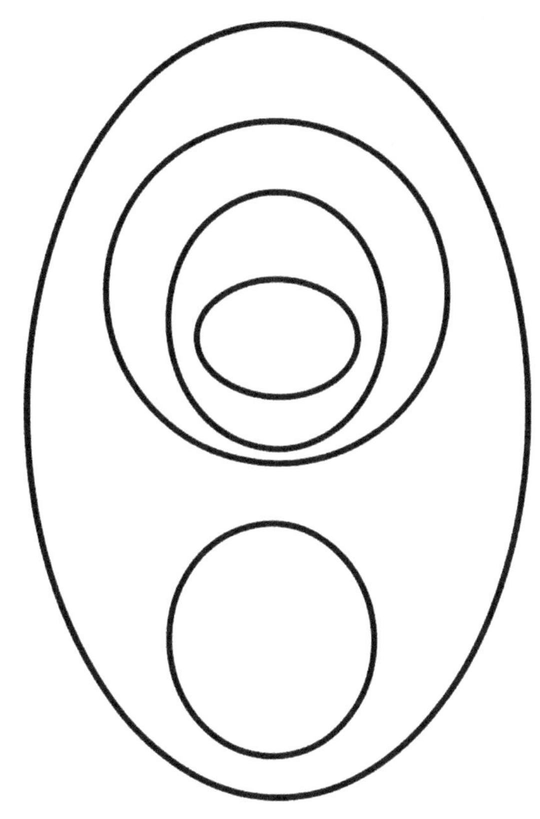

2) Use long division to convert the following fractions to decimals. Indicate repeating decimals like this: $0.\overline{34}$, which means $0.34\ldots$, which means the 34 repeats forever!

$\frac{4}{5}$

$\frac{5}{6}$

$\frac{6}{7}$

$\frac{7}{12}$

3) Below is a procedure for converting a number with a repeating decimal into a fraction! This procedure requires that you create and subtract two equations from each other, then solve for x. Every number with a repeating decimal can be converted like this, which means every number with a repeating decimal is a rational number! Put an asterisk next to the most important step you'll need to remember in each procedure.

$0.\overline{3}$ $0.\overline{3} = x$ (set the number equal to x) $3.\overline{3} = 10x$ (multiplying both sides by 10) $0.\overline{3} = x$ (re-written from above) $3 = 9x$ (subtracting the two equations) $x = \frac{3}{9} = \frac{1}{3}$ (solving for x)	$7.\overline{45}$ $0.\overline{45}$ (ignore the 7 for now) $0.\overline{45} = x$ $45.\overline{45} = 100x$ $0.\overline{45} = x$ $45 = 99$ $x = \frac{45}{99}$ $7\frac{45}{99}$ (put the 7 back in the answer)

4) Convert each of the following to fractions using the above method.

$0.\overline{5}$

$0.\overline{52}$

$0.\overline{12}$

$0.\overline{123}$

5) Compare each of the following pairs of numbers. Put a <, >, or = between them. You may need to convert between decimals and fractions in order to compare more effectively. Refer to the previous page to see how to do this, and convert when needed.

$0.5 \quad 0.\overline{5}$	$0.09 \quad 0.1$	$0.12 \quad 0.120$	$0.\overline{34} \quad 0.\overline{35}$
$\frac{1}{3} \quad 0.33$	$\frac{2}{3} \quad 0.\overline{6}$	$\frac{1}{5} \quad 0.\overline{2}$	$\frac{3}{10} \quad 0.33$
$\frac{6}{7} \quad 0.67$	$\frac{23}{100} \quad 0.23$	$\frac{456}{1000} \quad 0.\overline{456}$	$2 \quad 2.01$
$2 \quad 1.99$	$-2 \quad -2.01$	$-2 \quad -1.99$	$\frac{2}{3} \quad \frac{3}{4}$
$\frac{5}{7} \quad \frac{5}{8}$	$\frac{13}{14} \quad \frac{14}{14}$	$\frac{5}{3} \quad 2$	$\frac{17}{18} \quad \frac{18}{19}$
$\frac{48}{56} \quad \frac{6}{7}$	$0.\overline{18} \quad \frac{2}{11}$	$100 \quad 10^2$	$\frac{1}{100} \quad 10^{-2}$

6) Place the following numbers in the correct order and in the most precise location possible below the number line below.

$0.\overline{1} \quad \frac{1}{5} \quad -\frac{1}{5} \quad -\frac{2}{3} \quad 0.\overline{5} \quad -\frac{99}{100} \quad \frac{1}{100} \quad -0.\overline{3} \quad .\overline{34} \quad \frac{3}{5} \quad 0.9 \quad -\frac{4}{11} \quad \frac{78}{100} \quad 0.\overline{2}$

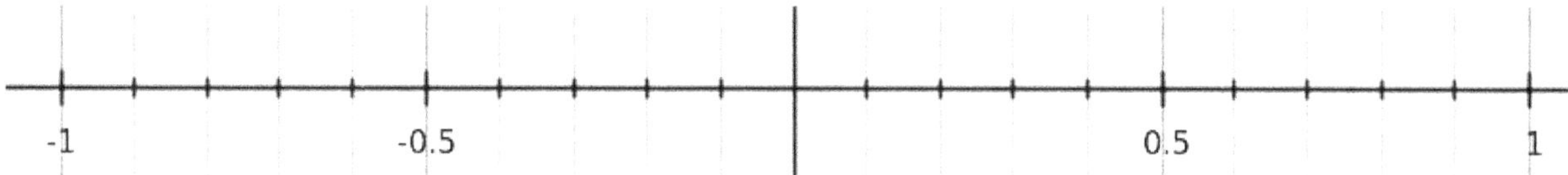

7) Add, subtract or multiply the following rational numbers:

$\frac{1}{2}+\frac{1}{3}$	$0.1-0.11$
$1.23-1.2\overline{3}$	$3.2-\frac{7}{3}$
$3.14+3.\overline{14}$	$0.\overline{1}+0.\overline{2}$
$\frac{11}{12}-\frac{12}{11}$	$\frac{11}{12}\cdot\frac{12}{11}$
$2\frac{2}{3}-2\frac{1}{7}$	$\frac{49}{25}\cdot\frac{100}{21}$
$\frac{1}{10}(0.1)$	$\left(\frac{1}{10}\right)0.01$

8) Write out sentences describing a procedure for adding two mixed numbers. Use the example problem below to support your explanation.

$2\frac{1}{3} + 7\frac{3}{5} =$

9) Write out sentences describing a procedure for multiplying two mixed numbers which requires reducing. You may give an example to support your explanation.

$\left(3\frac{1}{4}\right) \cdot \left(2\frac{5}{6}\right) =$

10) Try using your calculator to compute the decimal versions of the following numbers. If a repeating decimal is beyond the capacity of your calculator screen to show, use http://keisan.casio.com/calculator to compute to more decimal places. You can change the number of digits of the result by changing the number in the "digit" box, and you get the result by pressing the "execute" button. The first one is given to you.

$\frac{8}{7} = 1.\overline{142857}$	$\frac{22}{7}$
$\frac{5}{11}$	$\frac{5}{12}$
$\frac{5}{13}$	$\frac{1}{17}$
$\frac{1}{18}$	$\frac{1}{19}$
$\frac{1}{20}$	$\frac{1}{21}$
$\frac{1}{22}$	$\frac{1}{23}$
$\sqrt{2}$ (Do you think this has a **non-repeating** decimal?)	$\sqrt{3}$ (How about this one?)
π (You can find this number on your calculator and on the web calculator I gave you. Do you think it has a non-repeating decimal?)	e (Find this number on the web calculator, and make a guess if you think it repeats or not.)

11) True or False. Put an example or counter-example at right.

Statement	T/F	Example
Every number which has a repeating decimal is a rational number.		
Every fraction which is a ratio of two non-0 integers is a rational number.		
Every fraction which is a ratio of two non-0 integers has a repeating decimal.		
Some fractions which are the ratio of two integers have **terminating decimals.**		
All integers are rational numbers.		
All rational numbers are integers.		
All irrational numbers have non-repeating decimals.		
0 divided by any non-0 integer is a rational number.		
Any non-0 integer divided by 0 is 0.		
There are real numbers which aren't either rational or irrational.		

12) Define a rational number. Write three characteristics of rational numbers. Give examples as needed.

13) Define an irrational number. Write three characteristics of irrational numbers. Give examples as needed.

14) Can a number be both irrational and rational? Why or why not?

15) The rational numbers and the irrational numbers together are the real numbers. Are there any other kinds of real numbers? _____ Are there any numbers that aren't real? _____ Describe them!

16) Rewrite each of the following numbers in the circle for the most specific descriptor. Sometimes you may need to simplify first in order to determine correct placement.

$\frac{3}{10}$	0.1	$0.\overline{78}$	5	4.3	$2\frac{1}{3}$
$\frac{5}{1}$	$\frac{0}{6}$	$\sqrt{2}$	$\frac{\sqrt{2}}{3}$	$\sqrt{-3}$	$\sqrt{25}$
$\frac{27}{27}$	5^3	$(\sqrt{6})^2$	$\frac{5}{9}$	$\frac{45}{9}$	$\frac{\sqrt{36}}{2}$

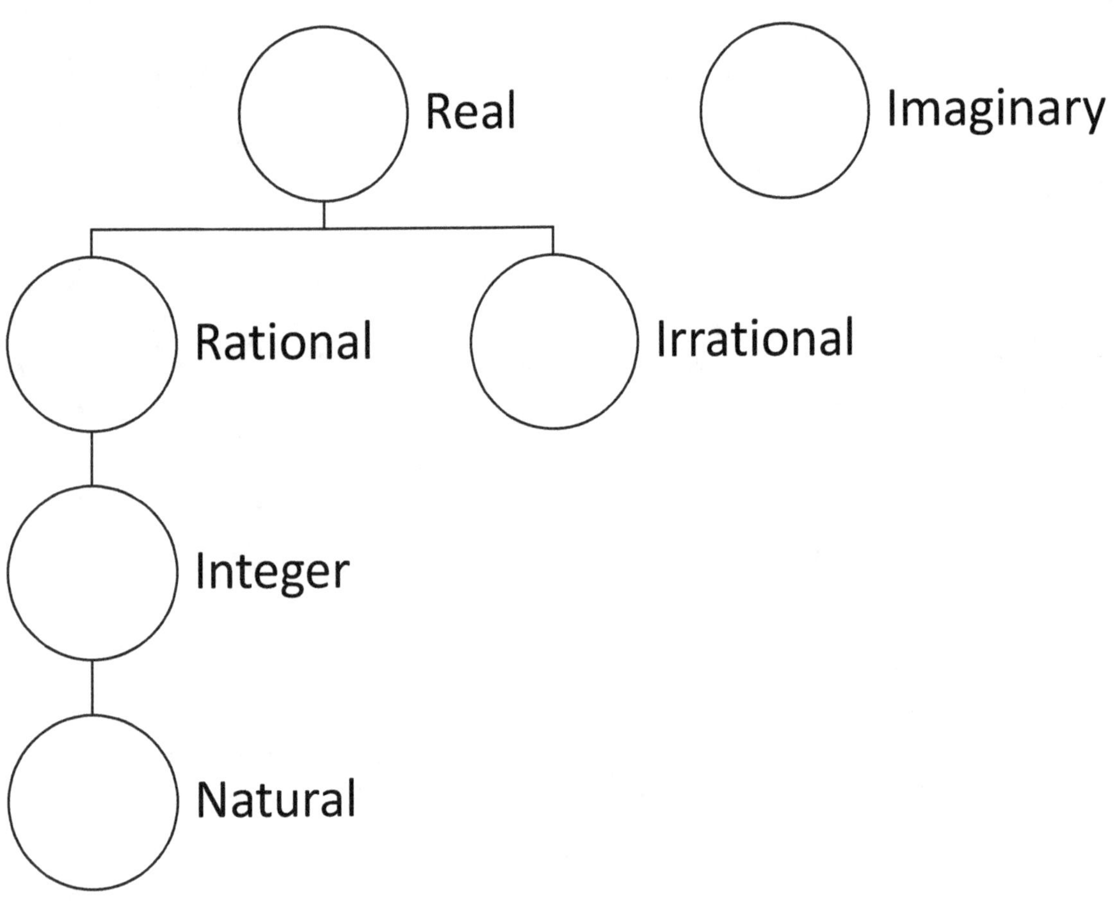

The Take-Home Message: Rational numbers are ratios of two integers, and as decimals they can be terminating or repeating. Irrational numbers like $\sqrt{2}$ and π have infinite non-repeating decimals.

NS5 Dividing Fractions Using Reciprocals

Name(s)

Date Class/Period/Group

1) Warm up by adding these fractions. What did you do to each fraction to create a common denominator? Does this step actually change each fraction or simply create an **equivalent fraction**?

$\frac{2}{12}+\frac{3}{7}=$	$\frac{15}{2}+\frac{10}{3}=$

2) **Reducing** or **expanding** fractions is simply multiplying the numerator and denominator of a fraction by the same number in order to create an equivalent fraction. Complete the following table to understand this. The first one is done for you.

Fraction	Multiply by factor of	What this looks like	Equivalent Result
$\frac{2}{3}$	5	$\frac{2}{3}\cdot\frac{5}{5}$	$\frac{10}{15}$ (expanded)
$\frac{3}{4}$	7		
$\frac{8}{10}$	$\frac{1}{2}$	$\frac{8}{10}\cdot\frac{\left(\frac{1}{2}\right)}{\left(\frac{1}{2}\right)}$	
$\frac{9}{12}$	$\frac{1}{3}$		
$\frac{\left(\frac{2}{3}\right)}{\left(\frac{4}{5}\right)}$	$\frac{5}{4}$	$\frac{\left(\frac{2}{3}\right)}{\left(\frac{4}{5}\right)}\cdot\frac{\left(\frac{5}{4}\right)}{\left(\frac{5}{4}\right)}$	
$\frac{2}{\left(\frac{6}{7}\right)}$	7	$\frac{2}{\left(\frac{6}{7}\right)}\cdot\frac{7}{7}=$	
$\frac{\left(\frac{10}{9}\right)}{5}$	$\frac{1}{5}$	$\frac{\left(\frac{10}{9}\right)}{5}\cdot\frac{\left(\frac{1}{5}\right)}{\left(\frac{1}{5}\right)}$	

3) Dividing fractions is accomplished by multiplying the numerator and denominator of the overall fraction in order to create a simpler equivalent fraction. We multiply by the **reciprocal** of the denominator. Complete the table below. The first one is completed for you as an example.

Fraction	Multiply by this reciprocal	What this looks like	Equivalent Result
$\frac{\left(\frac{2}{3}\right)}{\left(\frac{1}{3}\right)}$	$\frac{3}{1}$	$\frac{\left(\frac{2}{3}\right)\left(\frac{3}{1}\right)}{\left(\frac{1}{3}\right)\left(\frac{3}{1}\right)}$	2
$\frac{\left(\frac{4}{7}\right)}{\left(\frac{3}{7}\right)}$			
$\frac{\left(\frac{3}{5}\right)}{\left(\frac{1}{10}\right)}$			
$\frac{2}{\left(\frac{6}{7}\right)}$			
$\frac{\left(\frac{5}{8}\right)}{15}$			

4) Define reciprocal and give an example.

5) Simplify the following.

$\frac{\left(\frac{2}{3}\right)}{2}$	$\frac{\left(\frac{3}{4}\right)}{5}$
$\frac{5}{\left(\frac{6}{7}\right)}$	$\frac{\left(\frac{2}{3}\right)}{\left(\frac{\left(\frac{3}{4}\right)}{5}\right)}$

6) Simplify the following.

$\frac{0}{1}$	$\frac{0}{2}$
$(-3)(-2)$	$\left(-\frac{6}{2}\right)\cdot\left(-\frac{8}{4}\right)$
$\frac{\left(\frac{1}{2}\right)}{3}$	$\frac{1}{\left(\frac{2}{3}\right)}$
$-\frac{\left(\frac{3}{4}\right)}{5}$	$\frac{\left(\frac{-3}{4}\right)}{5}$
$\frac{-\left(\frac{5}{-6}\right)}{7}$	$-\frac{5}{\left(-\left(\frac{6}{-7}\right)\right)}$
$\frac{\left(\frac{7}{8}\right)}{\left(\frac{9}{10}\right)}$	$\frac{\left(\frac{12}{13}\right)}{14}$
$\frac{\left(\frac{12}{13}\right)}{12}$	$\frac{\left(\frac{1}{87}\right)}{\left(\frac{2}{87}\right)}$
$\frac{\left(\frac{100}{200}\right)}{\left(\frac{300}{400}\right)}$	$\frac{\left(\frac{6}{36}\right)}{360}$

7) Simplify the following. Hint: Start with the innermost fraction first. For example, to simplify the first fraction, write and simplify $\frac{3}{\left(\frac{4}{5}\right)}$ first, then re-write the overall fraction and continue.

$\dfrac{\frac{1}{2}}{\left(\frac{3}{\left(\frac{4}{5}\right)}\right)}$	$\dfrac{5}{\left(\frac{4}{\left(\frac{3}{\left(\frac{2}{1}\right)}\right)}\right)}$
$\dfrac{\left(\frac{\left(\frac{\left(\frac{1}{2}\right)}{3}\right)}{4}\right)}{5}$	$\dfrac{\left(\frac{\left(\frac{\left(\frac{5}{4}\right)}{3}\right)}{2}\right)}{1}$

The Take-Home Message: Dividing fractions is really just a fraction of fractions, expanded by the reciprocal of the denominator to convert the denominator to 1.

NS6 *Exponential and Scientific Notation*

Name(s)

Date Class/Period/Group

1) Imagine you are comparing the mass of the average adult and the mass of a 1 kg pizza. Which of the following statements is a more useful comparison, and why? Explain.

The average adult weighs 79 kg more than a medium pizza.
The average adult is 80 times as massive as a medium pizza.

2) The first statement above gives the **absolute** difference between the mass of the pizza and the person. The second gives the **relative** difference. How is each calculated? When is the absolute difference more useful and when is relative distance more useful and why? Give examples.

3) How many times as big is the second item compared to the first? The first answer is given to you in bold. Note that you are not expected to be sure about your answer—make some reasonable guesses.

Second compared to the first	How many times as big		
ant height—human height	65	**650**	6500
human mass—dinosaur mass	5.66	56.6	566
average 2 story house height—tallest skyscraper height	138	1380	13800
Usain Bolt top speed—cheetah top speed	2.13	21.3	213
housecat daily calories—human daily calories	1	10	100
fastest bicycle speed—passenger jet cruising speed	6.72	67.2	672
tallest tree in the world—tallest building in Madrid	2.15	21.5	215
population of Spain—population of US	0.68	6.8	68
total road length Spain—total road length US	0.96	9.6	96
American egg consumption per person—French egg consumption	0.97	9.7	97
walking speed in San Diego—walking speed in Dubai	0.729	7.29	72.9

4) An ichthyologist studies fish. The world's smallest fish is a carp which lives in the swamps of the island of Sumatra in Indonesia. It measures only 7.5 mm. The world's largest fish is a whale shark, whose mouth is bigger than you are and which measures 9.5 m in length. How long is the whale shark in mm?

5) What is the absolute difference between these two fish's lengths?

6) How would you calculate the relative difference in size between these two fish's lengths? Do so and round your answer to a whole number. Your answer for relative difference means that you would need to line up that many carp to reach from tail to nose of a whale shark! Can you visualize this? Which is a more useful comparison here: relative or absolute difference?

7) Before we move on, a little practice with exponents. After you complete these, define the terms **base** and **exponent**. Write 3 rules for simplifying exponential expressions like these.

$4 \cdot 4 \cdot 4 =$	$5^3 \cdot 5^4 =$	$\frac{6^5}{6^2} =$	$\frac{6^2}{6^5} =$	$250 \cdot 5^3 =$	$250 \cdot \frac{1}{5^3} =$

8) Consider the following story set in 2016. An ichthyologist named July discovers several species of fish in the Mediterranean. She first discovers fish A, which has a length of only 1 mm! Every fish she discovers after that, coincidentally, is 4 times as big as the previously discovered fish. At right is the table of her observations. Complete the table for her.

9) July wonders, how many times as big is fish F compared fish D?

If D is $1 \cdot 4^3$
and F is $1 \cdot 4^5$
then Fish D times 4^2 *gives you Fish F:* $1 \cdot 4^3 \cdot 4^2 = 1 \cdot 4^5$

So that means Fish F is 4^2 *times as big as Fish D.*

Fish	Length (mm)	Length (mm)
A	1	1
B	$1 \cdot 4$	4
C	$1 \cdot 4^2$	16
D	$1 \cdot 4^3$	64
E		
F		
G		

10) How many times as big is...
...fish D compared to fish A? ___________
...fish B compared to fish B? ___________
...fish G compared to fish D? ___________
...fish E compared to fish C? ___________
...fish C compared to fish F? ___________

11) On the previous page we used **exponential notation** to help us more easily represent how much bigger one fish was compared to another. You may have noticed in the last problem that you can use this notation to represent how much smaller a fish is compared with another as well by writing a fraction. For example, if Jenny is 4^3 times as tall as a penny standing on edge, the penny is $\frac{1}{4^3}$ times as tall as Jenny! Complete the following table by following the example in the first 3 rows. Explain how your answer to the last row.

Fish A (kg)	is ____ times as big as	Fish B (kg)
12	2	6
$1 \cdot 6^8$	6^4	$1 \cdot 6^2$
$3 \cdot 7^6$	$\frac{1}{7^4}$	$3 \cdot 7^{10}$
$2 \cdot 10^{11}$		$2 \cdot 10^6$
$3 \cdot 2^{20}$		$3 \cdot 2^5$
$6 \cdot 11^4$		$6 \cdot 11^8$

12) July discovers another fish. She names it the Golden Bubblefish, because it has a normal mass of 1 kg, but every time it is fed a candy it grows 6 times as big! She feeds it 4 candies. Write its new mass here using exponential notation: __________ Amazingly, if she feeds the Golden Bubblefish 1 peanut, it shrinks by the same factor! For every peanut it becomes $\frac{1}{6}$ times as big! She continues to experiment with the fish. Complete the table.

Stage	Feed	Previous mass multiplied by	New Mass (kg)
1	4 candies	6^4	$1 \cdot 6^4$
2	1 peanut	$\frac{1}{6}$	$1 \cdot 6^3$
3	3 candies		
4	3 peanuts		
5	4 candies		
6	6 peanuts		
7	1 candy		

13) Exponential notation is quite powerful! Here are some important concepts to understand:

Exponential Notation

If you want to multiply 7 by 6 forty times, you write $7 \cdot 6^{40}$. *The exponent counts the number of times you are multiplying by 6.*
If you want to divide 7 by 6 forty times, you write $7 \cdot \frac{1}{6^{40}}$. *The exponent in this case counts the number of times you are dividing by 6.*

14) We will use the Golden Bubblefish whose initial mass is 1 kg one more time to understand how to increase or decrease the relative size of a number. Complete the table for the new numbers of candies and peanuts.

Stage	Feed	Previous mass multiplied by	New Mass (kg)
1	5 candies	6^5	$1 \cdot 6^5$
2	2 peanuts		
3	2 candy		
4	1 peanuts		
5	5 candies		
6	5 peanuts		
7	1 candy		

15) If you look at the masses on the right of the table, you will see that when the fish ate a number of peanuts, the exponent decreased by that number. For example, in stage 2 the fish ate two peanuts and the mass went from $1 \cdot 6^5$ to $1 \cdot 6^3$. There are actually two ways to accomplish this change:

$1 \cdot 6^5 \cdot \frac{1}{6^2} = 1 \cdot 6^3$	*Here we divided by 6^2, removing two of the 6's multiplied in $1 \cdot 6^5$.*
$1 \cdot 6^5 \cdot 6^{-2} = 1 \cdot 6^3$	*Here we multiplied by 6^{-2}, which represents the same thing as dividing by 6^2 . You can confirm this by using the rule for multiplying exponential expressions with the same base like this example: $5^4 \cdot 5^7 = 5^{11}$. This works for negative exponents too: $3^7 \cdot 3^{-5} = 3^2$.*

16) A paper clip has a mass of about 1 gram. A liter of water has a mass of a kilogram. (For perspective, 3 cans of soda are about a liter.) How many times as massive is a kilogram than a gram? Write your answer in exponential notation.

17) Tricky question: How many times as massive is a kilogram than a kilogram? Write your answer as "10" to an exponent. (Hint: How many tens are you multiplying by?)

18) How many times as massive is a gram than a kilogram? (Write your answer as "10" to a negative exponent.)

19) Complete the table at right. What's bigger, and how many times bigger (answer in exponential form)?

1st	2nd	What's bigger?	By how many times?
$1 \cdot 6^3$	$1 \cdot 6^4$	2nd	6^1
$1 \cdot 10^2$	$1 \cdot 10^5$		
$3 \cdot 10^1$	$3 \cdot 10^4$		
4	$4 \cdot 10^9$		
$8 \cdot 7^{-2}$	$8 \cdot 7^3$		7^5
$9 \cdot \frac{1}{7^3}$	$9 \cdot \frac{1}{7^4}$		
$2 \cdot 9^{-1}$	$2 \cdot 9^{-3}$		
$6 \cdot \frac{1}{10}$	$6 \cdot 10$		
$7 \cdot 10^0$	$7 \cdot 10^{-1}$		
$3 \cdot 1000$	$3 \cdot 100$		
$3 \cdot 0.01$	$3 \cdot 0.1$		
$3 \cdot 10^4$	$3 \cdot 10^{-2}$		

20) You have practiced comparing two numbers represented with exponential notation. In many cases, the numbers were in a specific kind of exponential notation called **scientific notation**, which uses base 10. Sometimes you compared a number with a **base unit**, for example, when you compared a kilogram with a gram and observed that a kilogram is 10^3 times as big as a gram. The metric system makes it very easy for us not only to compare numbers, but also to convert from one unit to another. Complete the conversions below.

(Hint: When converting, ask yourself one question: Are you converting to units that are bigger or smaller? Converting to bigger units means you have *less* of them, and converting to something smaller means you have *more* of them!)

Amount	Units	Amount	Units	Why?
$4 \cdot 10^9$	m	$4 \cdot 10^{12}$	mm	Millimeters are smaller than meters. There are 10^3 times as many millimeters as meters here, so multiply by 10^3 .
$2 \cdot 10^7$	g	$2 \cdot 10^4$	kg	Kilograms are bigger than grams. There are $\frac{1}{10^3}$ times as many kilograms as grams here, so multiply by $\frac{1}{10^3}$.
$3 \cdot 10^1$	l		ml	
$7 \cdot 10^5$	kN		N	
8	cm		mm	
$3 \cdot 10^{12}$	nm		m	

21) Often, when working in scientific notation, we must pay attention to **significant figures**. In order to understand significant figures, we need to do some measuring. Take out your ruler and measure the width of another group member's open hand, from tip of pinky to tip of thumb. Have each person measure the hand of everyone else (unless you are cheating you should get slightly different answers!) and complete the chart.

Name	measure 1	measure 2	measure 3	measure 4	average

22) How **precise** are your measurements? In other words, are your measurements trustworthy to the nearest 1 cm? ______ the nearest 5 mm's? _______The nearest 2 mm? ______ The nearest mm? ________ Based on the average above and your assessment, give the most precise measure for your hand that you can below. Count the number of digits you wrote (the number of significant figures) and write it here: ______

23) Each of the different hand measurements below represents a different level of precision. Complete the table indicating the number of significant figures, representing the level of precision for the measurement. Some are completed for you as examples.

Measurement	Significant Figures	Measurement	Significant Figures
12 cm	2	11.3 cm	3
11.72 cm		12.0 cm	
12.561 cm		11.90 cm	

24) Below are five measurements of the speed of a car made with a radar gun. In field conditions (on highways), the gun is precise down to one-tenth of a km/hr but the display shows more digits. Round each of the following to tenths of a km/hr. After rounding, state the number of significant figures. Significant figures are the digits with trustworthy information. Some are completed for you a examples.

Car	Speed (km/hr)	Rounded Speed	Significant Figures
Fiat Panda	25.678	25.7	3
Seat Ibiza	4.25		
Mini	120.023	120.0	4
BMW	238.049		
Audi	87.000		
Porche	$2.3947 \cdot 10^2$		

25) The last problem on the previous page presented you with a number in scientific notation. In order to write a number in scientific notation, we write it with one number to the left of the decimal point. Convert each of the following to scientific notation with 2 significant figures. The first four are done for you as examples.

Number	Scientific Notation	Number	Scientific Notation
300	$3.0 \cdot 10^2$	349	$3.5 \cdot 10^2$
0.43	$4.3 \cdot 10^{-1}$	0.0043	$4.3 \cdot 10^{-3}$
54.6		7812	
0.067		6571.3	
$\frac{1}{10}$		$\frac{1}{100}$	
$\frac{2}{1000}$		$\frac{3}{10^5}$	

26) Compare the two numbers and put >, < or = as appropriate. In the cases where one number is larger than another, tell how many times as big the second number is as compared to the first. The first three problems are done for you. Sometimes it may be helpful to convert numbers to scientific notation.

$2 \cdot 7^2 < 2 \cdot 7^5$ The 2nd number is 7^3 times as big.	$4 \cdot 10^7 > 4 \cdot 10^4$ The 2nd number is $\frac{1}{10^3}$ times as big.	$0.0007 = 7 \cdot 10^{-4}$
$3 \cdot 10^4 \quad 3 \cdot 10^5$	$7 \cdot 12^{-7} \quad 7 \cdot 12^{-9}$	$6 \cdot 2^{-3} \quad 6 \cdot 2^{-7}$
$\frac{1}{1000} \quad \frac{1}{10000}$	$\frac{1}{100} \quad 10^{-2}$	$0.1 \quad \frac{1}{10}$
$\frac{1}{7^5} \quad 7^{-6}$	$3 \cdot 10^1 \quad 30$	$3 \cdot 10^{-1} \quad 0.3$
$17 \cdot 6^{-4} \quad 17 \cdot 6^{10}$	$6^{-2} \quad 6^2$	$0.003 \quad 3 \cdot 10^{-3}$
$0.03 \quad 3 \cdot 10^{-2}$	$0.3 \quad 3 \cdot 10^{-1}$	$3 \quad 3 \cdot 10^0$

27) Complete the following computations involving numbers with two significant figures, then give your answer in scientific notation to 2 significant figures. Note the use of the commutative and associative properties of multiplication to ease the simplification.

$0.10 \cdot 0.030 =$ $(1.0 \cdot 10^{-1}) \cdot (3.0 \cdot 10^{-2}) =$ $1.0 \cdot 3.0 \cdot 10^{-1} \cdot 10^{-2}$ $3.0 \cdot 10^{-3}$	$320 \cdot 1100 =$ $3.2 \cdot 10^{2} \cdot 1.1 \cdot 10^{3} =$ $3.52 \cdot 10^{5} =$ $3.5 \cdot 10^{5}$
$0.010 \cdot 0.010 =$	$0.12 \cdot 0.12 =$
$(1.5 \cdot 10^{4})(2.0 \cdot 10^{-3}) =$	$(2.1 \cdot 10^{3}) \cdot (0.00010) =$

28) Summarize how to compare the relative size of two numbers in exponential notation, with examples. Include an example which involves negative exponents.

29) Summarize how to convert numbers to scientific notation using significant figures, with examples.

The Take-Home Message: Expressing relative differences can be accomplished with exponential notation, or specifically with scientific notation, which is particularly helpful in unit conversion. Significant figures express the level of precision of a measurement.

Courage To Core

Share the Discovery

NS7 *Roots and Rationalizing Denominators*

Name(s)

Date Class/Period/Group

1) **Rooting** is simply the inverse of raising to an integer power. The $\sqrt{\ }$ symbol (also $\sqrt[3]{\ }$, $\sqrt[4]{\ }$, etc.) is called a **radical.** Simplify each row, comparing and contrasting the expressions.

$3^2 =$	$\sqrt{9} =$
$4^2 =$	$\sqrt{16} =$
$1^2 =$	$\sqrt{1} =$
$\left(\frac{1}{2}\right)^2 =$	$\sqrt{\frac{1}{4}} =$
$\left(\frac{1}{3}\right)^2 =$	$\sqrt{\frac{1}{9}} =$
$\left(\frac{2}{3}\right)^2 =$	$\sqrt{\frac{4}{9}} =$
$(2+3)^2 =$	$\sqrt{5} =$
$\left(\frac{1}{2+3}\right)^2 =$	$\sqrt{\frac{1}{25}} =$
$\left(\frac{1}{2}+\frac{1}{3}\right)^2 =$	$\sqrt{\frac{25}{36}} =$
$\left(\frac{1}{2}-\frac{1}{3}\right)^2 =$	$\sqrt{\frac{1}{36}} =$
$\left(\frac{1}{4}\cdot\frac{1}{3}\right)^2 =$	$\sqrt{\frac{1}{144}} =$
$(3\cdot 10^7)^2 =$	$\sqrt{9\cdot 10^{14}} =$

2) Gillian spent a year building a traditional sailboat and sailed it in the Gulf of Mexico. In preparation for sailing around the world, she decided to build another sailboat that would go faster. Her friend Janet suggested that she could build a longer boat, but she warned that while a longer boat will go faster, it doesn't increase the speed as much as you might expect. The main reason for this is that a longer (and thus heavier) boat generates a bigger wave in front (where the boat slices through the water) and a bigger wake in back (where the water fills in behind the boat). Essentially (though this is a simplified way of thinking about it), a longer boat "digs a deeper hole" in the water, so even though you go faster, the increase in speed isn't as much as you might expect.

In a perfect world it would be true that if you doubled the length of the sailboat, you would double the speed, but unfortunately it doesn't work out that way. Traditional boat designs meant that there is basically a maximum speed at which a boat was just digging a deeper hole for itself in the water without going much faster. Below is a table which shows the relationship between boat length and speed. The center column is how we wish it would be: note that when the length is doubled, the speed is doubled. The last column shows how it actually is!

Traditional Boat Length	"Perfect World" Top Speed (km/hr)	Real Top Speed (km/hr)
$9\,ft$	$22.32\frac{km}{hr}$	$7.44\frac{km}{hr}$
$18\,ft$	$44.64\frac{km}{hr}$	$10.52\frac{km}{hr}$
$25\,ft$	$62\frac{km}{hr}$	$12.4\frac{km}{hr}$
$50\,ft$	$124\frac{km}{hr}$	$17.54\frac{km}{hr}$

3) Take out your calculator! As boat length doubles from 9 to 18 feet, what happens to the speed in a "Perfect World"? ____ What about from 25 to 50? ____ Find a number to multiply by which converts boat length to "Perfect World" speed. ____ Does it work for every length above? ____

Write the **formula** for Perfect World Speed below, using "P" for Perfect World speed and "L" for Boat Length. Use your formula to find the Perfect World speed for a boat that is $120\,ft$ long.

4) At right is the table again. Write your formula for Perfect World Speed in the bottom of the middle row.

Length	Perfect World	Reality
$9\ ft$	$22.32\frac{km}{hr}$	$7.44\frac{km}{hr}$
$18\ ft$	$44.64\frac{km}{hr}$	$10.52\frac{km}{hr}$
$25\ ft$	$62\frac{km}{hr}$	$12.4\frac{km}{hr}$
$50\ ft$	$124\frac{km}{hr}$	$17.54\frac{km}{hr}$
L	$P =$	$R =$

5) In a Perfect World we would just use the formula you found, and voila, the top speed! Even better, in a Perfect World, when you doubled the boat length, the speed would double! In other words, increasing boat length would increase the top speed **proportionally**. Suppose you bought a new boat five times as big as your old one. Sweet, in a perfect world your new boat would be _____ times as fast! But does that seem realistic? _____ Why or why not? Support your argument with examples.

6) Bummer about how longer boats dig deeper holes in the ocean. Below is your Perfect World formula, and then the Real World formula. In the Real World, speed doesn't increase proportionally with an increase in boat length. How does the formula show you that? Support your argument by plugging in different value for L and calculating values for R.

	Formula for boat speed of boat with length L
Perfect World	$P = 2.48L$
Real World	$R = 2.48\sqrt{L}$

7) The Real World equation is a bummer—it means that when you go from a length of 9 to a length of 25 feet (which is almost a tripling!), the **square root** part of the formula just increases from _______ to _______ (not even a doubling!) Bummer. Well, I guess that's how it is in the Real World—things aren't always as cool as in a Perfect World. But wait, this whole problem has now been solved with modern naval technology! Guess which is the actual fastest speed achieved, and how long the boat was that achieved it:

	length of boat (feet)	Speed (km/hr)	year record set
fastest sailboat	40, 50, 60	71, 121, 171	1992, 2002, 2012
fastest motorized boat	7, 17 27	311, 411, 511	1978, 1993, 2008

8) Raising a number to a power is easy—you just multiply the number times itself multiple times. Rooting is a bit trickier, especially if the number you are rooting isn't a **perfect power**. However, we can still make really good guesses about roots, as well as calculating roots with our calculator to a high level of precision. Complete the table below. Note that this exercise requires you to make a guess without a calculator then use your calculator to find a more precise answer. Here is a suggested method for approximating a square root by using nearby perfect powers:

Approximate $\sqrt{2}$:

First off:	$\sqrt{1} = 1$
Secondly:	$\sqrt{4} = 2$
So that means:	$\sqrt{2}$ is between 1 and 2.
More precisely:	It should be closer to 1 than 2.
So I'm guessing:	1.3

Now it's your turn:

Natural Number	Square Root (guess)	Square Root (calculator, 3 sig figs.)
1		
2	1.3	
3		
4		
5		
6		
7		
8		
9		
10		

Same exercise as before, with **cube roots**:

Natural Number	Cube Root (guess)	Cube Root (calculator, 3 sig fig's)
1		
2		
5		
7		
8		
12		
15		
20		
26		
27		
28		

9) So if raising to a power makes the natural numbers bigger than 1 bigger while rooting makes them smaller, what happens with numbers between 0 and 1? Complete the table below to discover and write the answer! Use your calculator only when necessary.

Rational Number	Square Root	Square Root (calculator)	Rational Number	Square Root	Square Root (calculator)
$\frac{1}{25}$			$\frac{1}{4}$		
$\frac{1}{24}$			$\frac{9}{25}$		
$\frac{1}{20}$			$\frac{4}{9}$		
$\frac{1}{12}$			$\frac{1}{2}$		
$\frac{1}{9}$			$\frac{5}{6}$		
$\frac{4}{25}$			$\frac{24}{25}$		

10) So based on your experiments, what happens when you root a number that is between 0 and 1? What happens when you root a number that is greater than 1? Discuss results and provide examples to support your conclusions.

11) Put away your calculator! It turns out that we often have to find the square root of fractions like the ones in the last problem. They are particularly useful in **trigonometry**. One important numerical manipulation you need to know is how to **rationalize the denominator** in your fractions involving roots. Basically that means no irrational numbers in the denominator. Fortunately, this is pretty easy to do. It requires that you expand a fraction and understand how to multiply roots:

Multiplying Roots	Expanding a Fraction	Rationalizing a Denominator
$\sqrt{5} \cdot \sqrt{5} =$	$\frac{3}{5} \cdot \frac{2}{2} =$	$\frac{3}{\sqrt{2}} \cdot \frac{\sqrt{2}}{\sqrt{2}} =$
$\sqrt{7} \cdot \sqrt{7} =$	$\frac{6}{11} \cdot \frac{5}{5} =$	$\frac{7}{\sqrt{5}} \cdot \frac{\sqrt{5}}{\sqrt{5}} =$
$\sqrt{3} \cdot \sqrt{15} =$	$\frac{2}{5} \cdot \frac{35}{35} =$	$\frac{6}{\sqrt{2}} \cdot \frac{\sqrt{2}}{\sqrt{2}} =$

12) Let's practice a few more simplifications first. Simplify the following. (Some problems are worked as examples.)

$\sqrt{9} =$	$\sqrt{9} \cdot \sqrt{9} =$	$\sqrt{8} \cdot \sqrt{8} =$
$\sqrt{2} \cdot \sqrt{2} =$	$\sqrt{36} =$	$\sqrt{4 \cdot 9} =$
$\sqrt{4} \cdot \sqrt{9} =$	$\sqrt{2} \cdot \sqrt{3} =$	$\sqrt{5} \cdot \sqrt{7} =$
$\sqrt{4} \cdot \sqrt{3} = 2\sqrt{3}$	$\sqrt{4 \cdot 3} =$ $\sqrt{4} \cdot \sqrt{3} = 2\sqrt{3}$	$\sqrt{9} \cdot \sqrt{2} =$
$\sqrt{25 \cdot 7} =$	$\sqrt{49 \cdot 5} =$	$\sqrt{6} \cdot \sqrt{10} =$ $\sqrt{2} \cdot \sqrt{3} \cdot \sqrt{2} \cdot \sqrt{5} =$ $\sqrt{2} \cdot \sqrt{2} \cdot \sqrt{3} \cdot \sqrt{5} =$ $2\sqrt{15}$
$\sqrt{15} \cdot \sqrt{21} =$	$\sqrt{14} \cdot \sqrt{6} =$	$\sqrt{81} \cdot \sqrt{3} =$
$\sqrt{27} =$ $\sqrt{9} \cdot \sqrt{3} = 3\sqrt{3}$	$\sqrt{8} =$	$\sqrt{20} =$
$\frac{\sqrt{25}}{5} =$	$\frac{\sqrt{72}}{2\sqrt{9}} =$	$\frac{\sqrt{75}}{3\sqrt{25}} =$

13) Rationalizing the denominator means we can't have an irrational number in the denominator. This is a mathematical convention that we follow because often it makes further work with the numbers easier. Rationalize the denominator for each of the following fractions, being sure to simplify completely. Two problems are worked as examples. Note that often we simplify in advance of completing the step of rationalization in order to make the process easier.

$\sqrt{\frac{1}{2}} = \frac{\sqrt{1}}{\sqrt{2}} = \frac{1}{\sqrt{2}} =$ $\frac{1}{\sqrt{2}}\left(\frac{\sqrt{2}}{\sqrt{2}}\right) =$ $\frac{\sqrt{2}}{2}$	$\sqrt{\frac{1}{3}} =$
$\sqrt{\frac{2}{3}} = \frac{\sqrt{2}}{\sqrt{3}} =$ $\frac{\sqrt{2}}{\sqrt{3}}\left(\frac{\sqrt{3}}{\sqrt{3}}\right) =$ $\frac{\sqrt{6}}{3}$	$\sqrt{\frac{5}{7}} =$
$\sqrt{\frac{9}{10}} =$	$\sqrt{\frac{25}{7}} =$
$\frac{\sqrt{6}}{\sqrt{10}} =$	$\frac{\sqrt{1}}{\sqrt{12}} =$

14) Simplify each root as much as possible.

$\sqrt{9}$	$\sqrt{81}$	$\sqrt{\frac{4}{9}}$
$\sqrt{4} \cdot \sqrt{25}$	$\sqrt{100}$	$\sqrt{4 \cdot 7}$
$\sqrt{\frac{7}{4}}$	$\sqrt{4 \cdot 3}$	$\sqrt{12}$
$\sqrt{44}$	$\sqrt{\frac{1}{4}}$	$\frac{\sqrt{360}}{\sqrt{25}}$
$\sqrt{2} \cdot \sqrt{3} \cdot \sqrt{2}$	$\sqrt{3} \cdot \sqrt{25} \cdot \sqrt{3} \cdot \sqrt{7}$	$\sqrt{10} \cdot \sqrt{14}$
$\sqrt{49}$	$\sqrt{7^2}$	$(\sqrt{3})^2$

15) Rationalize the denominator for each of the following fractions and simplify as much as possible. One example is done for you.

$\frac{1}{\sqrt{2}}$	$\frac{1}{\sqrt{7}}$
$\frac{5}{\sqrt{11}}$	$\frac{3}{\sqrt{7}}$
$\sqrt{\frac{11}{13}}$	$\frac{\sqrt{4}}{\sqrt{5}}$
$\frac{\sqrt{30}}{\sqrt{20}} = \sqrt{\frac{30}{20}} = \sqrt{\frac{3}{2}} =$ (note that we reduced the fraction first) $\frac{\sqrt{3}}{\sqrt{2}} \cdot \frac{\sqrt{2}}{\sqrt{2}} = \frac{\sqrt{6}}{2}$	$\sqrt{\frac{8}{4}}$
$\frac{\sqrt{6}}{\sqrt{147}}$	$\sqrt{\frac{81}{21}}$

16) Summarize your understanding of the impact of rooting on different types of numbers. Give examples to support your conclusions.

17) Explain how to approximate roots without a calculator. Give examples to support your explanations.

18) Give at least three examples which demonstrate important tips and tricks for simplifying roots and rationalizing denominators, and explain the steps involved.

The Take-Home Message: Square rooting numbers has different impacts on numbers depending on if they are bigger or less than 1. Rooting of numbers that aren't perfect squares produces irrational numbers. Rationalizing denominators is accomplished by expanding the fraction.

NS8 Order of Operations

Name(s)

Date Class/Period/Group

1) In order to combine **operations** in **arithmetic**, we have to combine them in order! You can think of the numbers and symbols we use in arithmetic as the vocabulary, and the **order of operations** as the grammar of the language of arithmetic. Below are some arithmetic sentences, some of which involve **variables**. Some are silly and some make sense! Identify the silly ones with a ☺! Simplify the others as much as possible.

$2b \cap \sim 2b$ (Hint: ∩ means "or")	$\sqrt{-1}\ 2^3\ \sum \pi$ (Hint: $\sqrt{-1}$ is represented with i, and the $\sum$ symbol means "sum"
$\left(\frac{5}{2}-\frac{8}{3}\right)^2 =$	$2(3+4)^1 =$
$\left(\frac{2-7}{1+4}\right)^2 + \frac{1}{50} =$	$\left(\sqrt{4}+\sqrt{9}\right)^2$
$\left(-1+\frac{2}{4}-3\right)^2 =$	$\frac{3}{5-3}\cdot\frac{5}{3-2}-\frac{2}{3} =$

2) Create numerical expressions which require you to do the indicated steps in the indicated order. Use parenthesis to help you indicate the order of operations. Make sure it can be computed without a calculator. Then simplify the problem you created.

add, subtract, multiply, divide	multiply, raise to a power, add
add (by finding a common denominator), raise to a power, subtract	subtract (by finding a common denominator), multiply, subtract

3) Simplify each of the following as much as possible. Rationalize denominators where necessary.

$\left(\sqrt{2}+3\right)^2 =$	$\frac{10}{2+3}+\frac{2}{10}\cdot\frac{3}{10} =$
$\frac{1}{\sqrt{2}}+\frac{2}{\sqrt{2}} =$	$\left(\frac{1}{\sqrt{3}}-\frac{2}{\sqrt{3}}\right)^2 =$
$\left(-\frac{3}{2}-\frac{5}{3}\right)^3 =$	$\frac{2(7-6)}{4(2-5)}+\frac{1}{2} =$

4) Each pair of numerical expressions below is designed to help you avoid common mistakes with order of operations. Simplify and compare the two expressions in each row below. Discuss the similarities and differences. The first one is completed for you.

1st expression	2nd expression	compare and contrast
$(2+3)^2 = 25$	$2^2 + 3^2 = 13$	The sum must be completed first in the first expression.
$\sqrt{4+5} =$	$\sqrt{4} + \sqrt{5} =$	
$7(3\sqrt{2}+5) =$	$7(3\sqrt{2}) + 5 =$	
$\frac{5\sqrt{3}+10}{5} =$	$\frac{5\sqrt{3}}{5} + 10 =$	
$\frac{\sqrt{24}}{3} =$	$\frac{\sqrt{24}}{\sqrt{3}} =$	
$\left(\frac{1}{3}+\frac{1}{5}\right)^2 =$	$\left(\frac{1}{3}\right)^2 + \left(\frac{1}{5}\right)^2 =$	
$(3\sqrt{5})^2 + 7 =$	$(3\sqrt{5}+7)^2 =$	

5) The expressions below can be tricky to simplify. Simplify them if possible, then identify the pitfall which makes it tricky. The first one is identified for you.

expression	pitfall
$(5+10)^2 =$	It can be very tempting to square 5 and 10 idependently first. This is wrong, because powers or roots are not distributive over addition.
$5(4\sqrt{7}+2) =$	
$\sqrt{9+16} =$	
$\frac{\sqrt{15}}{5} =$	
$\frac{5+12\sqrt{3}}{5} =$	
$\frac{\sqrt{8+10}}{2} =$	
$\frac{1}{2}(12+7\sqrt{3}) =$	
$3(4+5)^2 =$	
$(3\sqrt{7})^2 =$	

6) Simplify as much as possible. Rationalize denominators where necessary.

$\pi + \pi =$	$\sqrt{3} + \sqrt{3} =$
$\left(\sqrt{3} + \sqrt{3}\right)^2 =$	$\left(\frac{1}{\sqrt{5}} + \frac{1}{\sqrt{5}}\right)^2 =$
$(2 \cdot 10^3)(2 \cdot 10^4) =$	$(2+3)(2+4) =$
$3(5+1)^2 + 1 =$	$3 \cdot 2^3 =$
$\left(\frac{3}{5}\right)^2 \left(\frac{1}{3}\right) + \left(\frac{4}{50}\right)^1 =$	$\left(\frac{1}{\sqrt{5}} + 2\right)^2 =$

The Take-Home Message: Simplifying numerical expressions using the order of operations correctly requires a correct interpretation of the expression and careful application of all the arithmetic procedures you've learned so far.

Courage To Core

Share the Discovery

NS9 Distribution and Factoring

Name(s)

Date Class/Period/Group

1) **Distribution** and **factoring** are two sides of the same coin! Imagine there is a genie who automatically converts the money you have in your pockets by multiplying it by a factor of 5. One day you meet the genie with $\$30$ in one of your pockets and $\$40$ in the other. Another day you show up with $\$70$ in just one pocket. Do you end up with the same result on both days? Why or why not? Use arithmetic to support your explanation.

2) Here is a table showing the genie's arithmetic of distribution. Complete the table. Is the numerical expression on the left equal to the one on the right in each case?

one pocket	two pockets	Are they equivalent?
$5(30+40)=5(70)=$	$5(30)+5(40)=$	
$6(10+20)=$	$6(10)+6(20)=$	
$3(50+60)=$	$3(50)+3(60)=$	

3) To summarize, we say that **multiplication is distributive over addition**. That means that if we multiply a number times a sum, we can multiply it times each part of the sum and get the same result. Create three of your own examples using the model below.

using the distributive property	checking the result by simplifying the original
$4(5+3)=4(5)+4(3)=20+12=32$	$4(5+3)=4(8)=32$ ☺

4) The distributive property works in both directions! The converse of distribution is called **factoring**. Here is an example. Afterwards, create three of your own examples using this model.

factoring	checking the result
$30+33=3(10+11)$	the left: $30+33=63$ the right: $3(21)=63$ ☺

5) So you might be thinking, it's pretty obvious that multiplication is distributive over addition. Clearly if a genie multiplies my money by 5, it doesn't matter if my money is in two pockets or one, I still end up with the same amount of money. What is the use of distribution or factoring? When we are working with irrational numbers, numbers in scientific notation or variables, distribution and factoring are essential for simplifying expressions. Here are a few examples:

	distribution	factoring
with irrational numbers	$3(5+\sqrt{2}) = 15 + 3\sqrt{2}$	$12 + 18\sqrt{5} = 6(2 + 3\sqrt{5})$
with numbers in scientific notation	$10^3(5+6) =$ $5 \cdot 10^3 + 6 \cdot 10^3$	$22 \cdot 10^8 + 7 \cdot 10^8 =$ $10^8(22+7) =$ $29 \cdot 10^8$
with variables	$3(4+5x) = 12 + 15x$	$2x + 3x = x(2+3) = 5x$

Which of the above examples was the most surprising, interesting or simply new for you? Why?

6) The last example above right uses factoring to **combine like terms.** Use factoring to combine like terms in each of the following.

$5y + 6y =$	$3\sqrt{2} + 5\sqrt{2} =$
$7\pi + 8\pi =$	$\frac{1}{2}e + \frac{2}{3}e =$

7) The key to factoring is finding the **greatest common factor**. Find the GCF in each of the following, then factor. The first is done for you. Note that because multiplication is commutative, you can factor "on the right" if you want!

expression	GCF	factored result, simplified
$2x^2 + 3x^2$	x^2	$(2+3)x^2 = 5x^2$
$4y^3 + 20y^4$		
$30\sqrt{2} + 35\sqrt{2}$		
$12\sqrt{5} + 18\sqrt{10}$		
$\sqrt{2} + \sqrt{6}$		
$\frac{1}{2} + \frac{1}{6}$		

8) Practice distribution on the left, and a similar factoring problem on the right. Simplify as much as possible after distributing or factoring. Some problems are worked as examples for you.

distribute, then simplify	factor, then simplify
$7(1+3) =$	$14+21 =$
$3(2\cdot 10^2 + 4\cdot 10^2) =$	$42\cdot 10^8 + 11\cdot 10^8 =$
$2(\sqrt{2}+\sqrt{3}) =$	$3\sqrt{5}+6\sqrt{7} =$
$2(4\sqrt{6}-3) =$	$10\sqrt{2}-45 =$
$2\sqrt{3}(\sqrt{3}+\sqrt{5}) =$	$2\sqrt{6}+\sqrt{12} =$
$2\sqrt{5}(1-3\sqrt{2}) =$	$2\sqrt{7}+3\sqrt{14} =$
$5(6+2\sqrt{2}) =$	$21+14\sqrt{3} =$
$\frac{1}{5}\left(1+\frac{1}{2}\right) =$	$\frac{1}{4}-\frac{1}{8} =$
$\frac{2}{3}\left(3+\frac{2}{3}\right) =$	$\frac{7}{9}-\frac{2}{3} =$
$\sqrt{3}(4\sqrt{3}+5\sqrt{6}) =$	$3\sqrt{7}+6\sqrt{35} =$

9) Distribute on the left, factor on the right, then simplify as much as possible.

distribute, then simplify	factor, then simplify
$2(5+7) =$	$18 - 30 =$
$4(6+\sqrt{2}) =$	$40 - 10\sqrt{2} =$
$5(1-\sqrt{3}) =$	$7 - 14\sqrt{5} =$
$2(\sqrt{5} - 9\sqrt{7}) =$	$12\sqrt{3} - 21\sqrt{5}$
$7(\sqrt{6} + \sqrt{3}) =$	$50\sqrt{6} - 25\sqrt{7} =$
$\sqrt{2}(\sqrt{2} + \sqrt{3}) =$	$\sqrt{35} - \sqrt{55} =$

10) A little more distribution and factoring practice, with fractions.

distribute, then simplify	factor, then simplify
$\frac{1}{3}(2+3) =$	$\frac{2}{3}+\frac{5}{3} =$
$\frac{3}{5}(7+8) =$	$\frac{2}{7}-\frac{5}{7} =$
$\sqrt{3}\left(\frac{1}{2}+\frac{1}{7}\right) =$	$\frac{\sqrt{6}}{5}+\frac{\sqrt{6}}{2} =$
$0.5\left(\frac{1}{3}+\frac{1}{5}\right) =$	$10^2+10^3 =$

11) The following expressions require you to use distribution or factoring in order to simplify the expression. Distribution and factoring are among the most powerful tools you have to help you simplify!

distribute, then simplify	factor, then simplify
$2(4+5x)+6x+4$	$\frac{(3\sqrt{2}+5\sqrt{2})}{\sqrt{2}}$
$4(2+3\sqrt{2})+5+\sqrt{2}$	$\frac{2\pi^2+3\pi}{\pi}$
$4\cdot 10^2(3+2\cdot 10^2)+5\cdot 10^2$	$2x+3x+4x$
$3(4e+3)+e+1$	$\frac{e^2+3e}{4e}$

12) Explain how factoring and distribution are related. Give two examples, with tips and tricks, of how to use factoring and distribution to simplify numerical or variable expressions.

The Take-Home Message: The fact that multiplication is distributive over addition makes simplifications a lot easier. Factoring is distribution in reverse, and is as useful as distribution.

Courage To Core

Share the Discovery

NS10 *Rational Exponents*

Name(s)

Date Class/Period/Group

1) Before we start, there are three rules of exponents which we need to review. They are shown below. Complete the table following the given examples as a guide. Then write sentences describing these rules are and why they work.

product/quotient of powers	power to a power	negative exponents
$2^2 \cdot 2^3 = 2^5$	$(7^3)^2 = 7^6$	$5^{-2} = \frac{1}{5^2}$
$4^5 4^3 =$	$(6^2)^5 =$	$3^{-6} =$
$3^5 3^{-3} =$	$(9^{-2})^4 =$	$(8^{-3})^2 =$
$\frac{4^7}{4^3} =$	$\left(\left(\frac{1}{2}\right)^2\right)^3 =$	$\frac{5^{-6}}{5^{-7}} =$

2) What is the biggest earthquake that you remember reading about or hearing about? Perhaps you thought of the same one I thought of the 2011 earthquake and resulting tsunami which devastated parts of coastal Japan and caused the meltdown of the Fukushima reactors. How powerful was that earthquake? We measure the **magnitude** of an earthquake using the **Richter Scale**, which essentially tells you the exponent of the **shaking amplitude** of the earthquake. (Magnitude is how big something is, and shaking amplitude is how far from **equilibrium** you shake.) Below are some simplified examples. Convert each shaking amplitude to scientific notation. Some are completed for you as examples. For some, you'll have to guess, using your calculator to narrow down to a more precise answer.

shaking amplitude	in exponential form	Richter Scale magnitude (2 significant figures)
10	10^1	1.0
100		2.0
1000		3.0
2000	$10^{3.3}$	3.3
4000		3.6
8000		3.9
10000		4.0

3) You might have noticed that you can have non-integer, rational exponents, like 3.3 and 3.6. That makes sense, since there could be earthquakes that measure something between 3 and 4 on the Richter Scale. But what does that actually mean? Complete the following chart featuring exponential notation. Some are completed for you as examples.

$3^4 \cdot 3^4 =$	$10^4 \cdot 10^4 = 10^8$
$3^{-4} \cdot 3^4 =$	$10^{-4} \cdot 10^4 = 10^0 = 1$
$3^2 \cdot 3^7 =$	$10^2 \cdot 10^7 =$
$3^{-7} \cdot 3^5 =$	$10^{-7} \cdot 10^5 =$
$3^3 \cdot 3^3 =$	$10^2 \cdot 10^2 =$
$3^1 \cdot 3^1 =$	$10^{\frac{1}{2}} \cdot 10^{\frac{1}{2}} =$
$3^{\frac{1}{2}} \cdot 3^{\frac{1}{2}} =$	$\sqrt{3} \cdot \sqrt{3} =$
$5^{\frac{1}{2}} \cdot 5^{\frac{1}{2}} =$	$\sqrt{5} \cdot \sqrt{5} =$

4) Look carefully at the two problems in the last line of the table. This is a demonstration of the fact that $5^{\frac{1}{2}} =$ ________. What is $6^{\frac{1}{2}}$? ____________ Describe what fractional exponents mean, with examples.

5) Complete the table below. Some are completed for you. Use your calculator when needed to complete the last column.

exponential notation	root or fractional version	decimal version (3 sig figs)
10^3		
10^2		
10^{-2}		
10^{-1}		
$10^{\frac{1}{2}}$		
$10^{\frac{1}{3}}$		
$10^{\frac{2}{3}}$	$\sqrt[3]{10^2}$ or $\left(\sqrt[3]{10}\right)^2$	
$10^{3.7}$	$10^{\frac{37}{10}} = \left(\sqrt[10]{10}\right)^{37}$ or $\sqrt[10]{(10)^{37}}$	
$10^{\frac{5}{4}}$		
$10^{1.1}$		

6) Use the table below to summarize results:

	this means:	example
Exponents can be natural numbers.		
Exponents can be negative integers.		
Exponents can be rational non-integers.		

7) Do you think that flocks of birds can cause earthquakes? True or false:

statement	T/F
March is earthquake month, because of a relationship between unstable weather and seismic activity.	
The 2004 earthquake in Indonesia made the earth more round.	
In the San Francisco earthquake of 1906, most buildings were lost to fire.	
The 2010 earthquake in Chile shortened the day.	
Oil extraction can cause earthquakes.	
The cumulative impact of falling leaves in autumn can cause earthquakes.	
There are $5 \cdot 10^5$ earthquakes every year.	
The German philosopher Immanuel Kant thought that the 1755 earthquake which destroyed Lisbon was caused by earth farts.	
Just as the sun and the moon pull on the water in the oceans to cause the tides, they can pull on the earth's plates to cause earthquakes.	
All the earthquakes described above were magnitude 10 or above.	
The 2004 Indian Ocean earthquake (subject of the movie, The Impossible) caused a tsunami that was 30 meters tall.	
In 2011 a magnitude 5.1 earthquake centered in Murcia was felt in Madrid.	
In 2013, Valencia was hit with a magnitude 8.7 earthquake which caused many forest fires.	
A 6.0 magnitude earthquake releases the energy equivalent of the Hiroshima atomic bomb.	
A magnitude 9.0 earthquake releases 10^5 times the energy of a magnitude 6.0 earthquake.	
The longer the geologic fault (where two plates meet), the more powerful the earthquake.	
A magnitude 10.5 earthquake is probable in your lifetime.	

8) Below are the magnitudes of some measured earthquakes. Without your calculator, give an approximation for the shaking amplitude of each earthquake. The first one was done for you. The guess was made based on the idea that $10^0 = 1$ and $10^1 = 1$. (After completing all the approximations, get out your calculator to complete the last column.)

earthquake	magnitude	approximation	calculator (3 sig figs)
Hand grenade	0.20	$10^{0.2} \approx 2$	1.58
Stick of dynamite	1.20		
2013 Valencia earthquake	3.20		
Chernobyl nuclear plant explosion	3.87		
2010 Haiti earthquake	7.00		
1908 Tunguska meteor explosion	7.90		
largest tested nuclear weapon	8.35		
1883 Krakatoa Island eruption	8.75		
1960 Chile earthquake	9.5		

9) Practice problems. Simplify as much as possible. One example is done for you.

$9^7 9^{-7} =$	$(3 \cdot 10^5) \cdot (4 \cdot 10^6) =$
$\left(10^5 \cdot \frac{1}{10^3}\right) = 10^2$	$\frac{3^7}{3^5} \cdot \frac{3^4}{3^8} = 3^{-2}$
$4^{-9} 4^8 =$	$\frac{4^{-4}}{4^{-6}} =$
$\frac{\left(\frac{1}{2^5}\right)}{\left(\frac{1}{2^7}\right)} =$	$4^6 \cdot \frac{1}{4^9} =$
$\frac{10^{-7}}{10^4} \cdot \frac{10^6}{10^4} =$	$6^0 6^{-1} 6^{-2} \cdot \frac{1}{6^2} \cdot \frac{1}{6^1} \cdot \frac{1}{6^0} =$
$\frac{2^3}{\left(\frac{2^4}{2^5}\right)} =$	$\frac{\left(\frac{8^3}{8^7}\right)}{8^2} =$
$\frac{5^7}{4^7} \cdot \frac{4^6}{5^6} =$	$\frac{10^7}{10^5} \cdot \frac{9^6}{9^8} =$

10) Practice working with exponents and fractions. Simplify as much as possible.

$12^5 12^{-6} 12^1 12^0 =$	$(5 \cdot 10^6)(3 \cdot 10^5) =$
$3^4 \cdot \frac{1}{3^4} =$	$\frac{5^9}{1} \cdot \frac{1}{5^4} =$
$\frac{7^1}{7^3} \cdot \frac{7^6}{7^0} =$	$\frac{7^{-1}}{7^4} =$
$\frac{\left(\frac{10^4}{10^5}\right)}{10^7} =$	$\frac{\left(\frac{3^7}{3^4}\right)}{3^{10}} =$

11) Practice working with exponents and fractions. Simplify as much as possible. One problem is worked as an example.

$\frac{2^4}{2^5} =$	$\frac{1}{\left(\frac{2^4}{2^5}\right)} =$
$\frac{2^6}{3^7} \cdot \frac{1}{3^5} \cdot \frac{2^1}{1} =$	$2^{\frac{1}{2}}2^{\frac{1}{2}} =$
$3^{\frac{1}{4}}3^{\frac{1}{4}}3^{\frac{1}{4}}3^{\frac{1}{4}} =$	$25^{\frac{1}{2}} =$
$27^{\frac{1}{3}} =$	$27^{\frac{1}{3}}27^{\frac{1}{3}} = 3 \cdot 3 = 9$
$125^{\frac{1}{3}}125^{\frac{1}{3}} =$	$10^{3.4}10^{3.6} =$
$(4 \cdot 10^4)(5 \cdot 10^5) =$	$\frac{(3 \cdot 10^{-5})(4 \cdot 10^3)}{7 \cdot 10^2} =$

12) Each row demonstrates a simplification using powers and then another similar simplification using roots. Compare and contrast the simplification process.

1st expression	2nd expression	compare and contrast
$(4+9)^2 =$	$\sqrt{4+9} =$	Add first in both cases.
$(5^2)^3 =$	$\left(5^{\frac{1}{2}}\right)^{\left(\frac{1}{3}\right)} =$	
$\left(\sqrt{5}\right)^2$	$(5^{\frac{1}{2}})$^2 =	
$\left(\sqrt{6}\right)^4$	$\left(6^{\frac{1}{2}}\right)^4 =$	
$\sqrt{7}\sqrt{7}\sqrt{7}$	$\left(7^{\frac{1}{2}}\right)^3 =$	
$\left(\sqrt[3]{10}\right)^6 =$	$\left(10^{\frac{1}{3}}\right)^6 =$	
$5\sqrt{3} \cdot 2\sqrt{3} =$	$5\left(7^{\frac{1}{3}}\right)(2)\left(7^{\frac{2}{3}}\right) =$	
$\sqrt{(11)^6} =$	$(7^6)^{\frac{1}{2}} =$	
$5^2 5^3 =$	$5^{\frac{1}{2}} 5^{\frac{1}{3}} =$	
$\sqrt[5]{7^{10}} =$	$(7^{10})^{\frac{1}{5}} =$	

13) Describe how the Richter Scale is used to indicate the magnitude of earthquakes. Give examples. Show the relationship between the magnitude of the earthquake and the shaking amplitude, and how exponents are involved.

14) Summarize the rules for working with exponents, including examples. Begin by reviewing positive integer exponents. Be sure to cover negative exponents, rational non-integer exponents, the 0 exponent.

15) What is $\left(\frac{1}{2}\right)^{\frac{1}{2}}$? Describe the various features of this number. Rationalize the denominator.

16) What is $\left(\frac{1}{3}\right)^{\frac{1}{3}}$? Describe various features of this number. Rationalize the denominator.

The Take-Home Message: Decimal or fractional exponents can be approximated by comparison with nearby integer exponents. Fractional exponents mean a root and integer exponent combined. The rules for fractional exponents are like those for integer exponents.

EE1 Evaluating and Simplifying Polynomials and Exponentials

Name(s)

Date Class/Period/Group

1) One big part of **algebra** is the idea that letters can **symbolically** represent numbers. We will talk more about algebra in the next mission—for now we want to practice **substituting** numbers in place of **variables** in **expressions** and **evaluating** the expressions. Write a definition for one of the bolded terms in this paragraph.

Term: ____________________ Definition: ____

____

2) Below are some expressions. Values are given for the variables. You substitute the given values in place of the variables in the expressions, and give the results, simplified. One example is worked for you.

x	y	z	expression	result
2	5	7	$3x$	
2	1	5	$3x + 4y = 3(2) + 4(1) = 10$	10
0	-2	3	xy	
-4	3	0	$xy + z$	
-5	3	4	$2x + 4y + z$	
1	2	3	$\frac{xy}{2z}$	
-7	2	4	$x^2 + y - z$	
-1	-1	5	$3x^2 + 5y^3 - z$	

3) In the last two expressions above you had to evaluate x^2 for negative values of x. When you square a negative number, is the result positive or negative? ______ Why?

4) In the last expression above you had to evaluate y^3 for a negative value. When you cube a negative number, is the result positive or negative? ______ Why?

5) The table on the previous page featured substitution of integers. You can also substitute non-integers into expressions. Evaluate the following expressions for the given values.

x	y	z	expression	result
$\frac{1}{2}$	$\frac{2}{3}$	3	$4x + 9y =$	
$\frac{3}{5}$	1	0	$\frac{1}{2}x + \frac{2}{3}y =$	
$\frac{4}{5}$	$\frac{6}{7}$	$\frac{0}{3}$	$xyz =$	
0.75	2	0.5	$xy + z =$	
5	3	4	$\frac{1}{x} + \frac{1}{y} =$	
$\frac{8}{27}$	$\frac{9}{4}$	5	$\frac{xy}{z} =$	

6) The following numerical expression is simplified as shown. Explain the steps involved, and why they work.

$$\frac{\left(\frac{5}{3}\right)}{10} = \frac{\left(\frac{5}{3}\right)}{\left(\frac{10}{1}\right)} = \frac{\left(\frac{5}{3}\right)}{\left(\frac{10}{1}\right)} \cdot \frac{\left(\frac{1}{10}\right)}{\left(\frac{1}{10}\right)} = \frac{5}{3} \cdot \frac{1}{10} = \frac{1}{3} \cdot \frac{1}{5} = \frac{1}{15}$$

7) Even though expressions contain variables, these variables still represent numbers. All the arithmetic you can perform with numbers you can also do with expressions. We often do this arithmetic in order to simplify an expression. One of the most important simplifications is shown below. Explain what was done.

$2x + 3x = 5x$

8) **Combining like terms** as we did above can be done because of the **distributive property of multiplication over addition**. Below are two examples of using the distributive property with numerical expressions. Explain the process for each.

Forwards	Backwards
$3(4+5) =$ $3(4) + 3(5) =$ $12 + 15 =$ 27	$2(7) + 2(9) =$ $2(7+9) =$ $2(16) =$ 32

9) How can you tell that the distributive property as shown above is mathematically valid? Simplify each side of the equation below to confirm distribution returns the same result.

$$5(2+7) = 5(2) + 5(7)$$

10) Now apply the distributive property backwards just like you did in the second column above to the following expressions. The first one is partially completed for you.

Backwards	Backwards
$3x + 4x = x(3+4) =$	$5x + 6x =$

11) The above application of the distributive property (backwards) is called **factoring**. You can factor on either side of the sum. Complete the simplifications below:

Factoring	Factoring
$7x + 8x = (7+8)x =$	$10x - 6x =$

12) Explain how factoring justifies combining like terms.

13) On the left is some arithmetic with numbers for you to complete, and on the right is a similar expression for you to simplify. The first row, and a few other problems are completed for you as examples.

operation	arithmetic with numbers	simplifying an expression
multiplication and addition (distributive property)	$4(5)+3(5)=$ $(4+3)(5)=$ $(7)(5)=35$	$2x+3x=$ $(2+3)x=$ $5x$
multiplication and subtraction	$6(7)-2(7)=$	$6x-2x=$
multiplication	$5\cdot 5\cdot 5=5^3$	$y\cdot y\cdot y$
division and addition	$\frac{3}{7}+\frac{5}{7}=$	$\frac{3}{z}+\frac{5}{z}=$
multiplication and addition	$2(3)+5(3)+7(10)=$	$2x+3x+7y=$
multiplication and division	$\frac{10(3)}{14}=$ $\frac{(5)(3)}{7}=\frac{15}{7}$	$\frac{xyz}{7x}=$
multiplication, exponentiation and addition	$3\cdot 2^3+5\cdot 2^3=$ $(3+5)\cdot 2^3=$ $7\cdot 2^3=56$	$3x^3+5x^3=$
division and division	$\frac{\frac{8}{3}}{\left(\frac{2}{7}\right)}=$	$\frac{\frac{x}{y}}{\left(\frac{y}{z}\right)}=$

14) Before we do more complex simplifying of expressions, let's review some previously learned properties of arithmetic, so that you have some vocabulary to talk about how you are simplifying. Under the given examples below, create your own using variables instead of numbers. The first one is done for you.

operation	**commutative**	**associative**
addition	$2+3=3+2$ $x+y=y+x$	$(1+2)+3=1+(2+3)$
subtraction (1st line changes it to addition)	$3-2=(-2)+3$	$(3-2)-1=3+(-2-1)$
multiplication	$3\cdot4=4\cdot3$	$3(4\cdot5)=(3\cdot4)5$

15) We've already observed that multiplication is distributive over addition. That means:

$2(3+7)=2(3)+2(7)$

Are other operations distributive over addition? Complete the table below. Use the examples to help you answer the questions.

operation	Is it distributive over addition?	Why or why not?
addition	$3+(4+5)=(3+4)+(3+5)$?	
subtraction	$3-(4+5)=(3-4)+(3-5)$?	
multiplication	$6(7+9)=6\cdot7+6\cdot9$	
division (1st line changes it to multiplication)	$\frac{7+4}{3}=\frac{1}{3}(7+4)$ $\frac{1}{3}(7+4)=\frac{1}{3}(7)+\frac{1}{3}(4)$?	

16) Simplify the following expressions as much as possible. Refer back to the properties of arithmetic on the previous page to assist you as needed.

$7x + 4x - 3x =$	$\frac{5x+3x}{16xy} =$
$2x(x + 5x) =$	$3 \cdot 2xyzyxy =$
$\frac{xx}{xxxx} =$	$\frac{x^2+x}{x} =$
$2x + 5y - 7x + 3z + 8y =$	$2x^2 + 5x^2 - 3y + 4y - 7 =$
$\frac{14x^2}{y} \cdot \frac{y^3}{7} =$	$\frac{3}{2x} + \frac{5y}{2x} =$
$6(x^2)(x^3) =$	$3(x^2)^3 =$
$xy^2 + 2xy^2 =$	$x^2 - 2x^2 + 3x + 5x - 7 + 4 =$

17) Evaluate each expression for the given values.

x	y	z	expression	result
1	5	-2	$2x + 3z$	
$\frac{1}{2}$	-1	5	$\frac{(x-y)}{z}$	
12	20	-1	$\frac{xy}{3z}$	
$-\frac{4}{3}$	3	0	$\frac{(3x-z)}{-y}$	
$-5\frac{1}{4}$	$3\frac{1}{2}$	2	$2x + 4y + z$	
1	2	$-\frac{2}{3}$	$\frac{x}{y} + \frac{z}{2}$	
-2	1	$\frac{2}{3}$	$x^2 + y - z$	

18) Simplify each expression. Use the associative, commutative, and distributive properties of arithmetic as needed.

$x + 2x - 3x =$	$\frac{4x+5x}{16xy} =$
$2(4x + x) =$	$3x \cdot 2x \cdot 5x =$
$\frac{x^5}{x^2} =$	$\frac{2x^2+5x^2}{x^3} =$
$5x + y - x - z + 5y =$	$\frac{1}{2}x^2 + 5x^2 - 3y + \frac{1}{2}y =$
$\frac{12x^2}{y^3} \cdot \frac{y^3}{6x^3} =$	$\frac{y}{2y} + \frac{5y}{5y} =$
$5 \cdot x \cdot y + 3 \cdot x \cdot y =$	$3(x^4)^5 =$
$x^2y^3 + 2x^2y^3 =$	$x^2 - 2x^2 + 3x + 5x - 7 + 4 =$

Take Home Message: Evaluating a variable expression means substituting a number for a variable then completing the arithmetic prescribed by the expression. Simplifying variable expressions means using same arithmetic you use with numerical expressions.

EE2 Equations

Name(s)

Date Class/Period/Group

1) Imagine you encounter a mischievous genie who will give you exactly the number of candies you desire, with one condition. He requires that you tell him the exact number you want, then give him a precise number of "seed" candies which he will magically transform through arithmetic into the number of candies you want. Specifically, here is the arithmetic he will perform:

seed number → multiply by 5 → add 7 → desired number of candies

The catch is, if you don't give him the correct seed number for the exact number you want from him, you won't get anything. Suppose you want 107 candies. How many seed candies do you have to give him?

____

2) Complete the table, finding the correct seed number. In the case of non-integer answers, round to nearest whole candy.

desired candies	seed # (exact)	seed # (rounded)
207		
325		
416		

3) How did you find the seed number each time? Explain your process in words, and explain why it worked.

4) To find the seed number, instead of guessing, you probably worked backwards, using **inverse operations** to find the seed number. And perhaps you were systematic, using the same process, or **algorithm**, each time. You used **algebra**! Guess what year algebra was created: ______

5) Before algebra, the Greeks used geometry to solve a great variety of problems, but equations and the procedures used to solve them were not yet developed. The word algebra comes from the Arabic, al-ǧabr, which was part of the title of a book written by the Persian mathematician Muḥammad ibn Mūsā al-Khwārizmī, in Baghdad in 830 AD. The entire title translates as "The Compendious Book on Calculation by Completion and Balancing." The Persians had figured out that there is more to mathematics than just solving one specific problem. They figured out that there are general types of equations, and each type has interesting properties, and each can be solved using a general algorithm, no matter what the specific numbers in the equation are. Highlight the sentence you want to remember most in this paragraph.

6) You used an algorithm to solve the candy problem on the previous page. Your algorithm was:

desired number of candies → subtract 7 → divide by 5 → seed number

That algorithm worked no matter what the desired candies were! You quickly generalized your algorithm and applied it to the equation even when the numbers changed. That's what the Persians did. Interestingly, instead of writing what we now call equations, the Persians did all the math work **rhetorically**, that is, with words, just like you wrote above. It wasn't until over 700 years later that two French mathematicians, Francois Viete and Rene Descarte, developed **symbolic algebra**, which allowed us to write and solve equations. Based on what you've read so far, write a two sentence summary of the history of algebra below. Include some specific pieces of information that are of interest to you.

7) Below are three columns. In the first one, you can see what the genie does to convert the seed number into the desired number of candies. In the second, we're working that process backwards, solving an equation to find the seed number if I want 107 candies. To the right, write the steps to solving this equation in words.

What the genie does to the seed number to give the desired number of candies.		Solving the equation to find the seed number. (Working backwards.)	Description of the steps to solving this equation.
seed number	20	$5x + 7 = 107$	write equation
multiply by 5	100	$-7 \quad -7$	
add 7	107	$\frac{5x}{5} = \frac{100}{5}$	
desired candies	107	$x = 20$	

8) In the above problem you used inverse operations in the reverse order from the genie. The genie multiplied by 5 then added 7. You subtracted 7 then divided by 5. This reversing is a great tool for solving equations and understanding why they work to find solutions to problems. Let's try this again with a different genie. In this case we don't know the seed number and we are going to write and solve an equation to find the seed number. Complete the table (some steps are completed for you. Afterwards, confirm your seed number works!

What the genie does to the seed number to give the desired number of candies.		Solving the equation to find the seed number. (Working backwards.)	Description of the steps to solving this equation.
seed number	x		write equation
multiply by 6	$6x$		
add 11			
desired candies			

9) If you know what the genie does, you can always create an equation to find the seed number you need to give him in order to get any desired number of candies. Creating an equation is as simple as performing arithmetic on a variable instead of a number. Create equations below but don't solve them. One example is completed for you. I've used the word **input** for seed number.

Input	multiply by 3	add 4	divide by 7	$= 10$
x	$3x$	$3x + 4$	$\frac{3x+4}{7}$	$\frac{3x+4}{7} = 10$
Input	multiply by 2	subtract 3	divide by 5	$= 11$
x				
Input	add 3	divide by 4	subtract 7	$= 12$
Input	add 5	multiply by -2	add 4	$= 13$

10) Now let's solve one more equation together before working some practice problems. We can solve equations by reversing the order of operations. We work through the original order of operations mentally, then do it again, working backwards. Here's an example:

Equation	The original order of operations	Working backwards	Solving the equation
$3x - 7 = 11$	Starting with x, first we multiplied by 3, then we subtracted 7.	Working backwards would be adding 7 then dividing by 3.	$3x - 7 = 11$ $+7 \quad +7$ $3x = 18$ $\frac{3x}{3} = \frac{18}{3}$ $x = 6$

11) Note that when we reverse an operation, we show this happening on both sides of the equation. This is a way of showing that we are eliminating that operation from the expression as well as altering the number on the right. For example, when we added 7 above it was to eliminate the -7, and that also turns our 11 into an 18. Does 18 have to be divisible by 3 in order for us to get a solution, or can we get non-integer solutions to equations?

12) Time to practice solving equations. Solve the equations below as shown in previous examples. Write a short verbal description of the process to the right of your solving. Be sure answers are simplified as much as possible. One problem is worked as an example for you.

$4x + 5 = 6$	$3x - 2 = 8$
$x - 54 = 7$ $\frac{4}{1} \cdot \frac{(x-5)}{4} = 7 \cdot \frac{4}{1}$ $x - 5 = 18$ $x = 23$	$\frac{x-3}{5} = -6$
$\frac{2x-4}{3} = 8$	$\frac{(-3x+8)}{7} = 4$
$-2(x + 5) = -8$	$5(3x + 4) - 7 = 9$

13) Continue to practice solving equations. Solve the equations below as shown in previous examples. Write a short verbal description of the process to the right of your solving. Be sure answers are simplified as much as possible. One problem is worked as an example for you.

$2x + \frac{1}{2} = 7$	$3x - \frac{2}{3} = 1$
$\frac{1}{3}x + 4 = \frac{1}{5}$	$\frac{3}{7}x - \frac{3}{4} = -\frac{5}{4}$
$\frac{3}{4}(x - 5) = -2$	$\frac{6}{7}(2x - 3) = 12$

14) Create equations below but don't solve them. I've used the word **input** for seed number.

Input	add 8	multiply by 4	subtract 3	$= 10$
x				
Input	subtract 4	divide by 3	add 4	$= 11$
y				
Input	multiply by 3	add 7	divide by 4	$= 12$
z				

15) Solve. Simplify answers.

$5x - 8 = 10$	$-2x + 12 = -3$
$3x - \frac{1}{2} = 4$	$4x + \frac{2}{3} = 2$

16) Solve. Simplify answers.

$2(x-5)=6$	$3(5+x)=4$
$4\left(\frac{1}{2}+x\right)=8$	$-2\left(x-\frac{1}{3}\right)=-\frac{16}{3}$
$-5(-2x+4)=7$	$3(2x-5)+7=8$
$\frac{3}{5}(x-2)-5=10$	$5(x)^2=80$

17) Solve. Simplify answers.

$\frac{2}{3}x - \frac{7}{9} = 8$	$\frac{2}{3}x - 8 = \frac{7}{9}$
$3(x - 5) = 9$	$-9 = -3(x - 5)$
$4x - 3x = \frac{5}{6} - \frac{4}{7}$	$1\frac{1}{7} + 2\frac{1}{3} = x - 2x$
$\frac{1}{2}x - \frac{1}{3}x = 5(2 + 1)$	$\frac{7}{14} - \frac{3}{21} = \frac{10}{20}x + \frac{40}{80}x$

The Take-Home Message: Algebra is built around expressions that take values as input and return values as output. Given a desired output, we can reverse the operations of the expression in order to find the input value.

Courage To Core

Share the Discovery

EE3 Polynomial Expressions

Name(s)

Date Class/Period/Group

1) Fill in the blanks using the words in the word bank below. Many ____________are **polynomials**. Some of the simplest polynomials are__________. Most __________ you have solved involved linear expressions. which are just a particular type of polynomial. When you solved those equations you used a predictable ___________. You did some of the same __________ from one problem to the next because all the equations were linear. So recognizing the sort of equation or expression you are dealing with is important, so that you can use the correct algorithm to manipulate the expression or solve an equation involving that type of expression.

steps	linear	expressions	algorithm	equations	pizza

2) Every expression on the left is a simplified polynomial in one variable. Every expression on the right isn't. Focus on the polynomials and use the non-polynomials for comparison. Then write 5 characteristics of polynomials below. Hint: Use the words "variable," "exponent," "**coefficient**," "**constant**," "multiply," "add," and "subtract."

Polynomials	Non-polynomials
$4x^2 + 3x - 5$	$\frac{1}{x}$
$4x - 3$	$\frac{3x}{x+3}$
$5x^3 - 2x^2$	$5x^{-1}$
$7x^4 + x + 5$	$3x^{-2} + x^{-1} + 7$
4	$5x^2 - \frac{3}{x}$
$\frac{1}{2}x^2 + \frac{1}{3}x - 6$	$\frac{1}{x^2} + \frac{2}{x} - 5$

3) In order to understand how to identify and classify polynomials, we can use the classification system used in biology as a metaphor. Fill in the blanks with words from the word bank below.

In ________________ we learn there are very simple animals like sponges, and more complex ones like jellyfish, and super-complex ones like ____________, and so on. But animals form a ________________ and every animal shares certain defining ______________ with every other animal. For example, all animals are multi-cellular, their cells have___________, and they have to eat to survive. Other characteristics were added or subtracted on top of this basic structure. For example, sometimes animals have added ____________ (like feathers). But no matter how simple or complex an animal is, it will still be an animal.

fish	kingdom	characteristics	biology	features	nuclei

4) The following paragraph is very similar to the above paragraph. Fill in the blanks using the word bank below.

In _____________ we learn there are simple polynomials expressions which we call linear, more complex ones we call **quadratic**, and even more complex ones we call ___________, and so on. But polynomials form a special _________________ of algebraic expression, and every polynomial shares certain defining ________________ with every other polynomial. For example, all polynomials are composed of **terms** that are a number called a **coefficient** times a variable to an exponent. These are ______________ by addition or subtraction, and the exponents can't be negative. In the course of _____________ a polynomial, we can add or eliminate as many terms as we want to make the polynomial simple or complex, but it will still be a _________________.

cubic	connected	creating	category	characteristics	algebra	polynomial

5) Below are examples of polynomials. When the largest exponent in a polynomial is a 3 we call the polynomial cubic. When the largest exponent in a polynomial is a 2 we call the polynomial quadratic. And when the largest exponent is a 1 we call it linear. Write an example of a linear expression here: ___________ You can also **classify** the polynomial based on the number of terms. A one term polynomial is a **monomial**. A two term polynomial is a **binomial**. A three term polynomial is a **trinomial**. Classify each of the following based on their highest exponent, as well as based on the number of terms. Some are completed for you as examples.

expression	description	expression	description
$5x^3 - 3x$	cubic, binomial	$4x^2 - 3x + 1$	
$x^2 - 4x$		$5x^3 + 1x - 7$	
$3x^2$		$4x - 3$	
$-2x^3$		$4x$	
11	**constant**, monomial	x^2	
$x^3 + 3x^2 - x + 4$		$6x^4 + 8x$	**degree** 4, binomial

6) Fill in the blanks.

The expression $6x + 7$ has ___ terms, which means it's a _____________. The exponent of the variable in the first term is ___, which means the expression is ___________. There is no variable in the second term, which means that term is constant. It's called constant because no matter what number we substitute for x the second term is always ___________.

The expression $5x^2 + 6x + 7$ has ___ terms, thus it's a ______________. The exponent of the variable in the first term is ___, which means the expression is ___________. The exponent of the variable in the second term is ____ which means that term is ______________. There is no variable in the third term, which means that term is ___________.

7) Summarize your understanding so far of polynomials and their different classifications. Give examples of polynomials and describe their parts. Give examples of non-polynomials to contrast with your polynomials. Use the bolded words you've seen so far in this mission in your summary.

8) Often we are given a polynomial that isn't simplified, or isn't in **standard form**. All the polynomials you've seen so far in this mission were in simplified standard form, with the term with the largest exponent first and then the next largest exponent second, and so on. Simplify and re-arrange each of the following polynomials so they are in standard form. Sometimes you'll have to **expand** part of the polynomial to simplify it. Expanding here will involve the distributive property.

$x^3 + x^4 + x - 2x^2$	$5(x + 2) - 3$
$3x^2 + 2(3x - 4) + 3x - 4$	$4(x - 2) + 3(x^2 - 7)$
$4x(x + 2) - 3x(x + 5)$	$5(x - 2) + 4x^2 - 3x$

9) Simplify and write in standard form.

$3x^2(x+2)-3x(4)$	$(3x-5)4$
$2x(x^2 \cdot 3)+7(x)$	$5x(x^3 \cdot 4 \cdot x)+x^2(8)$
$5 \cdot x(x-8)+3 \cdot 4$	$\frac{2}{3}x\left(\frac{x}{3}+7\right)$
$\frac{3}{5}x+\frac{2}{3}x$	$\frac{x}{5}+\frac{x}{7}+8x^2$

10) Polynomials are made of terms. Each term is made of a constant (which can be any real number), a variable and an exponent. Write three examples of terms below.

11) The exponent for each term in a polynomial has to be a non-negative integer. When the exponent is 0 that creates a constant term. Give three examples of constant terms below.

12) A polynomial is in simplified standard form when it is written as a sum or difference of terms of different degree, with a maximum of one term of each degree. Write three examples of polynomials in simplified standard form below.

13) Write three examples of polynomials which are not yet simplified and not yet in standard form.

14) Complete the table with examples of each type of polynomial.

cubic trinomial		quadratic binomial	
constant monomial		linear binomial	
degree 4 trinomial		degree 5 4 terms	
all negative coefficients		all fractional coefficients	
all coefficients of 1		no constant term	

15) Create a non-simplified polynomial which is not in standard form. Simplify it and put it in standard form. Classify the polynomial based on the number of terms as well as the degree.

16) Simplify each polynomial and write in standard form. Classify it based on degree and number of terms.

$5(x-3)$	$3x(2x+5)$
$4x(2x^2+3x)+8$	$2(x-4)+3(x-5)$
$-4x(2x-1)+3$	$(x^2)^3+5-3x$
$xxxxx+xxxx+xxx-xx-x-x^0$	$5x^3x^2-3x^1x+4x^0x^1x^2$
$5(x-3)+(4-x)x$	$5(-(-x))+3$
$x(-x)(-x)(-3)+6x^2$	$3(4x)+5x(4x)-1(4x)^0$

17) Simplify each polynomial and write in standard form. Classify it based on degree and number of terms.

$\frac{1}{2}(x-2)+\frac{3}{5}(x-5)$	$\frac{2}{3}x-\frac{5}{6}x$
$3x\left(\frac{1}{2}x+4\right)-x\left(x^2-\frac{2}{3}\right)$	$\frac{5}{6}\left(\frac{3}{10}\right)\left(\frac{x}{1}\right)+x$
$\frac{8}{9}x+\frac{7x}{3}-x$	$\frac{5}{3}(4x)\left(\frac{6}{25}\right)+3x$
$6\left(5x-\frac{4}{12}\right)-2x^2+x^2$	$-\left(-\frac{5}{7}\right)(14x-21)+x^0$

18) Create a linear equation below which has a binomial linear expression on the left and a constant on the right. Solve it.

19) Create a linear equation below which has a binomial linear expression on the left and a monomial linear expression on the right. Solve it.

20) Create a linear equation below which features a constant times a binomial linear expression in parenthesis on the left and a constant on the right. Solve it.

The Take-Home Message: Polynomials can be identified and classified by performing simplifications in accordance with the properties of arithmetic and by understanding a few essential terms. Once polynomials expressions and equations are correctly identified and classified, algorithms can be applied to make our work with them easier.

EE4 Linear Equations

Name(s)

Date Class/Period/Group

1) Linear equations are a lot like linear expressions—they have at most linear terms and constant terms, and nothing else. Solving equations often requires that we reverse the order of operations. Solve the following linear equations and write your answer as simplified as possible.

$5x - 3 = 7$	$9x + 2 = 8$
$\frac{3}{10}x - 7 = 8$	$\frac{2}{3}x - \frac{3}{4} = -1$
$0.5x + 0.7 = 1.2$	$x + 2 \cdot 10^4 = 5 \cdot 10^4$

2) Sometimes the linear expressions in equations are not simplified, and you should simplify first! Simplify the left and right side of each equation below first, then solve.

$x + x + 2 = 4 + 5$	$2(x - 3) = 5(2 + 4)$
$3(x - 5) = 4\left(\frac{2}{3} + \frac{1}{2}\right)$	$-2x + 3x^2 - 3x^2 + 5 = -10$
$\frac{3}{2}(x - 12) = 90x - 90x$	$4x^3 + 3x^2 - 2x - 3x^2 - 4x^3 = -1$
$-3(12x^2 - 5) + 36x^2 = 5x$	$(2x)^2 - 4x^2 + 5 = 4$

3) Fill in the blanks using the word bank below. So far we have explored two main components of ____________ equations. Our overall strategy is to reverse the order of operations in order to ______________ x. A useful tactic we just practiced is to simplify _______________ on the left or right side of the equation before isolating x.

One very basic concept is that equations are like a __________________ with equal weights on both sides. This means whatever you do to one side you have to do to the other. This allows us to add a second tactic of eliminating a ______________ when there are linear terms on both sides of the equation.

pizza	solving	popcorn	linear term	balance	expressions	**isolate**

4) Write an example of a linear equation with linear terms on both sides of the equation.

5) Each of the following linear equations has a linear term on each side. Eliminate the linear term on one side by adding or subtracting as needed, then solve the equation as you've done previously. One example is worked for you. Explain your steps as shown in the example.

$3x + 4 = 7x$ $-3x \quad -3x$ (subtract $3x$ both sides) $4 = 4x$ $\overline{4} \quad \overline{4}$ (divide both sides by 4) $x = 1$	$5x = 3x - 2$
$6x - 4 = 2x + 8$	$x + 2x = 5x - 4$
$3(x - 5) = -4(2x + 7)$	$4x^2 + 3x = 4x^2 + 5x - 9$

6) You may have noticed two equations in this mission which had **no solution**. Put a star next to those problems. "No solution" is a legitimate result for an equation! Sometimes, an equation will actually have **infinite solutions**. This means that no matter what you put in for x, the equation will always be true. Below are equations which have either one solution, no solution, or infinite solutions. State which! Some answers are given for you. Read the explanations carefully.

Equation	Number of solutions	Equation	Number of solutions
$2x = 6$ (divide by 2 both sides) $x = 3$ Only an x-value of 3 will make the equation true. No other x-values make the equation true. That means one solution!	1	$3x - 3x = 8$ (simplify left side) $0 = 8$ It is never true that $0 = 8$, so no matter what x is, the equation will never be true. That means no solution!	0
$2x = 2x$(subtract $2x$ both sides) $0 = 0$ It is always true that $0 = 0$, so no matter what x is, the equation will always be true. That means infinite solutions!	∞	$5x - 4x = 3 + 7$	
$6x + 7 - 6x = 3$		$2x + 2x - 4x = 0$	
$2x = 2x - 3$		$4(x - 5) = 2(2x - 10)$	

7) Create a linear equation which requires expansion to solve, and which requires adding or subtracting a linear term from both sides. Solve it.

8) Solve. Explain your steps.

$3(y+2)-7=9$	$2y+3y=10$
$\frac{12}{14}y+\frac{1}{7}y=3$	$\frac{3y}{5}+\frac{8y}{5}=22$
$-\frac{4}{7}(14y+2)=\frac{1}{7}$	$(3y-10)2=3(4-5)$

9) Solve. Explain your steps.

$2y = 3y + 4$	$5y - 6 = 8y + 2$
$-3(2y - 4) = -5(3y - 10)$	$5y - 7 = \frac{1}{2}(10y - 14)$
$y = -y$	$y = y$
$y - y = 2y$	$\frac{y}{3} = \frac{y}{2}$

The Take Home Message: Simplifying expressions is important when solving equations, and performing arithmetic operations on both sides of an equation produces an equivalent equation. Linear equations can have no solution, one solution or infinite solutions.

Courage To Core

Share the Discovery

EE5

Eliminating Fractions
Eliminating Negatives

Name(s)

Date Class/Period/Group

1) With a partner, play tic-tac-toe three times below. If you go first, can you always force a draw? How? If you go second can you always force a draw? How?

2) In tic-tac-toe there is an algorithm that you can use every time to make sure the game ends in a draw. There is an algorithm for the player who goes first and a separate one for the one who goes second. Hopefully you described those above! Play a 4x4 game of tic-tac-toe below (and try to get 4 in a row). Is there an algorithm for winning, or for forcing a draw? If so, describe.

3) In the game chopsticks, each player raises their fists in front of them within tapping distance of the opponent. Then, each player raises one finger on each hand. The first player taps either hand of the second player, and the tapped hand now has 2 (because 1 was added by tapping). Now, the second player can tap the first player with either hand. If, for example, she uses her 2 to tap a 1 on the first player, the first player will now have a 3 (because 2 were added by tapping). The game continues, with each player trying to force exactly 5 fingers onto the opponent's hand, at which point that hand is "dead" (a fist). If you "go over", for example by adding 4 to a 3, that hand ends up with 7, which then drops to 2 (the 5 get eliminated). Finally, you can use your move to "split," which means to re-distribute the fingers showing on your own two hands. For example, if you have 3, 2 you can change that to 4, 1. or if you have 4, 0 you can change that to 2, 2. The goal is to kill both hands of the opponent. Play this game with a partner and if you find an algorithm for win or draw, write it below.

4) We've been focusing on linear equations lately, and you've been implementing algorithms to solve them. You can think of algorithms as specific tactics and overall strategies in your game of solving equations. Here are the most important and essential tactics and strategies so far:

Essential Tools for Solving Equations

Tactic	Tactic	Strategy
Simplifying each side as much as possible.	*When there are linear terms on both sides of the equation, eliminate one on one side by adding or subtracting.*	*Reversing the order of operations in order to isolate x.*

Here is one more useful tactic: **eliminating fractions**. You won't want to try this all the time, but it works wonderfully in specific situations. As you become proficient with this tactic you won't need to write out every step because you'll complete them mentally. Write in the verbal explanations for the last 2 steps.

Eliminating Fractions	Steps
$\frac{1}{2}x + \frac{1}{4} = \frac{3}{4}x - \frac{1}{8}$	
$\frac{8}{1}\left(\frac{1}{2}x + \frac{1}{4}\right) = \left(\frac{3}{4}x - \frac{1}{8}\right)\frac{8}{1}$	Multiply both sides by the common denominator.
$\frac{8}{2}x + \frac{8}{4} = \frac{3 \cdot 8}{4}x - \frac{8}{8}$	Expand.
$4x + 2 = 6x + 1$	Reduce fractions.
$1 = 2x$	
$x = \frac{1}{2}$	

5) Now you try it:

$\frac{2}{3}x - \frac{1}{6} = \frac{4}{12}x - 2$

$-\frac{2}{5}x + \frac{1}{10} = \frac{3}{10}x - \frac{3}{20}$

6) You can use a similar tactic to **eliminate negatives**. By the second step below, how many negatives were eliminated? ________ How many were created? __________ Add explanations for the last three steps.

Eliminating Negatives	Steps
$-5x - 12 = -4x + 3$	
$-1(-5x - 12) = -1(-4x + 3)$	Multiply both sides by -1.
$5x + 12 = 4x - 3$	Expand.
$x + 12 = -3$	
$x = -15$	

7) Now you try it.

$-8x - 9 = -7x - 14$

$-\frac{1}{3}x - 2 = -\frac{2}{3}x - 5$

$-\frac{3}{5}x - 7 = -2$

$-x = -5$

8) Solve each equation. Remember to simplify your answers as much as possible, and rationalize denominators when necessary.

$\frac{2}{11}x - \frac{3}{22} = \frac{1}{11}$

$-\frac{5}{7}x - \frac{3}{14} = -\frac{2}{21}$

$\frac{1}{10^3}x + \frac{1}{10^2} = \frac{1}{10}$

$\frac{1}{6^4}x - \frac{1}{6^3} = \frac{1}{36}$

$\frac{1}{\sqrt{2}}x + \frac{5}{\sqrt{2}} = \frac{7}{\sqrt{2}}$

$-\frac{3x}{5} - \frac{1}{25} = \frac{3}{5^2}$

9) Solve each equation. Remember to simplify your answers as much as possible, and rationalize denominators when necessary.

$5(x-4) = 12\left(\frac{1}{2}x - 2\right)$

$\frac{2}{3}x - \frac{5}{3} = \frac{1}{3}$

$-3x = -5 - 2x$

$\frac{4x}{7} - \frac{3x}{7} = \frac{5}{14}$

$\frac{x-5}{10^3} = 10^{-2}$

$\frac{x}{\sqrt{3}} + \frac{2}{\sqrt{3}} = \frac{5}{\sqrt{3}}$

10) Solve each equation. Remember to simplify your answers as much as possible, and rationalize denominators when necessary.

$\frac{1}{10^4}(x-2) = \frac{3}{10^3}$

$x \cdot 3 + x \cdot 4 = 28$

$x\sqrt{3} = 5$

$(-\sqrt{5})x = -7$

$(2 \cdot 10^6)x = 4 \cdot 10^7$

$-\frac{7}{12}x - \frac{8}{12}x = -3$

The Take-Home Message: Eliminating fractions or negatives is based on the idea that arithmetic operations applied to both sides of an equation almost always result in equivalent equations.

Courage To Core
Share the Discovery

EE6 Equations with Multiple Variables

Name(s)

Date Class/Period/Group

1) Can you think of at least 2 famous equations and explain what they mean? Write them below with explanation at right. One is given for you.

Equation	Meaning
$A = \pi r^2$	The area (A) of a circle is pi times the **radius** squared.

2) Many of the basic behaviors of the observable universe can be described by equations. The falling of apples from trees, the way that it snows when the temperature, pressure and humidity are just right, the waves on the sea and the sound waves in your ear—all these and more are described by equations. You might say that mathematics is the **language** we use to describe the world. We have natural numbers like 4 and 7, and irrational numbers like $\sqrt{3}$ and π, and symbols like $=$, $+$ and x, and all those numbers and symbols are the **vocabulary** of our language. The order of operations, and the tactics and strategies for solving equations are the **grammar** of our language. And just like in English, if you just learn vocabulary and grammar and never read or write a story you are missing out. Now it's time to start telling stories. Order the equations below from 1 to 6, 1 being the one your group knows best, and 6 being the least well known to your group. Discuss their meaning.

$D = \frac{m}{V}$	$E = mc^2$
$F = ma$	$C = 2\pi r$
$s = \frac{1}{2}at^2$	$a^2 + b^2 = c^2$

3) Match each of the following equations with its description and units.

Equation	Description	Units
$D = \frac{m}{V}$	1) **Force** is **mass** times **acceleration**. When you swing a hammer towards a nail, you are accelerating a mass and thus it hits the nail with a force.	Newtons (N)
$E = mc^2$	2) **Power** is **work** over **time**. If it takes you a long time to do your work, clearly you are not that powerful yet!	Watts (W)
$F = ma$	3) The **circumference** (distance around) a circle is (π ≈3.14...) times the **diameter**. That means as the diameter of a circle increases, the circumference increases proportionally. That makes sense!	m
$C = \pi d$	4) The **final velocity** squared of an object is the square of the **initial velocity** plus twice the acceleration times the **displacement (s)**.	m/s
$s = \frac{1}{2}at^2 + v_i t$	5) **Einstein**'s famous equation which describes how a very small amount of matter (mass m) can be turned into a large amount of **energy** (E). The **speed of light** (c) is a big number! So in the equation, you multiply the mass times an even larger number (c^2), which results in a huge amount of energy.	Joules (J)
$v_f = v_i + at$	6) The **momentum** (p) of an object is the mass of the object times its velocity. For example, in a crash-up derby, if a robot-driven Mack truck collides head-on with a robot-driven Mini going at the same velocity, the truck clearly had more momentum and the crashed vehicles will keep moving (more slowly) in the direction of the truck's travel. Momentum keeps you moving even when lesser and/or slower objects push back!	N · s
$v_f^2 = v_i^2 + 2as$	7) **Density** is the mass divided by the **volume**. For example, an empty box has only a little mass and a lot of volume, which means it isn't very dense. A box of pencils has greater mass in the same volume—higher density!	kg/m^3
$a^2 + b^2 = c^2$	8) **Work** is defined as force times displacement (s). If you carry a book bag 100 meters up stairs you've got to apply more force to lift the bag, so that's a lot more work than carrying it 10 meters upstairs! And the heavier the bag, the more force, and therefore the more work.	Joules (J)
$p = mv$	9) The **gravitational force** between two objects in space is calculated by multiplying a constant times the mass of the 1st times the mass of the 2nd, then dividing by the square of the distance between them. This means that the bigger the masses, the bigger the force, but the bigger the distance the smaller the force since you divide by distance squared.	Newtons (N)
$W = Fs$	10) The displacement is half the acceleration times the square of the time, plus the initial velocity times the time.	m
$P = \frac{W}{t}$	11) The **Pythagorean Theorem**, which relates the length two **legs** (a and b) of a **right triangle** to the **hypotenuse** (c).	m
$F = \frac{Gm_1 m_2}{r^2}$	12) The final velocity of an object is the initial velocity plus the acceleration times the time that has passed. For example, a car going 30 km/hr which accelerates at 10 km/hr/hr for 2 hours will then be going 50 km/hr. (That was a very slow acceleration!)	m/s

4) Take the simple equation $F = ma$. Force equals mass times acceleration. Imagine that you know the mass of a skateboard with a talented dog riding it is 21 kg and it is being accelerating down the road at 12 m/s^2. That acceleration means that for every second that passes, the skating dog is going 12 m/s faster. Visualize 12 m/s by thinking about that dog skating across this room in about a second! And remember, he is accelerating 12 m/s every second! Good dog. What force does he need to apply with his little legs in order to keep accelerating like that? Use the equation to calculate the force, and write the answer with correct units (Newtons) written to 2 significant figures.

5) Now imagine your friend David riding a skateboard too. He has a mass of 55 kg with board but he is only able to apply the same force as the super dog above. What is his resulting acceleration? Pay attention here! I am giving you the mass and the force. Round to 2 significant figures and write your answer with the correct units.

6) When you think of equations or expressions, the first image that comes into your mind probably has one variable, x. It's true that we often work with equations and expressions in one variable, but equations in the real world often have multiple variables. It's not hard to work with them, but it is important to keep your work well organized. Often we substitute first, which is a procedure we frequently use with expressions, then we solve the equation once we have only one variable left to solve for. Let's practice. One example is given. Refer to page 2 and use correct units for your answer.

Equation	$v_f = v_i + at$	$s = \frac{1}{2}at^2 + v_i t$	$P = \frac{W}{t}$
Given values	$v_f = 42$ m/s $a = 4$ m/s^2 $t = 5$ s	$s = 2$ m $a = 3$ m/s^2 $t = 4$ s	$P = 6$ W $t = 5$ s
Equation with values substituted	$42 = v_i + 4(5)$ $v_i = 20$ m/s		
Solution	$v_i = 20$ m/s		

7) Find the correct formula on page 2 which involves the quantity you want to solve for and the given values. Solve for the desired quantity. Round to 3 significant figures and use the correct units for your answer.

Quantity to solve for	Density	Energy	Acceleration
Given values	$m = 5.00$ kg $V = 3.00$ m^3	$m = 1.00 \cdot 10^{-9}$ kg $c = 3.00 \cdot 10^{8}$ m/s	$F = 12$ N $m = 88.0$ kg
Equation with values substituted			
Solution			

Quantity to solve for	pi	Distance	Mass
Given values	measure C and d for a circle and compute	$a = 10.0$ m/s^2 $t = 4.00$ s $v_i = 5.00$ m/s	$G = 7.00 \cdot 10^{-11}$ $m_1 = 1.00 \cdot 10^{11}$ kg $r = 2$ m $F = 1.00$ N
Equation with values substituted			
Solution			

8) Complete the request, then fill in the blanks:

You may have noticed that in some of the previous equations we solved for a variable other than the variable that isolated on the left. Put a star by those. In another equation we solved for a constant, π. Pi is a constant because no matter how big or small the circle is, the ____________ of the circumference to the diameter is always π, and π is always approximately ___________ Remember that the decimal version of π has a non-repeating decimal, which means it's an ______________________ number.

9) Suppose you are quite interested in determining the acceleration which results when you apply a particular force to a particular mass. Which formula from page 2 relates these three variables? ________ Which variable would you solve for if you knew the force and the mass? ______ Below are several masses and the forces applied to them. Write an equation and solve for acceleration. Round to two significant figures.

Given values	Equation and solution	Given values	Equation and solution
$F = 2.0$ N $m = 3.0$ kg		$F = 0.50$ N $m = 1.5$ kg	
$F = 0.25$ N $m = 2.5$ kg		$F = 1.2$ N $m = 0.30$ kg	

10) Rather than repeatedly writing the original equation, substituting and solving above, we could have solved the general equation $F = ma$ for a first, then just used this modified equation. This is a very useful technique! Solve $F = ma$ for a below. Use this equation to find the acceleration when $F = 3.5$ N and $m = 6.25$ kg.

11) For each of the following, use algebraic strategies and tactics to isolate the indicated variable (in bold). Simplify the result as much as possible, but be aware that in many cases you won't be able to simplify much! Some examples are worked for you.

$F = m\boldsymbol{a}$	$D = \frac{\boldsymbol{m}}{V}$ $\left(\frac{V}{1}\right)(D) = \frac{m}{V}\left(\frac{V}{1}\right)$ $VD = m$ $m = VD$
$W = F\boldsymbol{s}$	$v_f = \boldsymbol{v_i} + at$
$v_f = v_i + a\boldsymbol{t}$ $v_f - v_i = at$ $\frac{v_f - v_i}{a} = t$ $t = \frac{v_f - v_i}{a}$	$F = \frac{G\boldsymbol{m_1}m_2}{r^2}$
$\boldsymbol{a}bc = d + 3$	$a\boldsymbol{b} + c = d - 4$

12) Solve each equation for the variable in bold.

$W = \boldsymbol{F}x$	$p = m\boldsymbol{v}$
$v_f^2 = v_i^2 + 2\boldsymbol{a}s$	$A = \boldsymbol{\pi}r^2$
$E = \boldsymbol{m}c^2$	$P = \frac{W}{\boldsymbol{t}}$
$\boldsymbol{a^2} + b^2 = c^2$ (Solve for a^2.)	$y = mx + \boldsymbol{b}$
$m = \frac{y_2 - y_1}{x_2 - x_1}$	$ab\boldsymbol{c} = 123$

13) Substitute the given values for the variables in the formula, and solve for the remaining variable. Give your answer to 3 significant figures with correct units.

$F = ma\ (a = 15.0, m = 24.0)$	$E = mc^2 \begin{pmatrix} m = 1.00 \cdot 10^{-12} \\ c = 3.00 \cdot 10^{8} \end{pmatrix}$
$p = mv\ (m = 70.0\ and\ v = 10.0)$	$C = \pi d\ \left(C = \frac{7\pi}{2}\right)$
$s = \frac{1}{2}at^2 + v_i t$ $(a = 10.0, v_i = 20.0, t = 5.00)$	$v_f = v_i + at$ $(a = 10.0, v_i = 10.0, t = 3.50)$
$v_f^2 = v_i^2 + 2as$ $(v_f = 30.0, v_i = 15.0, s = 2.00)$	$m = \frac{y_2 - y_1}{x_2 - x_1} \begin{pmatrix} x_1 = 5.00, y_1 = -6.00, \\ x_2 = -1.00, y_2 = 3.00 \end{pmatrix}$

14) Find one equation with a variable on page 2 that you think would be difficult to solve for. Write it here and attempt to solve for that variable. Explain your steps for solving the equation. You may need help from your teacher for this one eventually, but try your best with assistance from others in your group first.

15) Find three equations with variables on page 2 that you are fairly certain you can solve for. Write them here and solve for the variable you select. Explain your steps in solving.

16) Use Google to find a new equation like the ones on page 2. Write it below and describe the meaning of each variable. Select a variable to solve for and substitute reasonable values for each of the other variables and solve for your selected variable.

17) Type "Volume of Pyramid" into Google. An image of a pyramid and input boxes should come up. Fill these in with the values of the variables indicated in the first row below to find the Volume. To complete the second row, go to "Solve for" and select "base length" from the drop down menu. Notice that the equation was re-arranged so that l was isolated. Complete the table, with answers to 2 significant figures. Below the table, write the four different versions of the equation used.

base length (l)	Base width (h)	Pyramid height (h)	Volume (V)
2.0 m	3.4 m	5.6 m	
	3.5 m	5.2 m	45 m^3
4.5 m		3.1 m	37 m^3
3.8 m	2.4 m		41 m^3

The Take-Home Message: Real world formulas are equations for which we can substitute values to find quantities of interest. We can solve these equations in advance for quantities of interest so that the formulas are easier to use.

Courage To Core — Share the Discovery

LF1 *Linear Functions*

Name(s)

Date Class/Period/Group

1) Imagine a genie who converts the number of candies you give him by multiplying them by 2 then adding 9. What expression would represent this process? __________ Complete the table for the various numbers of candies that different people give the genie. (You make up that amount, between 0 and 20.) Everyone in your group should have the same values.

Name	Input (x)	$2x + 9$	Output (y)
Andy	3	$2(3) + 9$	15
Billy			
Carly			
Danny			
Geoff			

2) There are a lot of patterns which exist in the process you've displayed above. Describe them.

3) Fill in the blank using the words from below. As the **input** increases, the **output** __________. For every increase of 1 in input, the output increases by ____. This is because the first thing the genie does is __________ the input. Interestingly, even if you give the genie 0 candies, you still get ____ candies as output. $2x + 9$ is a polynomial, specifically, it is _________ and a ________. The doubling shows up in the expression $2x + 9$ as the ______________ of the _________ term. The 9 which gets added on shows up as the _________ term.

binomial	2	increases	linear	9	double	constant	linear	linear	coefficient

When you increase the input by one, does the output increases by two? Let's put in 6 candies and then put in $6 + 1$ candies and distribute, and compare. Complete the table.

6 candies	$6 + 1$ candies
$2(6) + 9 = 12 + 9 =$	$2(6 + 1) + 9 = 12 + 2 + 9 =$
Here you see $12 + 9$...	...but here you see $12 \mathbf{+2} + 9$.
$2x + 9$	$2(x + 1) + 9 =$

Based on what you see above, explain why increasing x by 1 always increases the output by 2.

4) We can visualize the different outputs for different inputs as a **graph**. The graph will show us the inputs on the x-axis and the outputs on the y-axis.

Fill in the table, then graph the points which represent the outputs for various inputs. Label the points A, B, C, etc. Use a ruler to connect the points with a straight line in order to show the regularity of the increase in output as the input increases. Label the x-axis "input" and label the y-axis "output."

Name	Input (x)	$2x + 9$	Output (y)
Alice	0		
Bob	1		
Chris	2		
David	3		
Gillian	4		
Harry	6		
Irene	9		
Janice	20		

5) You've already identified some patterns in this situation (see previous page). How do those patterns show up in the graph?

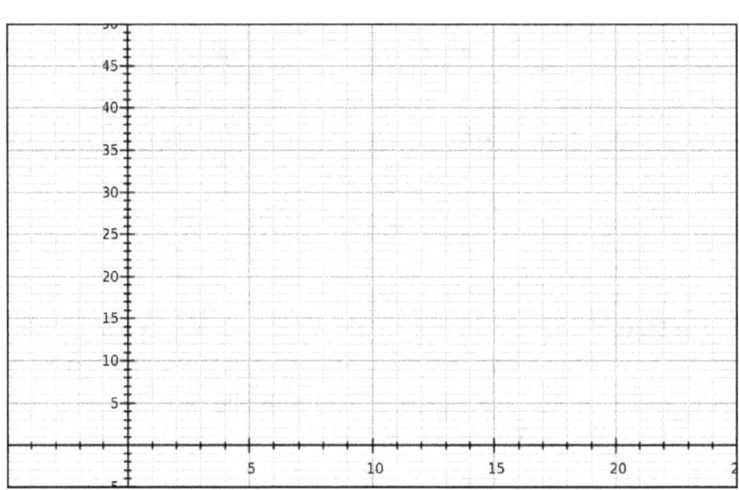

6) With a ruler, extend the line so it goes to the edge of the **graphing window**. Based on the graph, how many candies would you get if you gave the genie 15 candies as input? __________ Label this point Z on the graph. Now, evaluate the expression $2x + 9$ for an input of 15. ______ Was this the same output that you got based on looking at the graph? ______ Why or why not? Should they be the same? Why or why not?

7) Remember, the genie on the last two pages was doubling your candies and adding 9. Cool! Can you think of better genies than that one? How about worse ones? Give examples below and explain why they are better or worse.

8) Imagine you've got one genie who doubles your candies and then adds 50. And then imagine you've got another who multiplies your candies by 10 then adds 5. Which genie do you prefer? Discuss, debate, and write your conclusions below.

9) Now we are going to use a table to compare the two genies in #8. We are going to call the first genie f and the second genie g. And the output for f we are going to call y_1 and the output for g we are going to call y_2. Some people give their candies to genie f and others to genie g. Here are some tables, you fill them in!

Genie f — $2x + 50$

Name	x	y_1
Alice	0	
Bob	1	
Chris	2	
David	3	
Gillian	4	
Harry	6	
Irene	9	
Janice	10	

Genie g — $10x + 5$

Name	x	y_2
Adam	0	
Billy	1	
Cathy	3	
Dylan	5	
Greg	6	
Herman	8	
Irma	9	
John	10	

10) Using two different colors, graph the results for f and g and connect each set of points with a straight line. Label the x-axis "input" and the y-axis "output." Label the lines f and g.

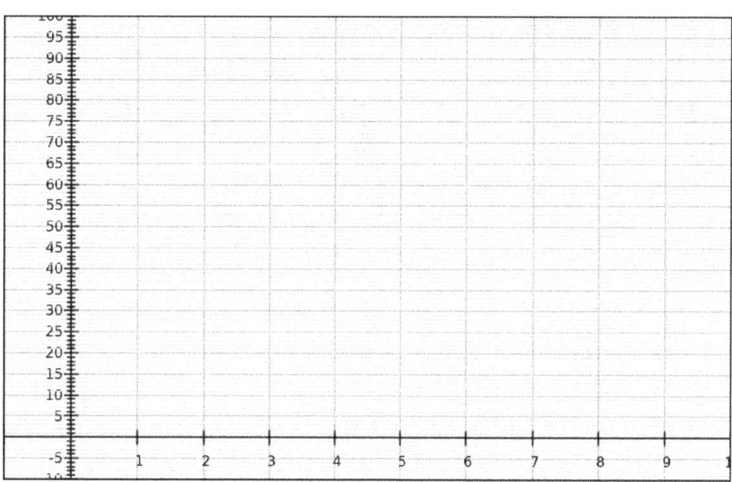

11) Following is one of the most important **essential questions** in algebra, so read it carefully:

Is there one single number of candies that two people could give the two different genies on the last page so the genies return the same result?____ What is that number, approximately?____

What output would both genies return? ____ Did you try to answer this question **graphically**, by looking at the graph, or by **trial and error** plugging in different values into the two expressions, or some other method? Why did you choose the method you chose?

12) In #8 I asked you which genie you preferred and why. Has your opinion changed? ___ Why or why not?

13) Suppose you want a genie that, if you give him 30 candies, he returns exactly 100. Can you think of different ways to make that happen? Create and write the **functions** for various genies in the table below. Some are provided for you. A function is an expression of how a genie functions.

Notice that in order to express a function we write, for example, $y = 10x - 200$, where y represents output and x represents input. That equation states that "output is ten times input, minus two-hundred."

Genie	his **function**	Good, bad, or just ok?	Why?
f	$y = 10x - 200$	IMO, bad...	...he returns negative candies if you give him 10.
g			
h			
i			
j	$y = \frac{1}{2}x + 85$		
k			
l			
m			
n	$y = -3x + 190$		
o			

14) Some of the genies are better than others! In the table above, rate the quality of each genie and give a brief explanation of your assessment.

15) Which genie above returns the most for an input of 40? ____ How much is returned? ____

16) Which genie above returns the most for an input of 0? ____ How much is returned? ____

17) There are lots of genies that function in lots of different ways. For example, some function by multiplying the input by a large positive integer first and then adding a positive integer. Describe some other combinations that are possible, and give examples, perhaps **linear functions** as you've seen so far or making up new types. Why are functions like $y = 2x - 6$ called linear functions?

18) Here are two more genies. Complete the tables.

Genie f $y_1 = 3x + 20$

Name	x	y_1
Alice	0	
Bob	5	
Chris	10	

Genie g $y_2 = x + 50$

Name	x	y_2
Adam	0	
Billy	4	
Cathy	8	

19) Graph the results for f and g and connect each set of points with a straight line. Label the x-axis "input" and the y-axis "output." Label the lines f and g. Use a ruler to extend the lines to the edges of the graphing window.

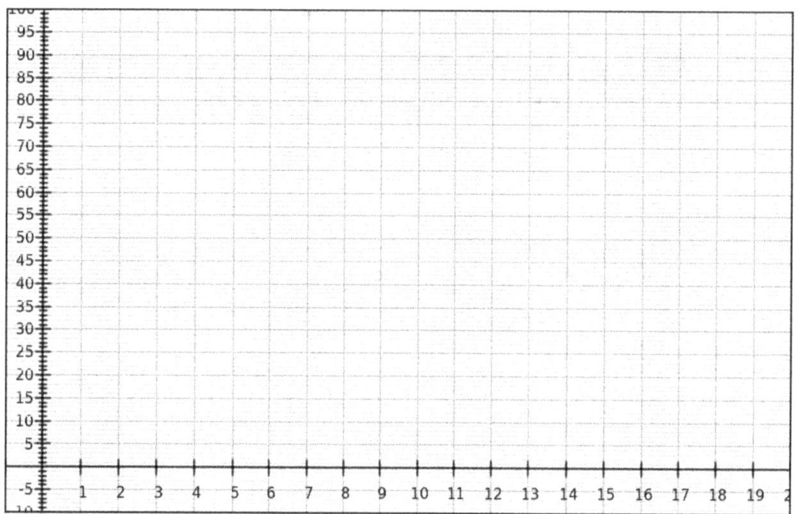

20) For every increase of 1 of input, how much does the output of f increase? ___

21) For every increase of 1 of input, how much does the output of g increase? ____

22) About what single input returns the same output for both? What output is returned? How did you determine this?

23) Fill in the blanks. Functions accept values as _______ and return values as _______. Functions are named using letters like ___ and ___, and each function can be written as an __________ like ___________. The relationship between the input and the output of a function can be shown in a __________ or in a ___________.

graph	f	output	$y = 2x + 3$	g	equation	table	input

24) Create two different functions which, if you input a value of 10 return a value of 40.

25) At right is a table representing a function. Find the equation for a function which produces these results.

Input (x)	Output (y)
0	7
1	10
2	13
3	16
4	19

26) Below is a graph representing the relationship between input and output for one of the three functions listed. (The horizontal axis is the x-axis representing input and the vertical axis is the y-axis representing output.) Which one? How do you know?

$y_1 = -3x + 10$

$y_2 = 2x + 14$

$y_3 = 2x + 5$

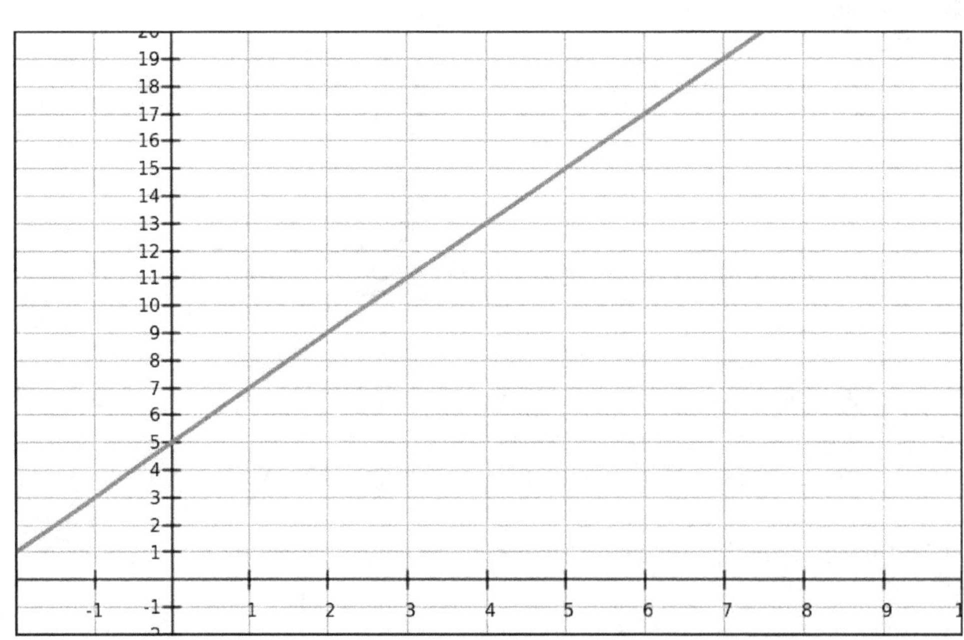

27) I'd now like to give you three more **essential questions** which can be asked about functions, for a total of four. A great many of the problems you'll see in algebra and beyond come down to these four essential questions. Here they are. The blanks represent any real number.

Essential Questions about Functions

Given one function:	Given two functions:
What input returns an output of___?	*What single input will produce the same output for two functions?*
What output is returned for an input of___?	*What output does the input above produce?*

These questions can be answered using the equations, the tables or the graphs for the function. What are the advantages and disadvantages of using each of these three methods?

Method	Advantages	Disadvantages
Equations		
Tables		
Graphs		

28) Below are two functions represented with tables. Answer the questions below the functions.

Function f

Point	x	y_1
A	0	8
B	1	12
C	2	16

Function g

Point	x	y_2
D	0	3
E	1	10
F	2	17

For function f:

Approximately what input returns an output of 14? (Decimals good.)	
What output is returned for an input of 2?	

Given two functions:

Approximately what single input will produce the same output for two functions? (Decimals good.)	
What output does the input above produce?	

29) Now that you've experimented a bit more, which questions were answered well by these tables and which questions were not answered so well, and why?

30) Below are two functions represented using graphs. One graph is steeper. Label the steeper graph f and the other one g. Label the x-axis "input," and the y-axis "output."

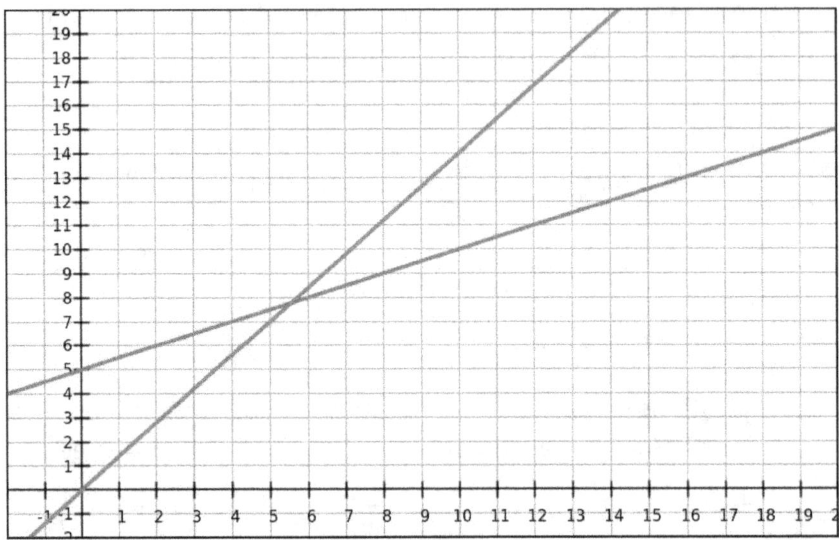

For function f:		Given two functions:	
Approximately what input returns an output of 11? (Decimals good.)		Approximately what single input will produce the same output for two functions? (Decimals good.)	
What output is returned for an input of 18?		What output does the input above produce?	

31) Which questions were answered well by the graph and which questions were not answered so well, and why?

32) Can you determine at least approximately, the equations for each of functions represented by the graphs above? Put your best guesses below.

33) Finally, let's answer some questions using the equations for the functions. Below are equations for two functions. I'm going to give you some useful initial steps to use these equations effectively. Pay close attention, as you'll want to use these steps often in the future. Complete the work as instructed.

Function f	$y_1 = 2x + 7$	Function g	$y_2 = x - 3$

For function f:	Given two functions:
Approximately what input returns an output of 15? (Decimals good.) $15 = 2x + 7$ (Substitute desired output for y.) (Now you solve the equation.)	Approximately what single input will produce the same output for two functions? (Decimals good.) $2x + 7 = x - 3$ (Set the expressions equal.) (Now you solve the equation.)
What output is returned for an input of 18? $y_1 = 2(18) + 7$ (Substitute input for x.) (Now you simplify.)	What output does the input above produce? $y = (-10) - 3$ (Substitute the above input for x in either equation.) (Now you simplify.)

34) Do you think that equations do a good job answering these questions? Why or why not?

35) The equations below involve one or both of the following genies:

Genie f: $y_1 = 3x + 20$ Genie g: $y_2 = x + 50$

What question is each equation trying to answer?

Equation	Question this equation is trying to answer
$3x + 20 = 100$	
$x + 50 = 100$	
$3x + 20 = x + 50$	

36) Create equations for two different genies who return 100 candies for an input of 12 candies.

37) Complete the table. Using two different colors, graph the results for f and g and connect each set of points with a straight line. Label the x-axis "x" and the y-axis "y." Label the lines f and g. Use a ruler to extend the lines to the edges of the graphing window.

Genie f $y_1 = \frac{5}{2}x + 20$

Name	x	y_1
Alice	0	
Bob	2	
Chris	10	

Genie g $y_2 = \frac{3}{2}x + 50$

Name	x	y_2
Adam	0	
Billy	2	
Cathy	8	

38) For every increase of input of 2, how much does f increase? _____ For every increase of input of 2, how much does g increase? _____ Which function increases as a faster rate as the input is increased? _____ Which function returns more candies just for running into the genie (giving him an input of 0? ____ Explain how you know the answer to these questions.

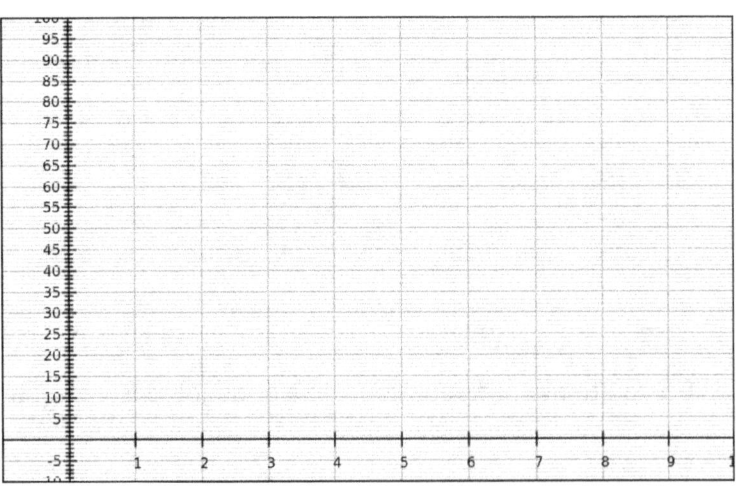

39) A genie f doubles your candies and then adds 60. How many candies does he give you if you give him 30? __________ How many candies do you have to give him in order to get 150? __________ (Write and solve an equation to answer the second question.)

40) A genie triples your candies then adds 48. How many do you have to give him in order to get 163? (Write and solve an equation to answer the second question.)

41) Below are two genies. Complete the tables and the graphs. Label the axes and the lines correctly.

Genie f $y_1 = \frac{10}{3}x + 45$

Name	x	y_1
Alice	0	
Bob	6	
Chris	9	

Genie g $y_2 = \frac{20}{3}x + 10$

Name	x	y_2
Adam	0	
Billy	3	
Cathy	6	

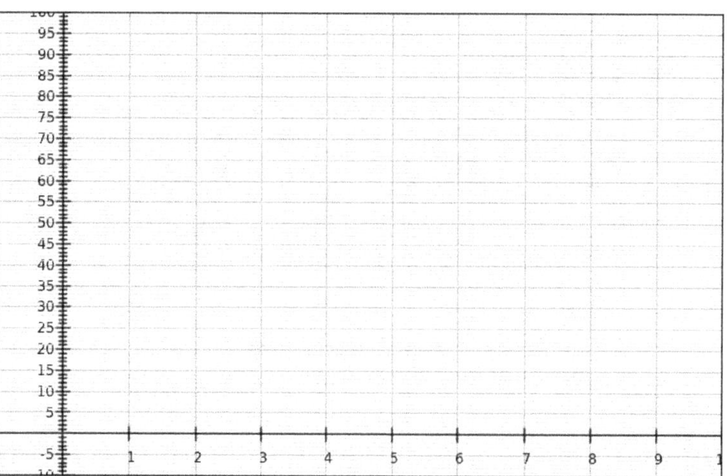

42) Here are the genies from the last problem:

Genie f $y_1 = \frac{10}{3}x + 45$ Genie g $y_2 = \frac{20}{3}x + 10$

Below are equations related to these genies. What questions is each equation trying to answer?

Equation	Question this equation is trying to answer
$\frac{10}{3}x + 45 = 100$	
$\frac{20}{3}x + 10 = 100$	
$\frac{10}{3}x + 45 = \frac{20}{3}x + 10$	

43) Solve each of the equations above.

44) Create two examples of linear functions and use them to explain what a function is. Explain how a function can be represented with an equation, table or graph. Create four essential questions about one or both of these functions and answer them.

The Take-Home Message: Functions are built from expressions which take values as input and return values as output. They can be represented using equations, tables and graphs. We can write and solve special equations to answer essential questions about functions.

LF2 Rate of Change

Name(s)

Date Class/Period/Group

1) Fill in the blanks with the words below. Put in your own numbers where necessary.

Imagine a genie who converts the number of candies you bring him into a new number of candies. The genie converts your candies based on the function $y = 2x + 5$. This means if you bring him 6 candies (that's the input, x), he returns ___ (that's the output, y).

First the genie _________ and then he __________. For the genie represented by $y = 2x + 5$, the output increases as the input increases because of the ___________ that the genie does. For example, an input of 3 produces an output of ___ and an input of 4 produces an output of ___. When the input increased by 1 the output increased by ___. In other words, the **rate** of increase of output is determined by the ___________ of the linear term.

coefficient adds multiplies multiplication

2) Consider each of the genies below. We are interested not just in how many candies they return; we are also interested in how much their output improves (or declines) as we increase the candies we give him. Is it worth it to bring each genie more candies next time? If input increases by one, how much does output increase for each? (One way to test this is to plug in **consecutive** numbers for x and then compare results. An example is done for you.)

Genie	Function	How much output increases when input increases by 1
f	$y = 3x + 20$	If $3(5) + 20 = 35$ and $3(6) + 20 = 38$, the increase $= 3$.
g	$y = 7x + 100$	
h	$y = x + 50$	

3) What is the relationship between the coefficient of the linear term and the increase in output? ___ Why? (Below, I've worked with $y = 3x + 20$ to help you with your explanation.)

Input of 5	Input of 6 (which is $5 + 1$)
$3(5) + 20 = 15 + 20 = 35$	$3(5 + 1) + 9 = 15 + 3 + 20 = 38$
Above, you see $15 + 20$...	...but above here you see $15 + 3 + 20$.
$3x + 20$ (in general)	$3(x + 1) + 20 = 3x + 3 + 20 = 3x + 23$ (3 more than $3x + 20$)

4) For some functions, it can be easier to find how much the output increases when the input increases by 3 (instead of 1). (One way to test this is to plug in two numbers that differ by 3 for x and then compare the results. An example is done for you.)

Genie	Function	How much output increases when input increases by 3.
f	$y = \frac{1}{3}x + 20$	$\frac{1}{3}(3) + 20 = 21$ and $\frac{1}{3}(6) + 20 = 22$, so increase $= 1$.
g	$y = \frac{2}{3}x + 100$	
h	$y = 5x + 200$	
i	$y = \frac{7}{3}x + 50$	
j	$y = \frac{11}{6}x + 40$	

5) For which questions was it easier to find how much the output increases when input increases by 3 above? Could there have been a better question to ask for functions h and j? What questions and why?

6) The output of the first function increases by 1 when input increases by 3. How much does output of the first function increase when the input increases by 1?

7) For the following, how much does output increase when the input increases by 5?

Genie	Function	How much output increases when input increases by 5.
f	$y = \frac{1}{5}x + 20$	
g	$y = -\frac{3}{5}x + 100$	
h	$y = \frac{17}{5}x + 200$	

8) When we describe the increase of output for a certain increase of input we are describing the **rate of change** of the function. Above, we described this as the change in output for every 5 units of change of input. That's change in candies out per change in candies in! In h, for example, that's an increase of ___ candies of output per every increase of ___ of input. Write 5 examples of rates of change in the real world and give the units for each.

9) You can answer questions about the rate of change of the genie's output by looking at a table too. For the genies below, what is the increase in output when the input increases by 1? Write your rate of change as a reduced fraction (sometimes it will be a whole number).

Genie f

Name	x	y_1
Alice	0	15
Bob	1	23
Chris	2	31

increase: ____________
rate of change: ________

Genie g

Name	x	y_2
Adam	0	12
Billy	2	42
Cathy	4	72

increase: ____________
rate of change: ________

10) For the genies below, what is the increase in output when the input increases by 2?

Genie f

Name	x	y_1
Alice	1	17
Bob	2	21.5
Chris	3	26

increase: ____________
rate of change: ________

Genie g

Name	x	y_2
Adam	3	16.5
Billy	4	20
Cathy	5	23.5

increase: ____________
rate of change: ________

11) You can also determine the rate of change from the graph. A good technique is to find two points on the graph whose coordinates are easy to approximate. Essentially we are using the graph to create a short table. Below I've selected two x values that work well—you find the corresponding y-values from the graph below.

x	y
2	
12	

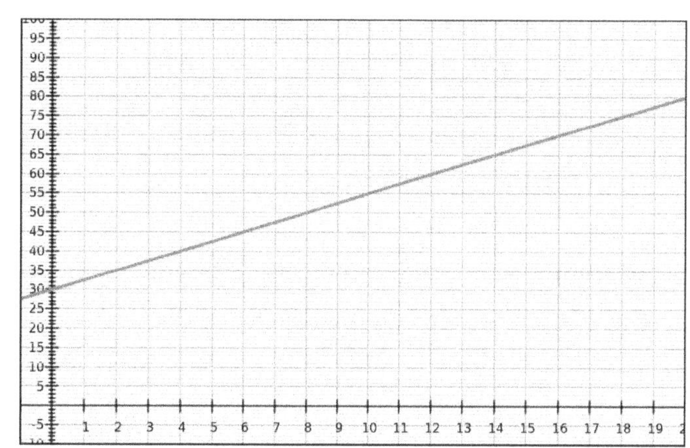

12) Based on the table, approximately how much does the output increase when the input increases by 10? ___ Rate of change: ___

13) Based on your answer to #11, how much does the output increase if the input increases by 20? ____ Rate of change: ____ How much does the output increase if input increases by 1? ____ Rate of change: ____

14) All the genies we **modeled** so far have a constant rate of change. No matter what two points you use to calculate the rate of change for a particular function, you always end up with the same value. where A genie is just one of many things we can model which have a constant rate of change. What are 3 real world examples of things which exhibit a constant rate of change? What are 3 real world examples of things which exhibit a changing rate of change?

15) Suppose you drive a car 200 kilometers in 2 hours. How fast are you driving? What are the units for your velocity? How did you calculate this value? For every increase of one hour of time, how much does your distance increase?

16) Mike is driving from Portland to San Diego. His distance from Portland is shown at different times in the graph below. Points A and B on the graph below represent the trip odometer readings in kilometers of his car at the 5 hour mark and the 10 hour mark. Label the x-axis "time (hours)" and the y-axis "distance (kilometers)." Write the lengths of the vertical and horizontal sides of the triangle with units. This is to make it easy for you to visualize the vertical (distance) and horizontal (time) change between points A and B.

17) Complete the table:

point	x	y
A	5	
B	10	

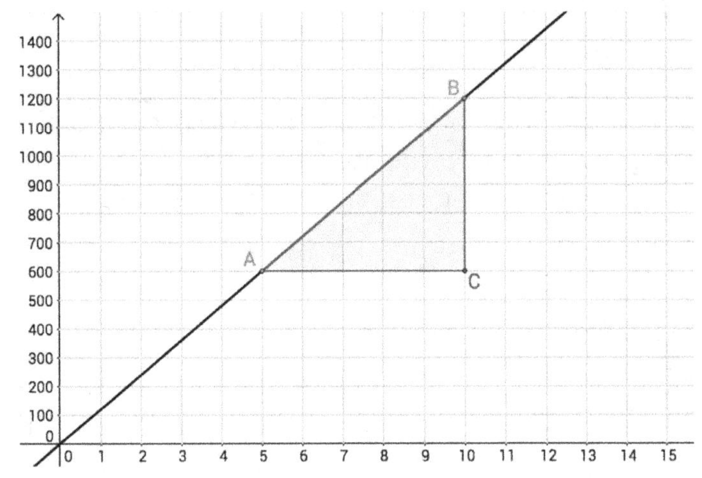

18) What is Δy (the change in y) between A and B? ____ What is Δx between A and B? ____

19) What is the rate of change?

20) Another way of stating the rate of change is $m = \frac{\Delta y}{\Delta x}$. Use this to find the rate of change. Is this the same as Mike's velocity? Why or why not?

21) On the above graph, what is the length of BC? ____ Does this correspond to Δy or Δx? ____ What is the length of AC? ____ Does this correspond to Δy or Δx? ____

22) The above triangle is called a **slope** triangle. Why is it called that?

23) Here is a function for a car which is traveling at 100 km/hr. Complete the table and graph the function. Label the x-axis "time (hrs)" and the y-axis "distance (km)" Extend your line, using a ruler, to the edges of the graphing window.

$y = 100x$

point	x	y
A	0	
B	1	
C	2	
D	3	
E	4	

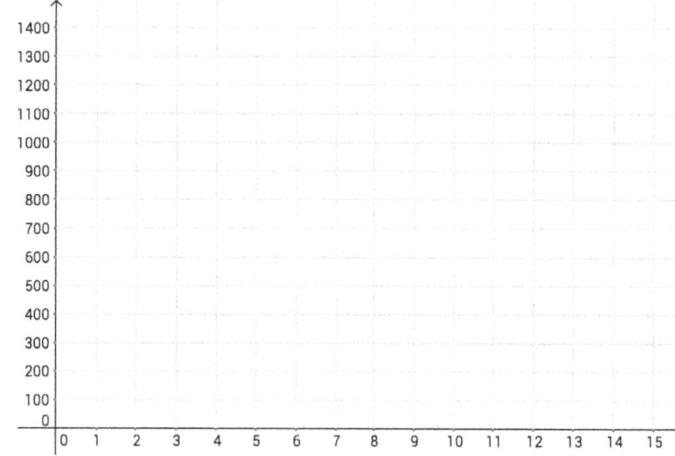

24) The slope of the line above is specifically how much the distance increases when the time increases by 1 hour. It's the rate of change. It is $\frac{\Delta y}{\Delta x}$ where Δ means "change in." What is the slope? ____________

Slope can be visualized as the steepness of the line. This can be visualized effectively by drawing a slope triangle between two points on the line. Pick two points from the table. Everyone in your group should pick a different two points. Make slope triangle between these two points (like in the graph on the last page). Write the length of each side next to each side, and write the units for these lengths (km and hr). Can you visualize the slope using this triangle? Does everyone in your group have the same slope? Why or why not?

25) Suppose you drive a car 300 kilometers in 5 hours. What is the rate of change? Write the function for the distance (y) this car has travelled after x hours.

26) Suppose you drive a car 400 km in 7 hours. What is the rate of change? Write the function for the distance this car has travelled.

27) Suppose you are driving a car from Iowa City to Jackson Hole. After you start driving you remember to reset your trip odometer to 0 km. At the 1 hr mark you stop at a gas station and you see your odometer shows 178 km. At the 4 hr mark you stop again and you notice that your odometer shows 478 km. What was your velocity for the 3 hr period between the odometer readings? How did you calculate that?

28) What two points represent the two stops described above? _________ and _________ Graph these two points on the axes below. Label the points A and B and draw a slope triangle between them. Write the side lengths next to the vertical and horizontal sides with units. Extend the line to the edges of the graphing window and label the axes appropriately. What is the rate of change (that is, the slope)? How did you calculate it?

point	x	y
A	1	178
B	4	478

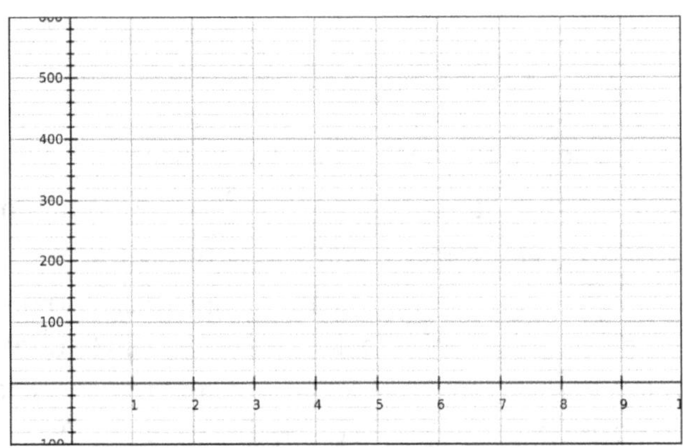

29) Where does your line cross the y-axis? This point is called the **y-intercept**. What do you think the exact value for this number is? Why? What does this number represent in the story?

30) Use the values you found in #28-29 to create a function for the distance your odometer shows at x hours. (Hint: the y-intercept is the constant term for the linear function. Why?)

$y =$ ____________

31) Use this function to find what your odometer reading will be at 10 hours. ____________ Is this pretty close to what your graph shows at 10 hours? ____________ Why or why not?

32) Suppose you want to drive from St. Louis to New Orleans. You reset your trip odometer to 0 and start your stopwatch, and start driving. You are in slow traffic for the first two hours, and you only travel 120 kilometers. What was your velocity for the first two hours of the journey? _____ The traffic subsides and after another 5 hours of driving your odometer reads 670 kilometers. What was your velocity for this faster 5 hour section of your journey? _____ How did you calculate that?

33) You are riding a bicycle from Asheville to the Outer Banks. On the first day of your ride you set your odometer to 0 and start riding. Your timer shows that after 3 hours of difficult trail riding you are at the 32 kilometer mark. But the next section is easier road riding and you finish your day in 10 hours total, at the 224 kilometer mark. How fast were you traveling for the second section of easier road riding? ______ How did you calculate that?

34) Slope is possibly the most important concept related to functions. The idea of slope will be central to your understanding of functions, particularly when you reach calculus. Here is an important table showing the different ways to think about slope.

Slope

Verbally	Algebraically	Algebraically Again	From a table	From a graph
Slope Rate of change Velocity	$m = \frac{\Delta y}{\Delta x}$	$m = \frac{y_2 - y_1}{x_2 - x_1}$	$\Delta x \leftarrow$ [x, y ; x_1, y_1 ; x_2, y_2] $\rightarrow \Delta y$	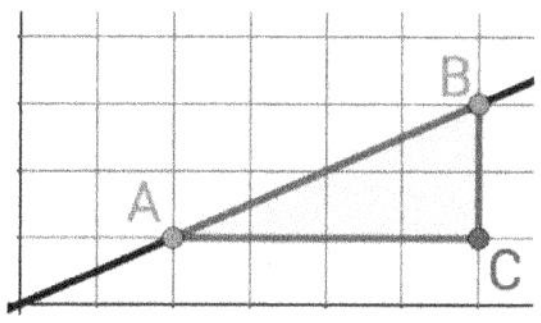 A, B, C $m = \frac{vertical\ change}{horizontal\ change}$ $m = \frac{BC}{AC}$
Example	a car travels $120\ km$ in $3\ hrs$ $m = \frac{120\ km}{3\ hr}$ $m = 40 \frac{km}{hr}$	Point A: $(20{,}130)$ Point B: $(30{,}170)$ $m = \frac{40}{10} = 4$	$2 \leftarrow$ [x, y ; 5, 12 ; 7, 18] $\rightarrow 6$ $m = \frac{6}{2} = 3$	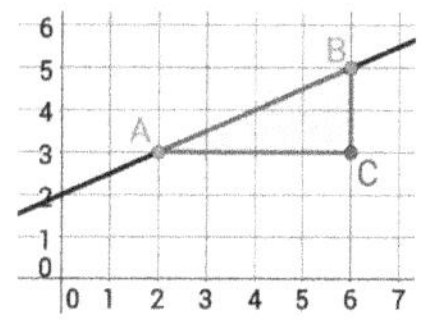 A, B, C $m = \frac{2}{4} = \frac{1}{2}$

35) You ride your bike from Albany to Hershey. At the 3 hour mark your odometer reads 93 kilometers. At the 11 hour mark your odometer reads 235 kilometers. How fast were you going for this section of your ride? ___________ Complete the tables. Graph the points and label them A and B. Draw a slope triangle and write the lengths of the vertical and horizontal sides with units. Extend the line to the edges of the graphing window. Determine the y-intercept approximately. Find the function for your odometer reading at x hours.

point	x	y
A	3	
B	11	

Δy	
Δx	
$\frac{\Delta y}{\Delta x}$	

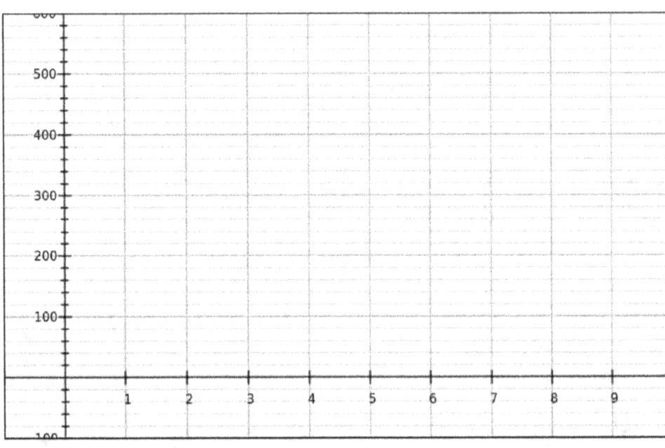

36) You ride a motorcycle from city A to city E. Below is a table which shows your timer and odometer reading at different cities. Note that your speed varies from city to city! Complete the table at right to show your **average velocity** for different sections of your journey.

City	Time (hours)	Odometer (kilometers)
A	0	1230
B	1	1332
C	3.5	1542
D	7.5	1850
E	9	2015

Section	Velocity
A to B	
B to C	
C to D	
D to E	
A to E (overall average velocity)	

37) What was your fastest section of riding? ____ What was your slowest section of riding? ____ On which section of your ride did you travel closest to your overall average speed? _____

38) How did you calculate the velocities? It was probably obvious to you that you needed to subtract the smaller number from the bigger number in both cases. Why do we do that? Would it work if we subtracted the bigger from the smaller in both cases? Why or why not?

39) In fact, it would work if we subtract the bigger number from the smaller number in both cases. Here's how that works for the section from B to C. Complete the arithmetic here.

$$velocity = slope = \frac{\Delta y}{\Delta x} = \frac{1332-1542}{1-3.5} = \frac{\quad}{\quad} =$$

40) Consider the following two points which represent the hours elapsed and odometer readings of a bike at two different moments: $(4{,}245)$ and $(7{,}380)$. How fast is this car going? ________ Sketch a graph, with slope triangle, labeled. Extend the graph to the edges of the graphing window. Approximate where the line crosses the y-axis. Determine a function. Use your function to determine the kilometers traveled at 5 hours. Confirm that this matches with your graph.

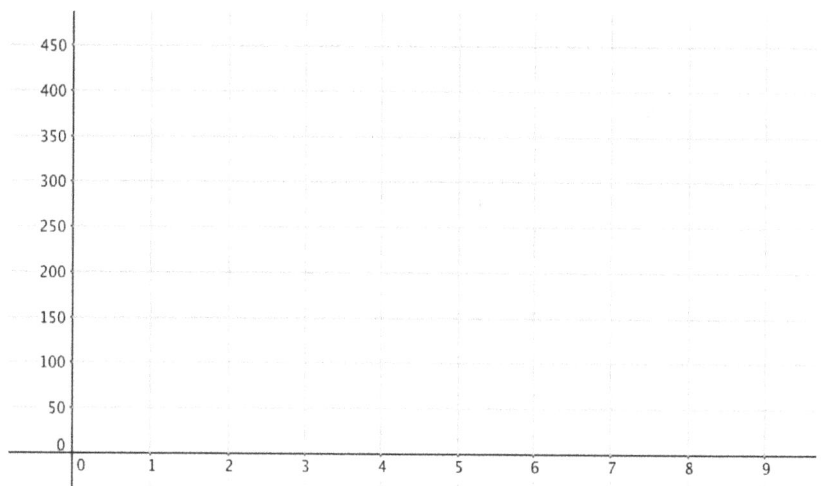

41) Below are three representations of the same function. What is the rate of change for this function? _____ Explain how you find it from the equation, from the table, and from the graph.

$y = 2x + 3$

x	y
0	3
1	5
2	7
3	9

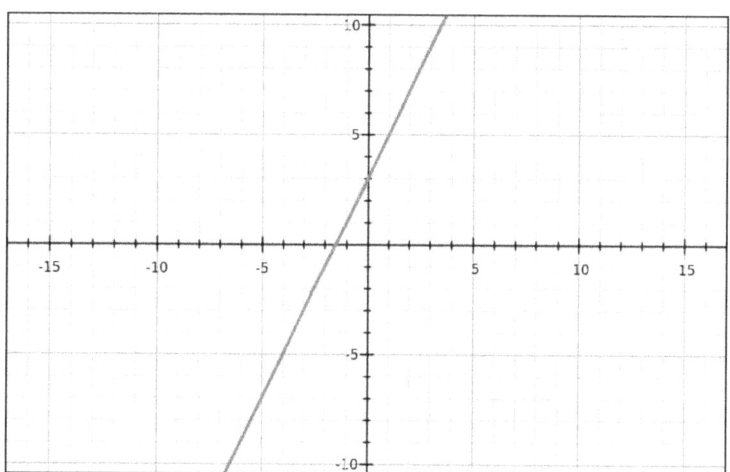

42) Here is a table representing a function. Find the equation for the function.

x	y
0	7
5	18
10	29

$y =$ ______________

43) Here is a graph representing a function. Complete the table and write the equation for the function.

x	y
0	
2	
7	

$y =$ ____________

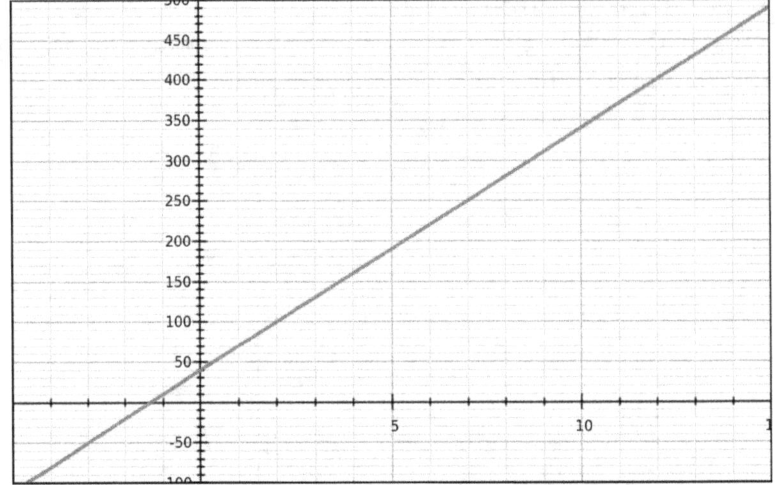

44) Given the points $(2,3)$ and $(7,10)$, create the graph for this linear function. Label axes. Be sure to extend the line to the edges of the graphing window. Write the points in the table below. Find the slope by drawing a slope triangle between these two points. Write the vertical and horizontal distances next to those sides. Confirm this slope correct by finding the slope from the points in the table. Approximate where the graph intersects the y-axis. Add this point to the table. Find the function for the line.

x	y

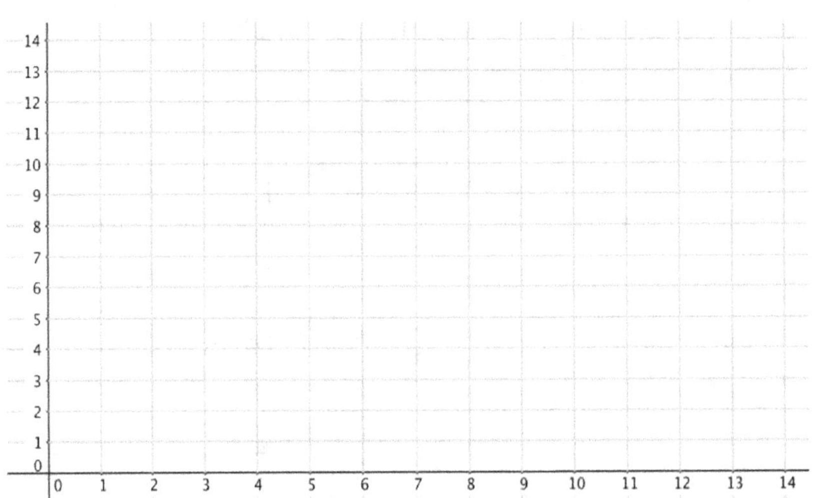

45) Given the points $(2,50)$ and $(8,350)$, create the graph for this linear function. Be sure to extend the line to the edges of the graphing window. Write the points in the table below. Find the slope by drawing a slope triangle between these two points. Write the vertical and horizontal distances next to those sides. Confirm this slope correct by finding the slope from the points in the table. Approximate where the graph intersects the y-axis. Add this point to the table. Find the function for the line.

x	y

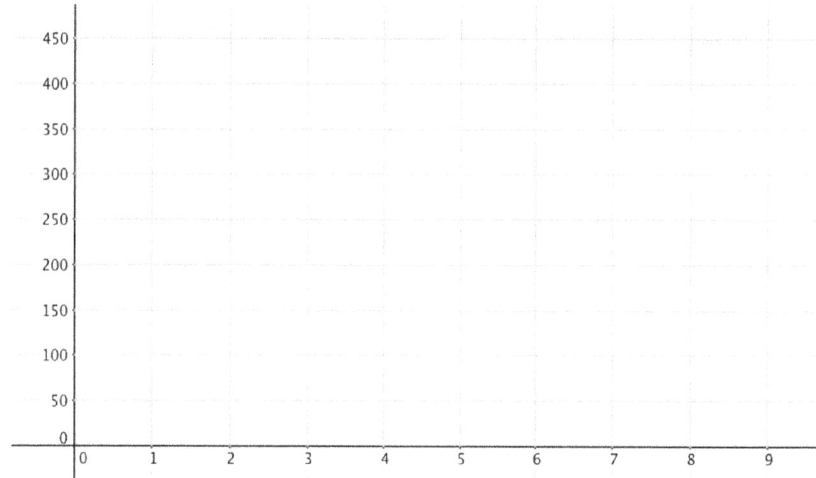

46) Here is a graph representing a function. Write the equation for the function.

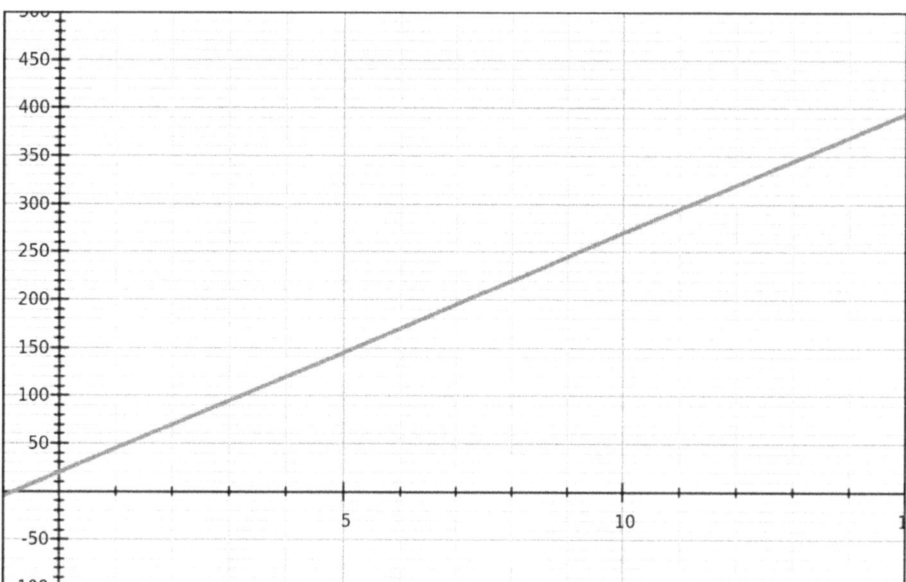

47) Here is a table representing a function. Graph the function using the points in the table. Write the equation for the function.

x	y
0	12
1	22
2	32

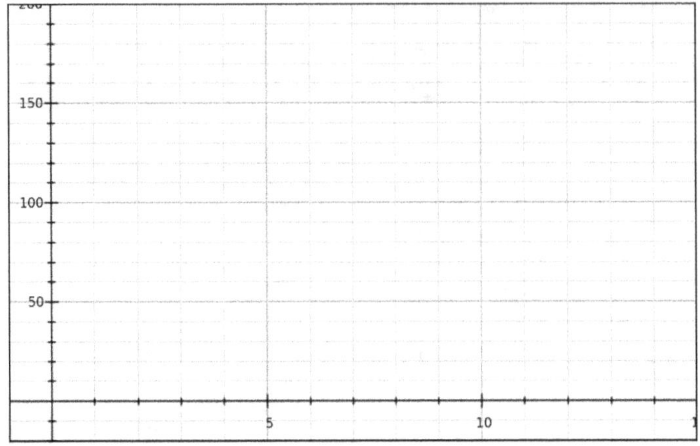

48) Use the above equation for your function to find the y-value for an x-value of 9. Confirm that the graph gives approximately the same value.

49) Below is an equation for a function. Based on this equation, what is the slope? ________Use your function to complete the table. Use the table to make a graph. Use a slope triangle to show the slope in the graph.

$y = 15x + 45$

x	y
0	
5	
10	

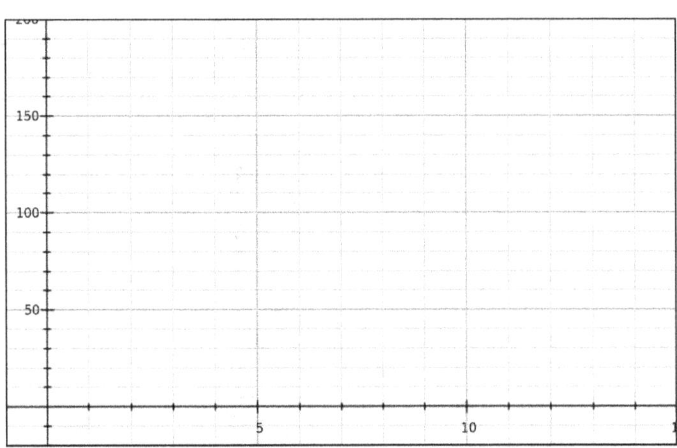

50) Below is an equation for a function. Based on this equation, what is the slope? _________Use your function to complete the table. Use the table to make a graph. Use a slope triangle to show the slope in the graph.

$$y = \frac{15}{2}x + 35$$

x	y
0	
6	
10	

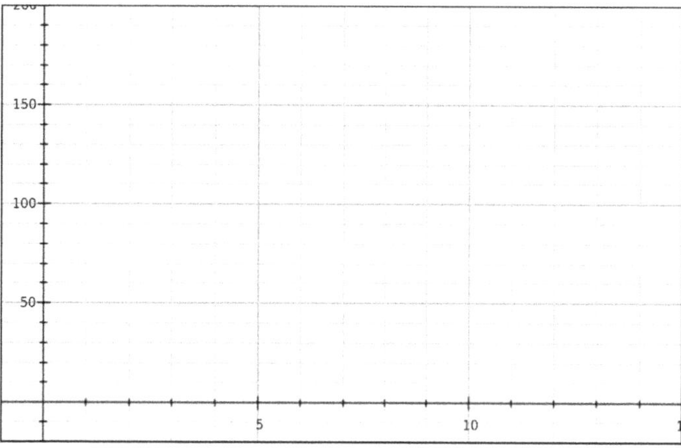

51) What are the coordinates of the y-intercept above? ________ Where is the y-intercept represented in the equation? Why?

52) Summarize in your own words the meaning of slope and how to find it. Use equations, tables and graphs to support your explanation. Look on page 7 as a source for your summary, but enrich it with your own language and perspective.

The Take-Home Message: The rate of change of a linear function is its slope. it is the ratio of $\frac{\Delta y}{\Delta x}$ for two points on the line, or the vertical change over the horizontal change in the graph. It shows up as the coefficient of the linear term in the equation, while the y-intercept shows up as the constant term.

LF3 *Finding Functions*

Name(s)

Date Class/Period/Group

1) Imagine a genie who converts the number of candies you bring him into a new number of candies. The genie converts your candies based on the function $y = 2x + 5$. This means if you bring him 6 candies (that's the input, x), he returns 17. (that's the output, y).

It is great to run into a genie, but sometimes you might not know the function for the genie. However, if we see a few people visit a certain genie and give candies as input and receive candies as output, we can figure out the function and decide which one is best for you. Consider each genie below. What is the function for each? Write your answer as a linear function. The first one is given for you as an example.

Genie f $y_1 = 5x + 15$

Name	x	y_1
Alice	0	15
Bob	1	20
Chris	2	25

Genie g

Name	x	y_2
Adam	0	50
Billy	1	65
Cathy	2	80

2) Here are a couple more. Be careful here and make sure you got it right! you can check by substituting each x-value into the function and confirming that it produces the expected y-value.

Genie f

Name	x	y_1
Alice	0	0
Bob	1	20
Chris	2	40

Genie g

Name	x	y_2
Adam	0	10
Billy	2	50
Cathy	4	90

3) And a couple more…be careful!

Genie f

Name	x	y_1
Alice	0	12
Bob	2	32
Chris	10	112

Genie g

Name	x	y_2
Adam	0	16
Billy	3	61
Cathy	10	166

4) Explain in words the various techniques you used to figure out the problems above.

5) Sometimes we might be given a graph of input and output instead of a table. Below is a graph of one genie's output vs the input. Label the y-axis "output" and the x-axis "input." What is the approximate output for an input of 5? ____ Approximately how many candies would you need to give him in order to get 80 candies? ____

6) Use the graph to determine the function for this genie. Write it below in the blank below:

$y =$ ______________

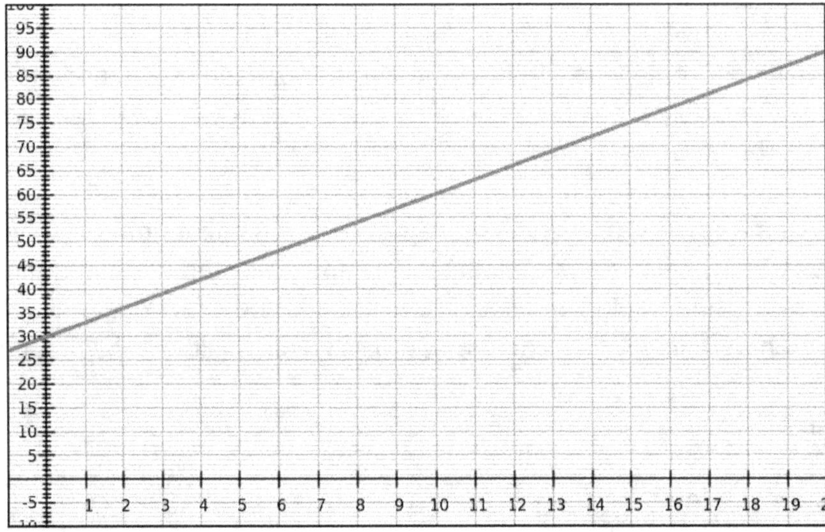

7) Using the function you wrote above, determine the amount of candies returned for an input of 30. ____ Write and solve an equation to determine the number of candies needed to get 80. Check to see that this answer matches up well with the graph.

8) Use the graph to determine the function for this genie. Write it below in the blank below.

$y =$ ______________

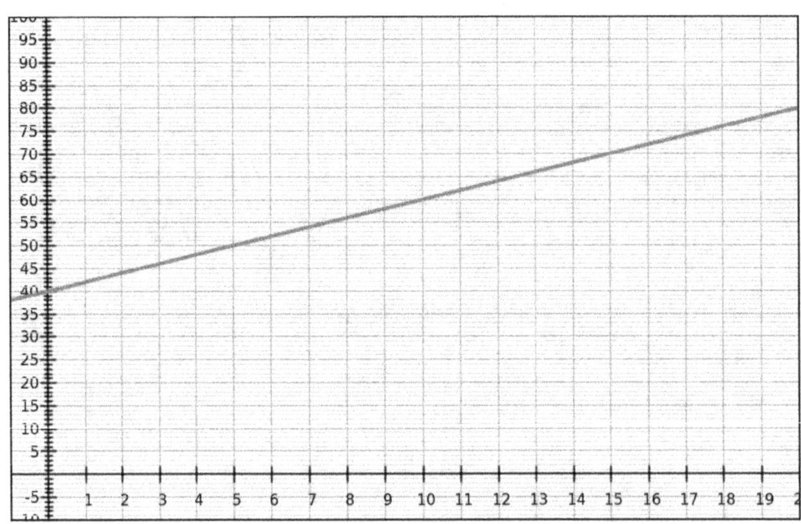

9) The first two pages of this mission challenged you to develop tools for finding the equations for functions given a table or a graph. Summarize the algorithms (procedures) you developed so far for finding equations from a table or from a graph. Give examples to support your summary.

10) Previously in the course you explored different ways to think about slope. Here is a table showing a summary of the various places slope shows up. Review this carefully and refer back to it frequently as we refine our algorithms for finding the equation for a linear function.

Slope

Verbally	Algebraically	Algebraically Again	From a table	From a graph
Slope Rate of change Velocity	$m = \frac{\Delta y}{\Delta x}$	$m = \frac{y_2 - y_1}{x_2 - x_1}$	$\Delta x \leftarrow$ [table: x, y; x_1, y_1; x_2, y_2] $\rightarrow \Delta y$	A, B, C $m = \frac{vertical\ change}{horizontal\ change}$ $m = \frac{BC}{AC}$
Example	a car travels $120\ km$ in $3\ hrs$ $m = \frac{120\ km}{3\ hr}$ $m = 40 \frac{km}{hr}$	Point A: $(20,130)$ Point B: $(30,170)$ $m = \frac{40}{10} = 4$	$2 \leftarrow$ [table: x, y; 5, 12; 7, 18] $\rightarrow 6$ $m = \frac{6}{2} = 3$	A, B, C; 0 1 2 3 4 5 6 7; 0 1 2 3 4 5 6 $m = \frac{2}{4} = \frac{1}{2}$

11) You have also seen previously that the slope shows up in a linear function as the coefficient of the linear term, and the y-intercept shows up in a linear function as the constant term. Let's briefly review this by graphing a linear function. Make a table and graph. What is the slope, and what is the y-intercept?

$y = 20x + 50$

x	y
0	
1	
5	
7	

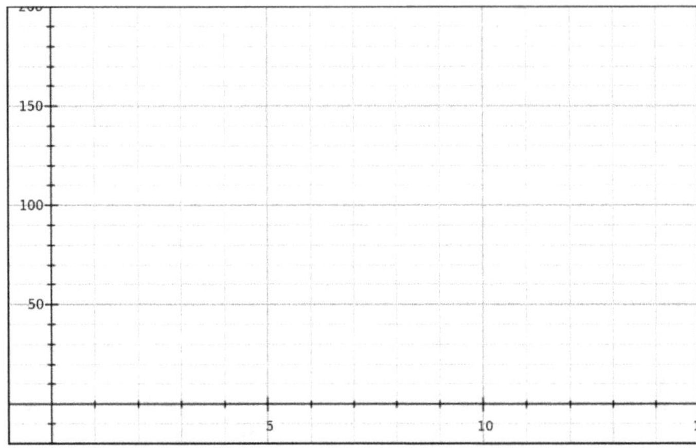

12) To summarize, you know how to find the slope from a table or graph or description using the tools on page 3. You can also find the y-intercept easily when it is visible in the graph or the table. And you know where to put the slope and the y-intercept when we are writing an equation. Specifically, we write a linear equation like this:

Slope-Intercept Form For a Line

$$y = mx + b$$

m is the slope, b is the y-intercept

The point of writing this general form for a line is to remind you that the coefficient of the linear term is the slope and that the constant term is the y-intercept. It also allows you to solve for m or b when needed, as you'll see in some examples soon. Quiz each other to ensure you've remembered the name, structure and meaning of the parts of the general equation above.

13) The table below represents a linear function. Use the tools on page 3 and 4 to find the equation for the function. Confirm your function is correct by plugging in the x-value that you didn't use during your process and confirming that it produces the indicated y-value.

x	y
0	7
1	10
5	13

14) When the y-intercept is clearly visible in the table or graph it's fairly easy to find the equation for the line. What can we do when the y-intercept isn't obvious? Look at the example below. It shows a procedure using our knowledge of slope combined with slope-intercept form for a line.

How to Find the Equation for A Line

	x	y
A	2	9
B	5	21
C	10	41

Example	Algorithm
$m = \frac{\Delta y}{\Delta x} = \frac{12}{3} = 4$	1) Find slope
$y = mx + b$	2) Write general slope-intercept form for a line.
$9 = 4(2) + b$	3) Substitute known information.
$b = 1$	4) Solve for b.
$y = 4x + 1$	5) Write specific function.
$y = 4(10) + 1 = 41$ ☺	6) Confirm function is correct using other point.

In step 1 above I used points A and B to find the slope. Could I have used points A and C? How about B and C? Why or why not? In step 3 above I substituted point A into the function. Could I have used B or C? Why or why not?

15) Find the equation for the line represented in the table below.

	x	y
A	3	7
B	8	22
C	10	28

Show work here	Algorithm
	1) Find slope
	2) Write general slope-intercept form for a line.
	3) Substitute known information.
	4) Solve for b.
	5) Write specific function.
	6) Confirm function is correct using other point.

16) Given the following results from genie encounters, find the function for each.

Genie f

Name	x	y_1
Alice	0	9
Bob	1	29
Chris	2	49

Genie g

Name	x	y_2
Adam	1	12
Billy	2	17
Cathy	3	22

$y_1 =$ ____________ $y_2 =$ ____________

17) Find and write the functions as we did above.

Genie f

Name	x	y_1
Alice	3	0
Bob	5	14
Chris	7	28

Genie g

Name	x	y_2
Adam	0	4
Billy	1	55
Cathy	2	106

18) Find and write the functions.

Genie f

Name	x	y_1
Alice	2	0
Bob	5	14
Chris	8	28

Genie g

Name	x	y_2
Adam	0	4
Billy	1	55
Cathy	2	126

19) Find and write the functions.

Genie f

Name	x	y_1
Alice	1	0
Bob	3	20
Chris	5	40

Genie g

Name	x	y_2
Adam	0	11
Billy	2	41
Cathy	4	71

20) Find and write the functions.

Genie f

Name	x	y_1
Alice	2	17
Bob	5	47
Chris	8	77

Genie g

Name	x	y_2
Adam	1	12
Billy	2	22
Cathy	3	32

21) From the graph below, approximately how many candies are returned for an input of 8 candies? ______________ Approximately how many candies input return 70? ___________ What is the function?

$y =$ 

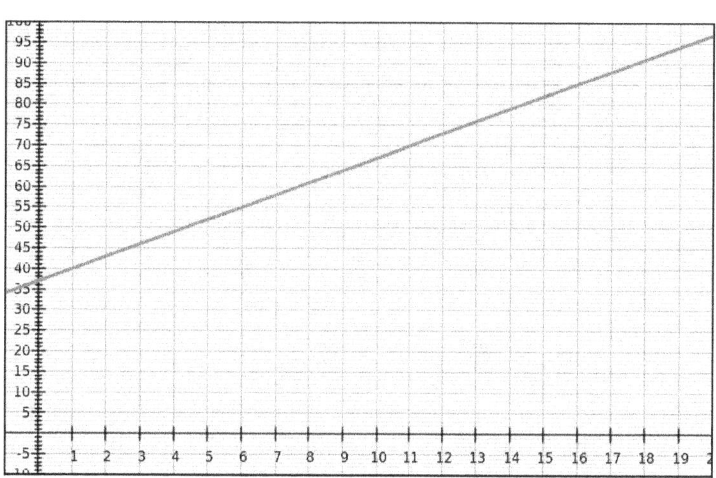

22) Summarize the relationship between the table, the graph and the equation for a line. Explain how the slope and the y-intercept are expressed in each. Use an example to support your explanation.

23) Demonstrate and explain the algorithm for determining the equation for a line from a table or from a graph. Use an example to support your explanation.

The Take-Home Message: The equation for a line includes a slope ($\boldsymbol{m}$) and a y-intercept ($\boldsymbol{b}$). Slope-intercept form for this equation is $\boldsymbol{y = mx + b}$. The equation can be determined from any two points on a line. These points can be given in a table or determined from a graph.

LF4 *Linear and Non-linear Models*

Name(s)

Date Class/Period/Group

1) Match each of the following equations with its meaning. (You may have done this in a prior mission.)

Equation	Description	Units
$D = \frac{m}{V}$	1) **Force** is **mass** times **acceleration**. When you swing a hammer towards a nail, you are accelerating a mass and thus it hits the nail with a force.	Newtons (N)
$E = mc^2$	2) **Power** is **work** over **time**. If it takes you a long time to do your work, clearly you are not that powerful yet!	Watts (W)
$F = ma$	3) The **circumference** (distance around) a circle is ($\pi \approx 3.14...$) times the **diameter**. That means as the diameter of a circle increases, the circumference increases proportionally. That makes sense!	m
$C = \pi d$	4) The **final velocity** squared of an object is the square of the **initial velocity** plus twice the acceleration times the **distance** travelled.	m/s
$s = \frac{1}{2}at^2 + v_i t$	5) **Einstein**'s famous equation which describes how a very small amount of matter (mass m) can be turned into a large amount of **energy** (E). The **speed of light** (c) is a big number! So in the equation, you multiply the mass times an even larger number (c^2), which results in a huge amount of energy.	Joules (J)
$v_f = v_i + at$	6) The **momentum** (p) of an object is the mass of the object times its velocity. For example, in a crash-up derby, if a robot-driven Mack truck collides head-on with a robot-driven Mini going at the same velocity, the truck clearly had more momentum and the crashed vehicles will keep moving (more slowly) in the direction of the truck's travel. Momentum keeps you moving even when lesser and/or slower objects push back!	N · s
$v_f^2 = v_i^2 + 2as$	7) **Density** is the mass divided by the **volume**. For example, an empty box has only a little mass and a lot of volume, which means it isn't very dense. A box of pencils has greater mass in the same volume—higher density!	kg/m^3
$a^2 + b^2 = c^2$	8) **Work** is defined as force times **displacement** (s). If you carry a book bag 100 meters up stairs you've got to apply more force to lift the bag, so that's a lot more work than carrying it 10 meters upstairs! And the heavier the bag, the more force, and therefore the more work.	Joules (J)
$p = mv$	9) The **gravitational force** between two objects in space is calculated by multiplying a constant times the mass of the 1st times the mass of the 2nd, then dividing by the square of the distance between them. This means that the bigger the masses, the bigger the force, but the bigger the distance the smaller the force since you divide by distance squared.	Newtons (N)
$W = Fs$	10) The displacement is half the acceleration times the square of the time, plus the initial velocity times the time.	m
$P = \frac{W}{t}$	11) The **Pythagorean Theorem**, which relates the length two **legs** (a and b) of a **right triangle** to the **hypotenuse** (c).	m
$F = \frac{Gm_1m_2}{r^2}$	12) The final velocity of an object is the initial velocity plus the acceleration times the time that has passed. For example, a car going $30\ km/hr$ which accelerates at $10\ km/hr/hr$ for 2 hours will then be going $50\ km/hr$ at that time. (That was a very slow acceleration!)	m/s

2) Imagine a genie that takes the number of candies you bring him and magically transforms them into a new number of candies using a function, like $y = 4x + 5$, where x is the input and y is the output. But not every genie is linear as in that example. Some genies are so generous that they square your input! Complete the table for the function below and graph the function. Do NOT connect the points with lines. Why not?

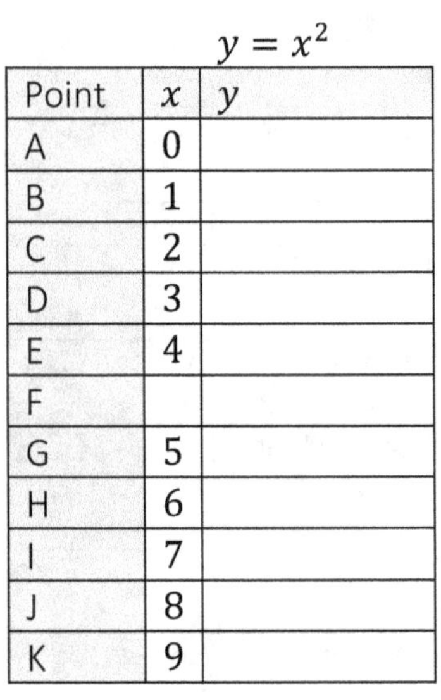

$y = x^2$

Point	x	y
A	0	
B	1	
C	2	
D	3	
E	4	
F		
G	5	
H	6	
I	7	
J	8	
K	9	

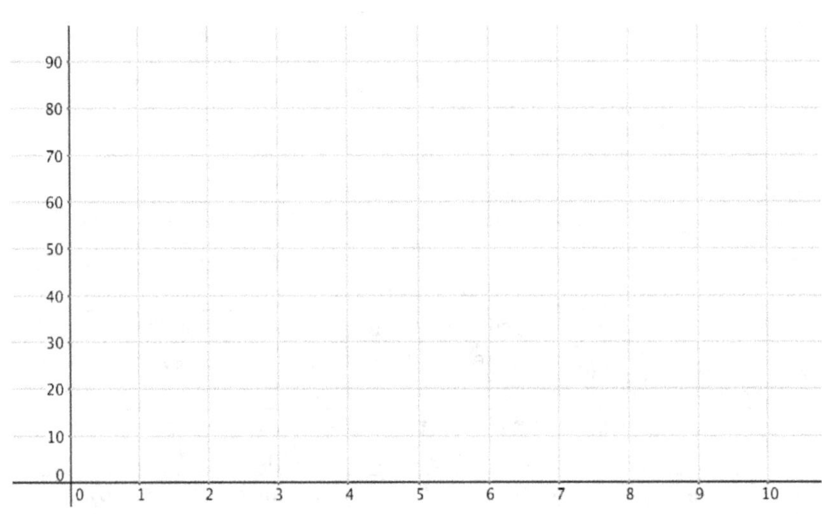

3) Connect the points above with a smooth curve. Based on your graph, what would be the output for an input of $4\frac{1}{2}$? ______ Put this in the table as point F and label it point F on your graph.

4) When the input is increased from 3 to 4, how much does the output increase? ______ When the input is increased from 7 to 8, how much does the output increase? ______ Is that the same or different from what happens with linear functions? Why or why not?

5) In a different color, graph a slope triangle between points D and E. Put the lengths of the sides next to the sides. $\frac{\Delta y}{\Delta x} =$ ______. Do the same for points H and I. $\frac{\Delta y}{\Delta x} =$ ______. Are the slopes the same or different? Why or why not?

6) Is the function you graphed in #3 linear? ______ Is it a polynomial function? ______ Is it a quadratic function? ______ Give two examples of equations for linear functions below. Give two examples of equations for quadratic functions below.

7) Read these carefully and discuss them until you understand them. Rate them based on your level of understanding, with 5 being the highest.

Quantity	Meaning	Units	Example	1-5
Force	mass accelerated	Newtons (N)	The typical apple on earth applies 1 Newton of force to your hand.	
Energy	Force times distance	Joules (J)	Lifting the apple 1 meter requires 1 Joule of energy.	
Work	Force times distance	Joules (J)	Lifting the apple 1 meter requires 1 Joule of energy. (Yep, same units as energy, same description.)	
Power	Work over time	Watts (W)	Powering your phone requires about 1 Watt of power. (Put another way, you have to lift an apple 1 meter to power your phone for one second.)	
Momentum	mass times velocity	$N \cdot s$	An apple traveling at 10 m/s (36 km/hr) across the room has 1 $N \cdot s$ of momentum. This also means it would take a force of 1 N applied for 1 second to stop it.	
Density	mass over volume	$\frac{kg}{m^3}$	Air density in Salt Lake City is about 1 kg/m^3 A kg per every cubic meter, wow!	

8) Imagine a cubic meter of air in Salt Lake City. That's 1 kg of air. Sketch a large cubic meter below on the isometric grid paper where each unit is 1 decimeter. Make sure that each edge is 10 decimeters to make each edge one meter. What is the density of this air? What are the units for this density? Visualize a cubic meter in the room. Why is it difficult to notice that there is a kilogram of air in each cubic meter of space around you?

9) Give five examples of things that are denser than air below, with approximate densities.

10) As we increase in elevation, does air become denser or less dense? ______ Guess the density of air at 10000 meters. ______

11) Do you think the relationship between elevation and density is linear? _______ Why or why not?

12) Below is a table showing the different densities of air at different elevations. Graph these on the axes provided. Do NOT connect the dots with a straight line. Instead, use a smooth curve. Label the x-axis "elevation (km)" and the y-axis "density (kg/m^3)."

x (km)	y (kg/m^3)
0	1.2
1.5	1
3	0.9
7	0.7
10	0.4
30	0.1
40	0.03

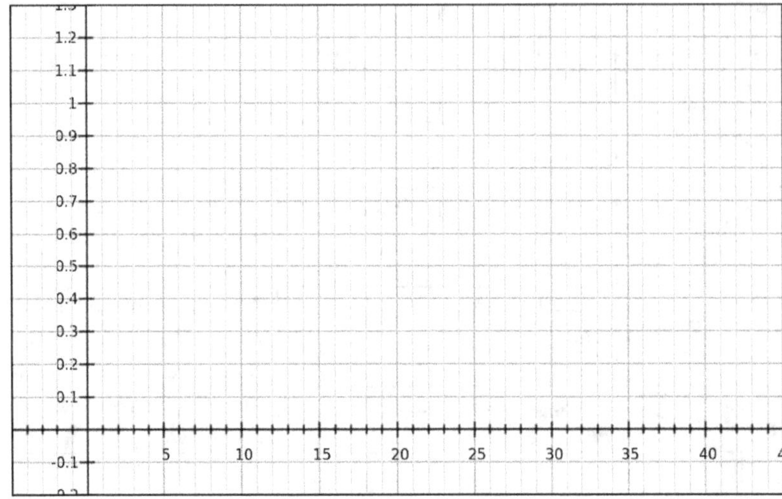

13) So, one more time, would you describe this relationship as linear? ______ Do you think it is possible to have an air density of 0? ______ Do you think it is possible to have an air density below 0? ______ Why or why not? Based on your graph, what is the approximate density of air 20 km above the surface of the earth? ______ Does this fit in your table as well? ______ Add it there. Do you think this function is linear, quadratic, or something else? Why? Describe the **behavior** of the function as we get close to sea level. Describe the behavior of the function as we go higher and higher.

14) Rank the materials at right from less dense to denser, guessing the density in kg/m^3 for each.

material	rank	density
Styrofoam		
fresh water		
cork		
salt water		
balsa		
oak		
ice		
helium		
space		

15) Supposing an apple with a mass of 0.2 kg falls from a cliff. Does it fall at a constant velocity or does it accelerate? _______ Suppose that a rock with a mass of 5 kg falls from the same height. Ignoring air resistance, does it accelerate at the same rate as the apple? _______ Acceleration due to gravity is a constant on the surface of the earth. Specifically, it is about 10 m/s^2, which means that an object dropped accelerates at a rate of 10 m/s^2 every second. How fast would that apple be falling after 3 seconds?

16) What is the equation for force? ______ If I am holding an apple whose mass is 0.1 kg, how much force is being applied to my hand? _______ How about if I am holding a 2 kg chocolate bar? ______ How about if you are holding your friend whose mass is 40 kg? ______ As the mass increases, does the acceleration increase? _______ As mass increases, does the force increase? _______ Why or why not?

17) So, on earth we could say that the force applied to your hand is **given** by the equation:

$$F = m(10)$$

This is because the acceleration (a) due to gravity on earth is 10 m/s^2. The a never changes for us here on earth, so for us it is a constant. Re-arranging so the coefficient is before the m gives:

$$F = 10m$$

Is this equation linear? How do you know?

18) Complete the table and graph Force **as a function of** mass. Label axes and extend the graph to the edges of the graphing window.

$F = 10m$

m	F
0	
1	
2	
5	
10	
20	
50	

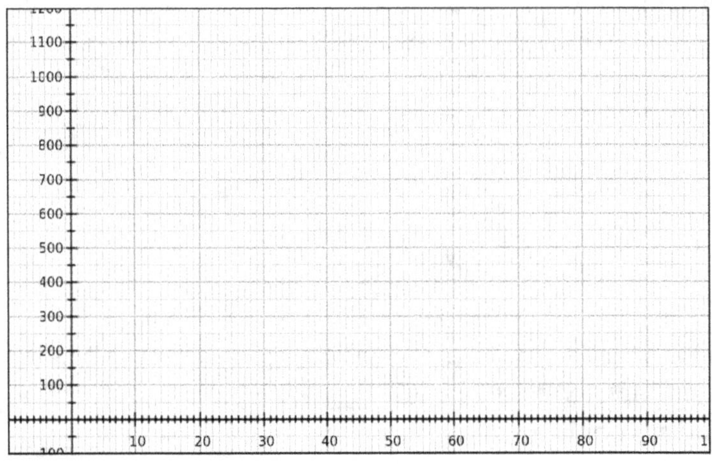

19) What is the force applied to your hands when you are holding 30 kg? Identify this as point A on the graph. Write and solve an equation to determine what mass applies a force of 850 N. Identify this as point B on the graph.

20) Of course, on the moon it's totally a different story! Acceleration due to gravity on the moon is only 1.6 m/s^2. That means an object dropped on the moon accelerates ______ m/s every second. Which means after 3 seconds of falling it would only be going ______ m/s.

21) What is the function for the amount of force applied to your arms when you carry mass on the moon? ____________ Find the force required to carry 100 kg on the moon. Is this more or less than the force required to carry 30 kg on the earth that you found in #3? Using the force required to carry 30 kg on earth, how much could you carry on the moon?

22) Graph force as a function **of** mass on the moon below. Note that there is already one line graphed. It is the graph of force on earth as a function of mass. Label axes. Label the given line by writing "$F = 10m$" along the line.

$F = 1.6m$

m	F
0	
1	
2	
5	
10	
20	
50	

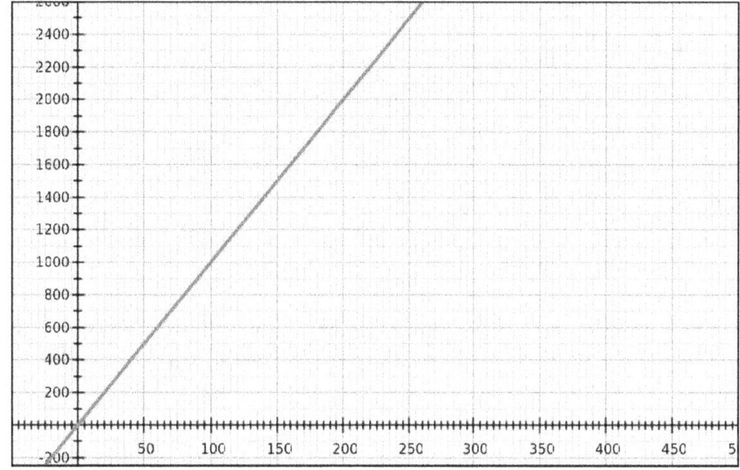

23) What is the maximum mass your entire group working together could carry on the earth? Find the force required to carry this mass on earth. Graph and label this point E on the earth graph. Use this force to determine the mass your group could carry on the moon. Graph and label this point M on the moon graph. Explain any interesting relationships between these two points. The mass of a smart car is 750 kg. Could your group carry a smart car on the moon? Why or why not?

24) On earth, the world record for dead lifting weights is a nearly unfathomable 460 kg. How much force does that require on earth? How much could that record holder lift on the moon?

25) Look at #1 and #7 to help you with this problem. What is the equation for power? ______ How much power is required to lift an apple 1 meter in 1 second? ______ How much power would be required to lift the same apple 1 meter in 2 seconds? ______

26) Usain Bolt ran the 100 meter dash in less than 10 seconds and broke the world record. In order to achieve this feat, Bolt produced roughly 2600 Watts of power every second to move his body during his 10 second run. That means he converted 2600 Joules of energy into motion and overcoming air resistance every second, or 26000 Joules per 10 seconds. (Remember, a Joule is lifting an apple a meter in a second.) Bolt's 2600 Watts of power is about what you need to sustain the amplifiers at an outdoor rock concert. He rocked, basically. Let's suppose my friend Carl has half the power of Usain Bolt. Super roughly, it would take him 20 seconds to run the dash. And the more time it took someone, the less power they would use. Here is a table showing this idea. Label the axes, graph and label the points, connect with a smooth curve to the edges of the graphing window.

$$P = \frac{26000}{t}$$

	t (sec.)	P (Watts)
Bolt	10	2600
Carl	20	1300
Earl	40	650
Frank	80	325

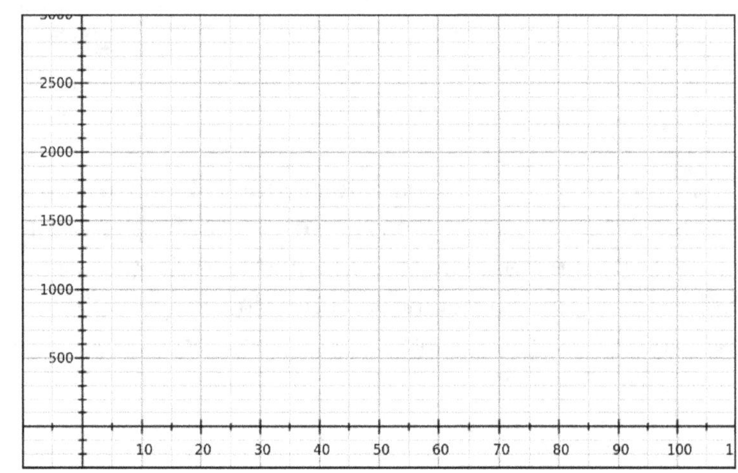

27) Is this function linear? _____ How do you know?

28) How much power would be required to run the dash in 30 seconds? _____ Put this point in the table, then graph and label this point D. How fast could you run the dash if you used 3000 Watts of power?

29) Graph the following two functions by completing the tables and then graphing. (Note that the first genie is a bummer for Alice, Bob and Chris because just by encountering them they end up with a candy deficit!)

Genie f $y_1 = x^2 - 8$

Name	x	y_1
Alice	0	
Bob	1	
Chris	2	
	3	
	4	
	5	

Genie g $y_2 = \frac{1}{2}x + 2$

Name	x	y_2
Adam	0	
Billy	1	
Cathy	2	
	3	
	4	
	5	

30) Which of these functions is linear? ________ For f, what x-value produces a y-value of 7? ______ Table, graph and label this point G.

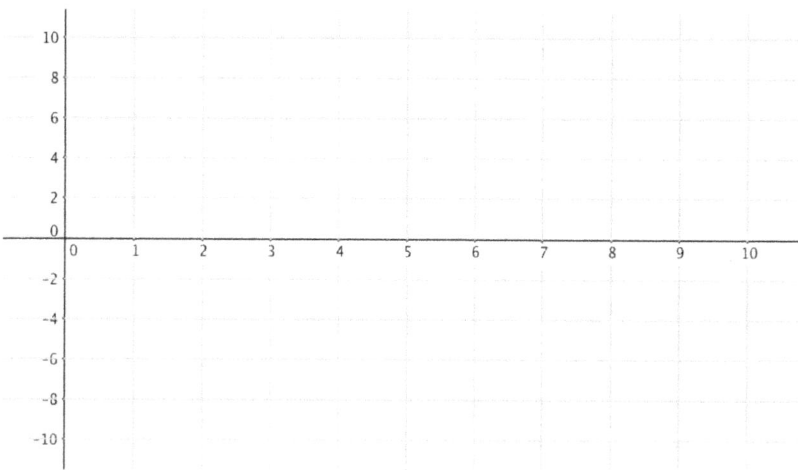

31) For g, based on the graph, approximately what x-value produces a y-value of 4.5? ______ Label this point H.

32) At approximately what point do the graphs of the two functions cross? ___________ Graph and label this point I for intersection. This **intersection point** shows us an x-value produces the SAME y-value in BOTH functions. Now let's see if that's true! Take the x-value for this point and evaluate the function f at this x-value, then do the same for g. Are the resulting outputs pretty close? Are the resulting outputs close to the y-value for point I on the graph? Do these results match up with point I on the graph? If not, review your work, starting with the tables in #29.

33) What is the equation for Force? ________ What are the units for Force? ________ What is acceleration due to gravity on the earth? ________ What is the maximum kilograms you can carry? _______ How many Newtons of force must you apply in order to carry that mass?

34) Now that we know how much force you've got in you, let's see if you can use the force to stop a car. Yes, a car. It's a convertible smart car with a mass of 750 kilograms. Go back to #1 and #7 and review momentum. What is the equation for momentum? ________ What are the units? ________ It probably seems a little strange that momentum is "mass times velocity" and yet the units are "force times time." Here is one way to explain that. Give a reason for each step below.

Statement	Reason
$p = m \cdot v$	
$kg \cdot \frac{m}{s}$	
$kg \cdot \frac{m}{s^2} \cdot s$	
"mass · acceleration · time"	
"Force · time"	
$N \cdot s$	

35) The above reasoning gives us a second, different equation for momentum: $p = Ft$

So imagine that 750 kg smart car is going at 10 m/s. You are a passenger in the car, wearing skate shoes. Suddenly the brakes and steering give out. You jump out of the car and land in front of the car, pushing against the hood with your skate shoes sliding on the asphalt. What is the momentum of the car? _______ So the question is how long will it take you to stop the car? Based on #33, how much force can you apply to the hood of the car? _______ So how long will it take you to generate the equivalent momentum to effectively stop the car?

36) Following are several linear functions which represent the momentum of cars as a function of velocity. Complete the tables and graph each of them in different colors, accurately (on the same axes) to the edges of the graphing window. Label the x-axis "velocity (m/s)" and the y-axis "momentum (N· s). Label the lines using their function name (f, g, etc.). Find the equation for each function and write it above p in the corresponding table. Find the mass of each car.

Car f

Point	v	p
A	0	0
B	10	7500

Car g

Point	v	p
C	0	0
D	20	7500

Car h

Point	v	p
E	0	0
F	1	1000

Car i

Point	v	p
G	0	0
H	2	4000

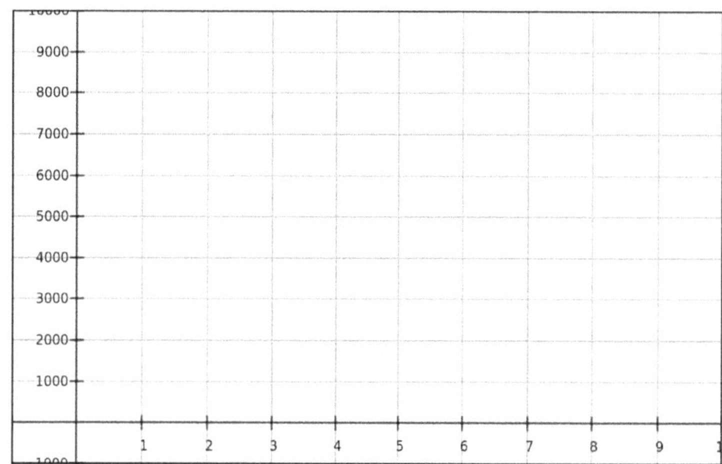

37) Complete the tables below by creating examples of linear functions, and examples of non-linear functions, using the equations in #1. Explain the meaning of each as shown in the examples below.

Linear

Function	Explanation
$F = 10m$	The force of an object on earth as a function of its mass. Acceleration (a) is a constant, $10\ m/s^2$.

Non-linear

Function	Explanation
$s = 3t^2 + 4t$	The displacement of an object as a function of time. The object is accelerating at $6\ m/s^2$ with an initial velocity of 4 m/s.

38) How can you tell the difference between linear and non-linear functions?

The Take-Home Message: Functions are models of quantities like density and force which exist in the real world. Sometimes these functions are linear and sometimes not. Recognizing linear and other types of functions makes it easier to work with them.

Courage To Core

Share the Discovery

LF5 Linear Models and Problem Solving

Name(s)

Date Class/Period/Group

1) A gym membership costs $50 to open, and then $30 a month thereafter. Another gym membership costs $100 to open, then $25 a month thereafter. Different people enroll for different amounts of time. Complete the tables for each gym to show the cumulative costs for the indicated months. Write the function for each gym above each table. Label the axes and graph and label the functions. Be sure to extend your lines to the edges of the graphing window.

Gym f

Gym Member	x	y_1
Alice	0	50
Bob	1	
C	2	
	3	
	5	
	8	

Gym g

Gym Member	x	y_2
Arnie	0	100
Brad	1	
C	2	
	3	
	5	
	8	

3) Which gym would you join? Why?

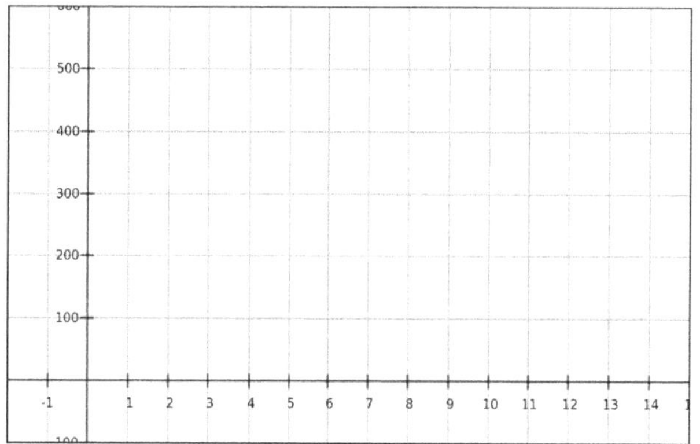

4) Write and solve an equation to determine the month the gym memberships have the same cumulative cost. Find the corresponding cost for each gym. Graph this as point I. Does it match up with where the lines cross? Why or why not?

5) Base jumping is a sport which involves leaping from building, cliffs or towers, then pulling a parachute and gliding to safety. More recently, base jumpers have decided that simply jumping isn´t enough. They want to fly. They use special wingsuits to glide rather than fall, and in this way they are able to jump off of cliffs where the ground is very close, and are able to glide close to the cliff face, or through narrow canyons. It's dangerous, and first off they need to jump from high enough that the chute has time to open. What do you think is the lowest elevation a person can jump from and still have time for their chute to open?

6) Frank wants to jump from 1200 meters. He will be gliding at exactly a $45°$ angle. What is the slope of a line that is descending at a $45°$ angle? _____ Is $-\frac{100}{100}$ the same or different from this number? _____ Label the x-axis "horizontal distance (m)" and the y-axis "vertical height (m). Fill in the table, then write and graph the function for Frank's flight, extending the line to the edges and labeling it y_1.

horizontal distance (m) x	vertical distance (m) y_1
0	1200

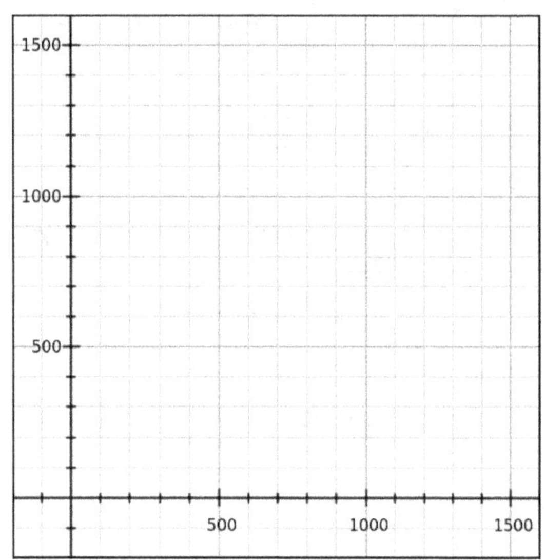

$y_1 =$ ____________

7) Jim is jumping from only 1000 meters up the cliff, and has the same **angle of depression**. Write his function y_2 then graph and label.

$y_2 =$ ____________

8) Assuming they pull their chutes and continue at the same angle (just moving more slowly!), based on the graph, how far away from the cliff base does Frank land? _______ How far away from the cliff base does Jim land? _________ Label these points A and B. These two points are called **x-intercepts**. Why?

9) Write and solve two equations to answer the questions in #8.

10) Now Frank wants to jump from a cliff that is 650 meters high. The slope of his glide path this time will be $-\frac{1}{2}$. Write a function for Frank's glide path and graph it below. Jim is jumping from 500 meters with a slope of $-\frac{1}{3}$. Write his function and graph. Label axes as before.

$y_1 =$ ______________

$y_2 =$ ______________

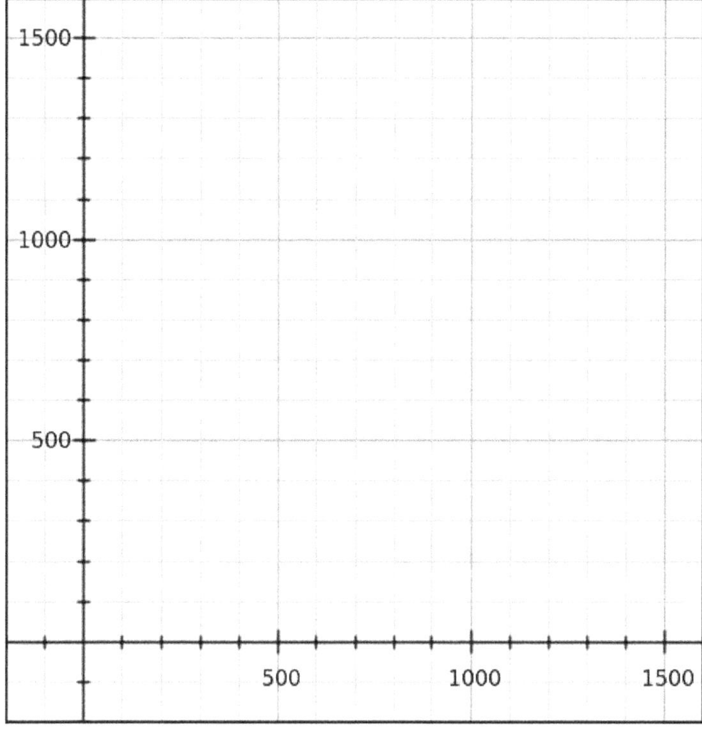

11) The heights they are jumping from are called y-intercepts. Why are they called that?

12) Write and solve equations to determine where they land. Label those points A and B.

13) In the above situation, Frank and Jim jump at the same time. Write and solve an equation to determine the exact point when they are both the same distance from the cliff face and at the same height. Label this point I.

14) Using a protractor, measure then write the angle of depression for each of the fliers and write them here: $\angle 1 =$ ______ $\angle 2 =$ _______ This angle is measured from a horizontal at the y-intercept, rotating clockwise.

15) Now Frank wants to jump from a cliff that is 1150 meters high. The angle of depression of his glide path will be $65°$. Jim is jumping from 700 meters with an angle of depression of 25 °. Use a protractor to measure the angle of depression then graph the glide path for each.

16) Draw slope triangles to find the slope for each flier. Use your calculator and write your answer as a decimal rounded to tenths place.

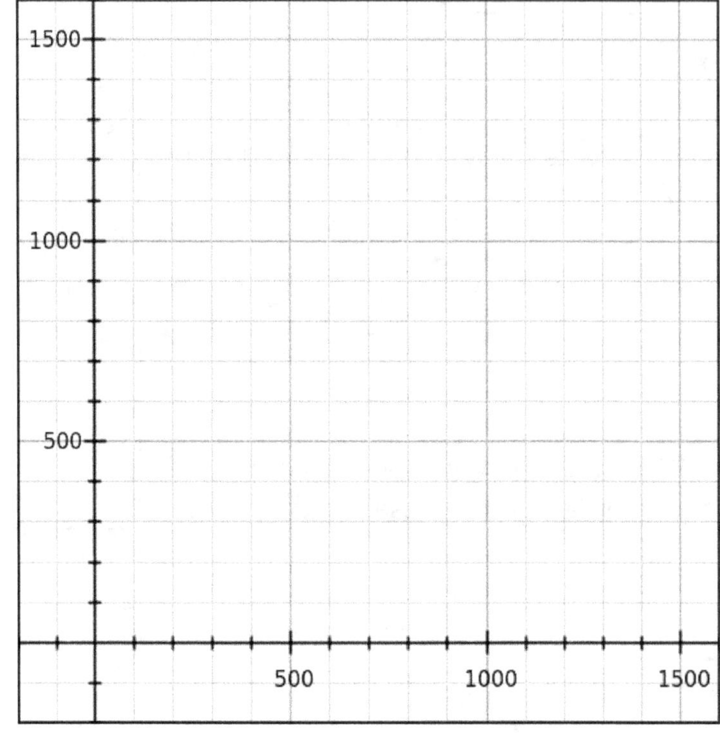

$m_1 =$ ____________

$m_2 =$ ____________

Now write the equation for each line:

$y_1 =$ ____________

$y_2 =$ ____________

17) Write and solve equations to determine where they each hit the ground. Label these points A and B.

18) Write and solve an equation to determine the point where the two fliers cross paths.

19) Using your protractor and a ruler, carefully draw a slope triangle below with an angle of depression of $15°$. Measure and write the side lengths with a ruler and calculate the slope with your calculator, rounding results to hundredths place. A horizontal line is provided for you to use as as a side (or part of a side).

____

20) Create a flier with a $40°$ angle of depression who jumps from high enough on a cliff to hit a target on the ground that is 1000 meters out from the base of the cliff. What is your y-intercept? ____________ What is your x-intercept? ____________ Label these points A and B on your graph. Find the slope using these points and your calculator rounding to tenths place, then find the function. Write the function here:

$y =$ ________________

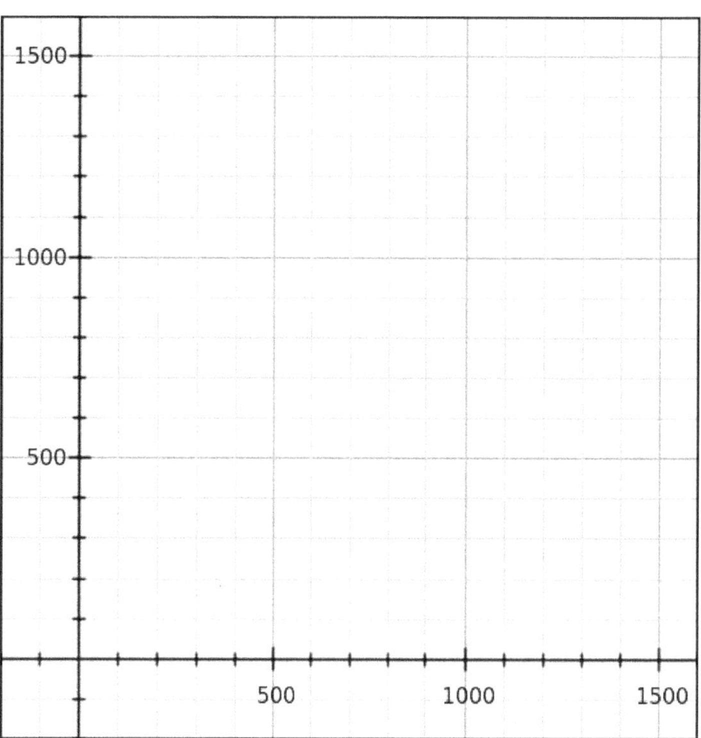

21) Now Frank wants to jump from a cliff that is 1050 meters high. There is a thin rock tower at 500 meters out, which is 700 meters tall. Draw this tower on the graph as a thin vertical segment. Use a ruler to draw a glide path for Frank which makes him just barely clear the top of the tower. passing through the point $(500,700)$ which you label point T.

22) Find Frank's slope. Then find the equation for the glide path.

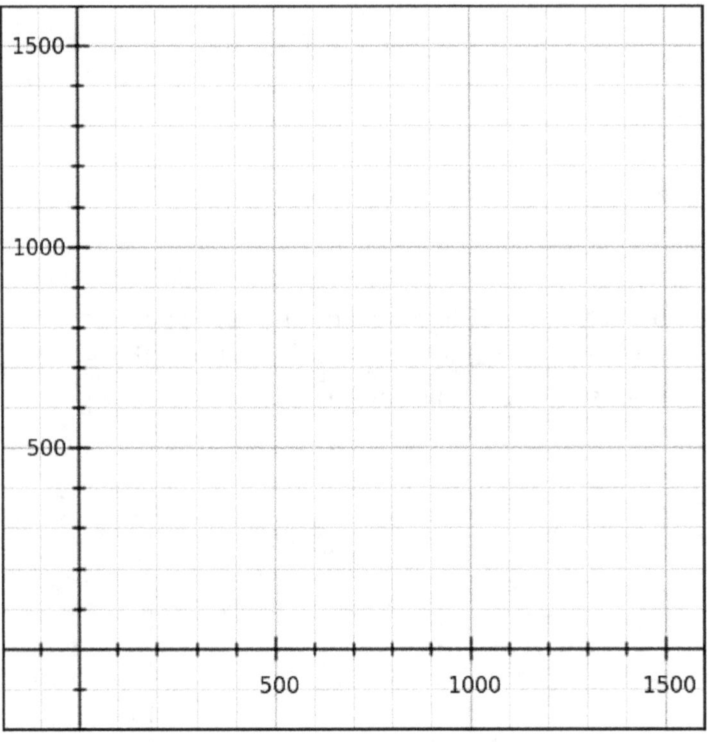

23) Using the function, how high is Frank at 250 horizontal meters? _________ At 700 meters out? _____ At 1000 meters out? ______ Label these points A, B and C.

24) Write and solve an equation to determine where Frank lands. If it's in the graphing window, label it point D.

25) Understanding functions requires that you imagine a relationship between two changing quantities in the real world. For example, the more time you walk, the farther the distance you achieve. The more you study the better your grade is. The higher the mass of the object the more force it takes to hold it up. Write 6 more phrases like these below:

26) For some functions there is a **one-way causal relationship** between one quantity and the other, that is, the causality runs in one direction and not the other. Although there may be exceptions, it seems clear that increased studying causes better grades, and not necessarily the other way around. And it seems natural, for example, to say that first comes an increase in power and that causes a lower time in the 100 meter dash, not the other way around. Below, create 3 examples of a relationship which seems like a one-way causal relationship:

27) All of these relationships between quantities can probably be modeled with functions. We just need to decide, which of the two quantities would be the logical input for the function and which would be the logical output? Below is a table showing two quantities which change with respect to one another. Decide which would be the logical input for the function then, in between, draw an arrow towards the other quantity which would be the output.

quantity	arrow	quantity
time to get home		amount of traffic
quality of college you get into		grades in high school
number of rainy days in Fort Collins		number of umbrellas sold there
# people on summer vacation		# people at beach
sun exposure		plant growth rate
mass of cat		amount of food eaten by a cat
probability of snow		air temperature

28) The number of hot dogs sold in the ballpark is a function of the number of people who attended the baseball game and not the other way around. But can you think of any relationships which are **bi-directional**, that is, the causality runs both ways? Write 3 below.

29) When my parents gave me candies when I was a kid, I always had to give my brother $\frac{1}{4}$ of my candies. "What" was a function of "what" in this situation? Explain below:

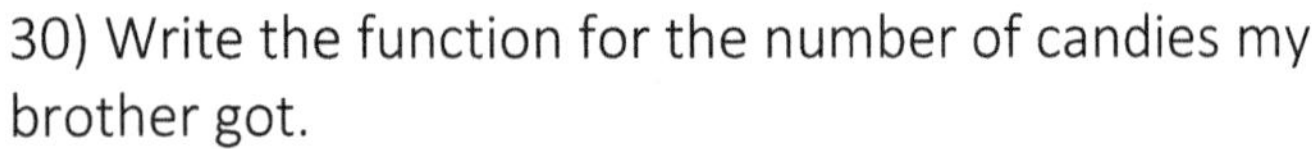

30) Write the function for the number of candies my brother got.

$y =$ __________________

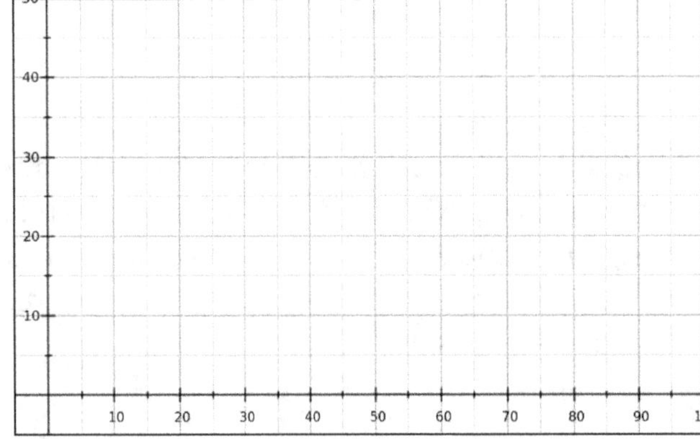

31) Sketch the graph and label the axes appropriately.

32) Write each linear function in this mission in the table below. Then write an explanation for the meaning of each function. The first one is given for you.

Function	Meaning
$y = 30x + 50$	The cumulative cost of a gym as a function of the # months of membership.

The Take-Home Message: Functions can be used to model relationships in the real world. Often, we create functions with one-way causality in mind, expressing this by saying that one variable (y) is a function of another variable (x).

LF6 *Vertical and Horizontal Lines*

Name(s)

Date Class/Period/Group

1) Imagine a glider thrown from a 100 meter tall cliff. It glides completely horizontally. Sketch a graph of this function below. Create and label axes "horizontal distance (x)" and "vertical height (y)." What is the slope of this line? What is the equation for this function?

2) Imagine a genie that takes the number of candies you bring him and magically transforms them into a new number of candies using a function. Let's imagine encountering a genie whose function is $y = 6x + 4$, where x is the input and y is the output. How many candies would this genie return you if you bring him 3 candies? _____ Write and solve an equation to determine how many candies you would need to bring him in order to receive 100 in return.

3) Now imagine a genie who returns you 124 candies no matter if you bring him 0, 12, 78 or any other number of candies. Complete the table below for this function. Sketch a graph. Find the slope and write the equation for the function.

x	y
0	
12	
78	

4) Sketch a graph of the following points: $(0,12)$ and $(5,12)$. Find the slope of the line between these points. Find the equation for the function.

5) All the functions on the previous page are linear functions. Some of them can be specifically defined as **constant** functions. Which ones, and why? How can you recognize a constant function from a table, graph or equation?

6) Complete tables for and sketch graphs of the following functions, labeling each one.

f $\qquad y_1 = 2x + 3$

	x	y_1
A	0	
B	5	
C	10	

g $\qquad y_2 = \frac{1}{2}x + 3$

	x	y_2
D	0	
E	4	
F	8	

h $\qquad y_3 = 3$

	x	y_3
G	0	
H	5	
I	10	

7) For function g, why did I choose x values which were even numbers?

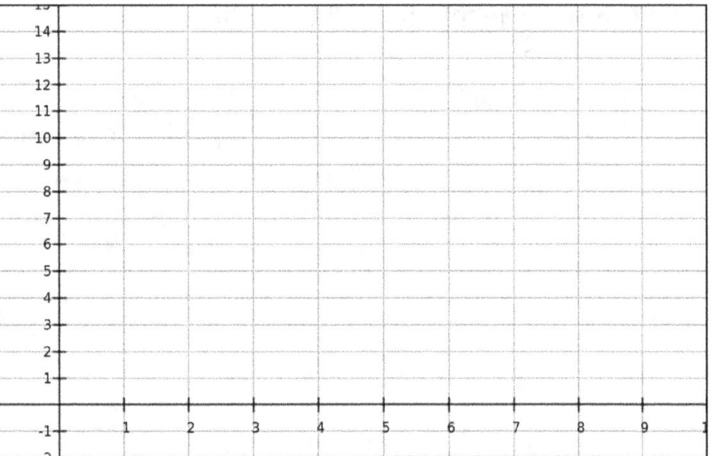

8) Draw slope triangles between the first two points for each function. Put the lengths next to the vertical and horizontal sides of each. Is it possible to draw a slope triangle between the first two points for function h? _____ Why or why not?

9) Write and solve equations to determine the x-intercept for each of the above functions. What happened in the case of h? Why?

10) Complete the tables for the functions below, then graph, using only the two points in the table. Label each line.

f $y_1 = \frac{1}{8}x + 5$

	x	y_1
A	0	
B	1	

g $y_2 = 5$

	x	y_2
C	0	
D	10	

11) Would you say your graph of f is of high or low **precision**? Why or why not? Find an additional x-value which you could use to help improve the precision of this graph. Explain why this improves the precision of the graph.

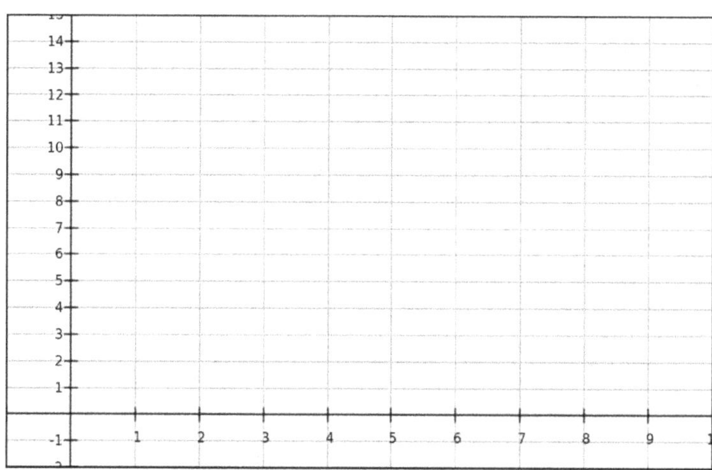

12) Suppose you want to graph the line $y = 10x - 20$. How would you scale the x-axis and the y-axis, and what would be the limits of the graphing window? What points would you choose to graph for maximum precision, and why? Use a ruler and sketch below. Try to be as precise as possible making the axes perpendicular, scaling equally and graphing points.

13) Below is a table comparing two different lines. Sometimes the lines are indicated with a function, sometimes they are indicated by two points. Compare the steepness of the two lines. Highlight or circle the line which is steeper. If they have the same slope, highlight or circle both.

1st	2nd
$y = 2x + 40$	$y = x - 1000$
$y = \frac{1}{2}x - 48$	$y = \frac{1}{3}x - 48$
$y = \frac{3}{5}x$	$y = \frac{1}{2}x + 10$
$y = \frac{5}{9}x$	$y = \frac{2}{3}x$
$(0,10)$ and $(1,15)$	$(0,0)$ and $(1,6)$
$(5,9)$ and $(15,9)$	$(4,12)$ and $(5,12)$
$(3,4)$ and $(5,10)$	$(2,1)$ and $(7,20)$

1st	2nd
$y = 5x$	$(0,11)$ and $(2,17)$
$y = 3$	$(4,5)$ and $(5,5)$
$y = 0.2x$	$y = 0.19x$
$y = \frac{\sqrt{2}}{2}x$	$y = \frac{\sqrt{3}}{2}x$
$(5,3)$ and $(7,15)$	$y = \frac{13}{2}x$
$(-2,7)$ and $(5,8)$	$y = \frac{1}{7}x + 10$
$(-4,-2)$ and $(3,2)$	$y = \frac{2}{7}x - 56$

14) What are the x-intercepts and y-intercepts for each of the following lines? Do work in boxes as needed.

line	y-intercept	x-intercept
$y = 3x - 5$		
$y = \frac{4}{3}x - 8$		
$(0,7)$ and $(4,0)$		
$y = -9$		

15) Imagine a genie who doesn't accept any candies unless you bring him exactly 7 candies. Anyone who tries to bring him a different number of candies gets sent away with their candies in hand—he doesn't even take them, which means of course he doesn't return anything to them. Meanwhile, the genie takes your 7 candies. He returns 10 to you. Then he returns 15, then he returns 0, and -6, and $\frac{2}{3}$, and $\sqrt{2}$ and 100, and then he continues to return literally all the numbers of candies you can imagine. You get all possible numbers of candies. Which is cool except for the fact that on a practical level, when he gives you 100, he also gives you -100, which cancels out the 100 he just gave you. But anyway, complete the table below for this genie. Then sketch a graph at right.

x	y
7	
7	
7	
7	
7	

16) Is the line you graphed above is horizontal or vertical? What is the equation for this line? What is the slope of the line? Explain your answers.

17) Here is a summary of important features of horizontal and vertical lines:

type	description	example	slope	x-intercept	y-intercept
horizontal	constant function	$y = 7$	0	none, unless it's the line $y = 0$	same as the value of y in the equation
vertical	not a function	$x = 5$	none	same as value of x in equation	none, unless the line is $x = 0$

There is only one horizontal line that has x-intercepts and one vertical line that has y-intercepts. Which lines, and why?

18) Create the equations for a vertical line below. Give two points that are on the line. Sketch the graph. State the x-intercept.

19) What are the slopes, x-intercepts and y-intercepts for the following lines? A sketch may be useful but is optional.

line	slope	y-intercept	x-intercept
$y = 6$			
$y = \frac{3}{4}x - 12$			
$(0,2)$and $(7,0)$			
$x = -4$			

20) Give the point of intersection between the two lines. Sketch a graph showing the lines and the intersection point. Find the intersection point algebraically in the last column.

lines	sketch	point of intersection
$y = 2x - 5$ $x = 3$		
$y = 8$ $y = \frac{1}{2}x + 5$		

21) Use the given information to find the specific equation for the line. Sketching a small graph may be helpful but is optional.

$(0,5)$ and $(2,13)$	$(0,7)$ $y = 2x + b$
$(2,4)$ and $(5,16)$	$(2,8)$ $y = 3x + b$
$(3,6)$ and $(7,-6)$	$(-2,4)$ $y = mx + 7$

22) Use the given information to find the specific equation for the line. Sketching a small graph may be helpful but is optional.

$(0,5)$ and $(2,5)$	$(5,7)$ $\quad y = -\frac{1}{2}x + b$
$(2,4)$ and $(2,15)$	$(9,-4)$ $\quad m = 0$
$m = 0$ $\quad b = 9$	$(-2,7)$ $\quad y = mx + 7$

23) Create lines which are consistent with the description on the left.

Description	Rough Sketch	Equation for Line
A vertical line with an x-intercept of 4.		
A horizontal line with a y-intercept of 7.		
A line with a slope of 0.		
A line with no slope.		
A line with a slope of $\frac{2}{5}$.		
A line with a slope of $\frac{2}{3}$ and a y-intercept of 4.		
A line with a negative slope and a y-intercept of 0.		
A line with a positive slope and an x-intercept of 0.		
A line with a slope of 0 and a y-intercept of 0.		
A line with no slope and an x-intercept of 0.		
A line with more than one x-intercept.		
A line with more than one y-intercept.		

24) Explain how you can recognize vertical or horizontal lines from a table, graph or equation. Give examples to support your explanation. Describe slope, the y-intercept and the x-intercept for each type. Organize your work carefully.

The Take-Home Message: The graph of a constant function with a slope of $\mathbf{0}$ is a horizontal line. The graph of a line with no slope is a vertical line.

Courage To Core

Share the Discovery

LF7 Domain and Range

Name

Date Class/Period/Group

1) Functions come in many different types and model many different things. Below are some examples of functions. Name them correctly using the word bank at right. Highlight or circle the polynomials. Put an asterisk next to the ones you haven't seen before.

function	type	word bank
$y = -\frac{2}{3}x + 7$		exponential
$y = 5$		trigonometric
$y = x^2 + 5x - 9$		constant
$y = 3\sqrt{x}$		cubic
$y = \frac{10}{x}$		quadratic
$y = sin(x)$		linear
$y = -x^3 + 8x$		rational
$y = 2^x$		square root

2) Write equations for three constant functions below.

3) Write equations for three rational functions below.

4) Write equations for three square root functions below.

5) Which of the functions in the table does not accept negative numbers as input? Why not?

6) Which of the functions does not accept 0 as input? Why not?

7) Complete the tables and graph each of the following functions on the same axes. Connect the points for g and h with a smooth curve. Graph h in two sections: one on the left of the y-axis and one on the right, separated by an imaginary vertical line. Label each function.

f $\quad y_1 = \frac{1}{3}x - 6$

	x	y_1

g $\quad y_2 = x^2$

	x	y_2
	-3	
	-2	
	-1	
	0	
	1	
	2	
	3	

h $\quad y_3 = \frac{10}{x}$

	x	y_3
	-10	
	-5	
	-2	
	-1	
	0	
	1	
	2	
	5	
	10	

8) Which function doesn't accept 0 as input?

9) Which function never returns negative y-values?

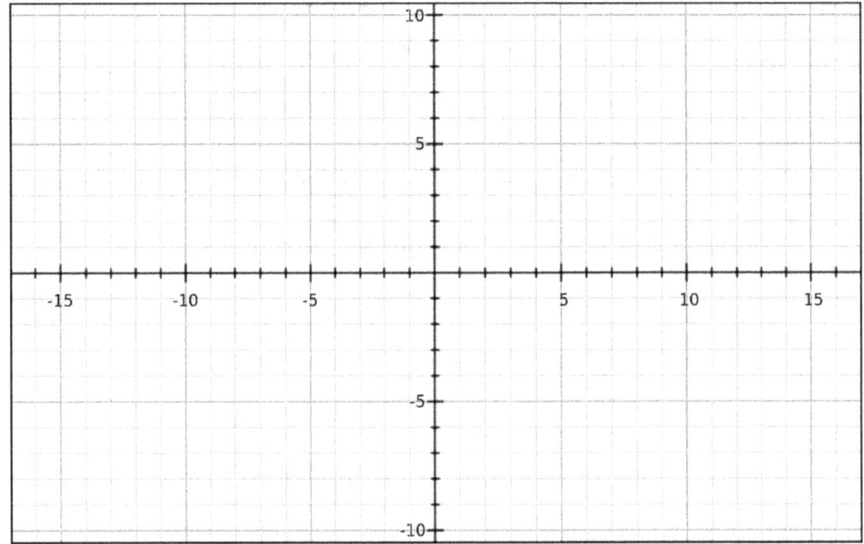

10) What are the x-intercepts for the first two functions?

11) Does the 3rd function appear to have an x-intercept? Why or why not?

12) Functions sometimes have rules for what they accept as input. All the functions below have a rule restricting what they accept. Describe the rule in words and explain why that rule exists.

function	rule restricting input:
$y = \frac{4}{x}$	
$y = \sqrt{x}$	
$y = 6\sqrt{x}$	
$y = \frac{12}{x}$	

13) Different functions also return different values after they perform their function. The function $y = x^2 + 7$ never return y-values below 7, for example. Below are functions that never produce a y-value below a certain number. Identify that number and explain how you know that's the lowest value the function can return as output.

function	This function never produces a y-value below:
$y = x^2$	
$y = x^2 + 5$	
$y = \sqrt{x}$	
$y = x^4$	

14) To understand a function completely we need to know what the rules for input are (if any), and we need to know the range of possible values it can return. Every function has a **domain**, which is the set of values which can serve as input, and a **range**, the set of values the function returns as output. Below is a function with the domain and range described in words, then an explanation. Read this carefully to prepare for the subsequent problems.

function: $y = x^2 + 7$

domain	range
all Real numbers	all Real numbers ≥ 7
The domain is the input. I can put all Real numbers in for x and get a result. For example, I can square 4, -5, and even 0.	*The range is the output. When I input 0 I get 7, and that's the lowest output I can get. That's because if, for example, I input 5 I get 25 and if I input -5 I also get 25. So an input of 0 returns the lowest y-value, 7. After that, the output gets bigger.*

15) Suggestions for identifying domain and range:

	tactic one	tactic two	function types to look out for	examples
domain	*Try to input various values and determine if there are any which the function doesn't accept as input.*	*Identify the function type*	square root functions	$y = 5\sqrt{x}$ all Real #'s ≥ 0
			rational functions	$y = \frac{8}{x}$ all Real #'s $\neq 0$
range	*Input various values and observe how the output changes as you change the input.*	*Identify the function type and visualize its graph to help you.*	quadratic functions	$y = x^2 + 11$ all Real #'s ≥ 11
			square root functions	$y = 2\sqrt{x}$ all Real #'s≥ 0
			rational functions	$y = \frac{7}{x}$ all Real #'s $\neq 0$

Explain the reasoning behind each of the domains and ranges identified in the examples above. They are re-written below for you:

function	domain	range	explanation
$y = 5\sqrt{x}$	all Real #'s ≥ 0		
$y = \frac{8}{x}$	all Real #'s $\neq 0$		
$y = x^2 + 11$		all Real #'s ≥ 11	
$y = 2\sqrt{x}$		all Real #'s≥ 0	
$y = \frac{7}{x}$		all Real #'s $\neq 0$	

16) It can sometimes be useful to graph a function to help you visualize the domain and range. Graph each of the following functions and then give the domain and range. Explain how the graph helps you visualize the domain and range.

function	$y = x^2 + 3$	$y = \frac{10}{x}$
graph	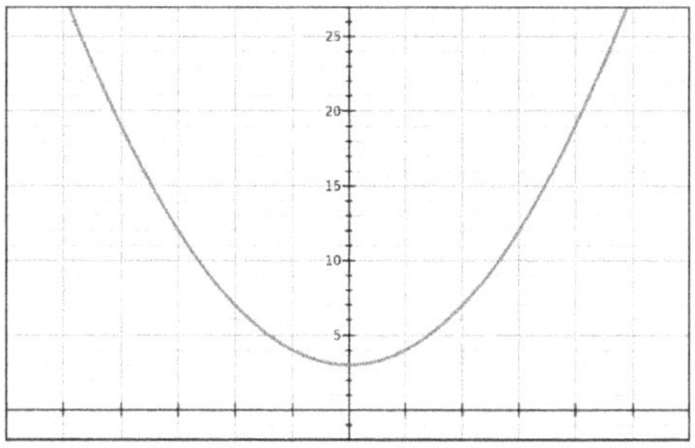	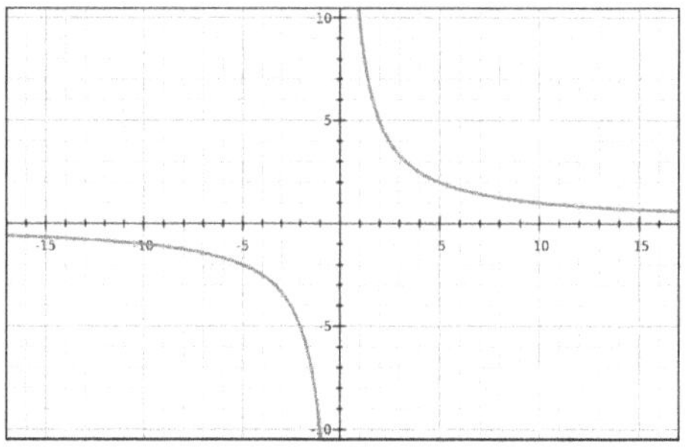
domain and how it can be visualized in the graph		
range and how it can be visualized in the graph		

17) Give the domain and range for each of the following.

function	domain	range
$y = 2x$		
$y = x^2$		
$y = \sqrt{x}$		
$y = 7$		
$y = -\frac{3}{7}x + 8$		
$y = x^3$		
$y = \frac{10}{x}$		
$y = x^2 + 12$		
$y = \frac{11}{x}$		

18) Imagine a genie who accepts the number of candies you have as input and returns a number of candies as output. He converts your candies into a new number according to a function, like $y = 6x + 4$, where x is the input and y is the output. You can imagine all sorts of genies like $y = x^2 + 7$ or $y = \frac{6}{x}$, all of which return a single number of candies to you based on the amount you give him. Here are graphs of two genies. Use the graphs to complete the table for each.

	f	g
graph	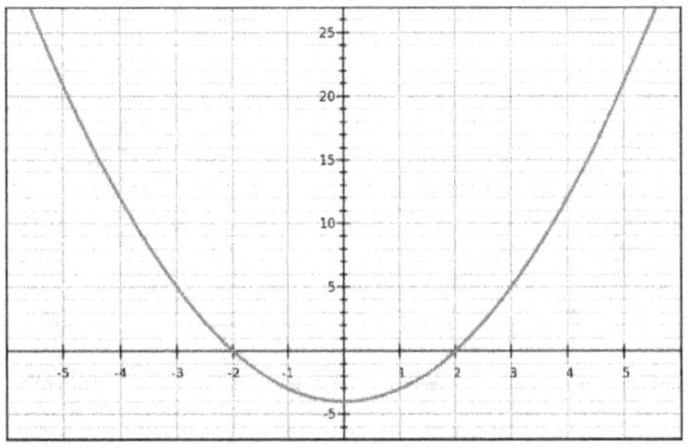	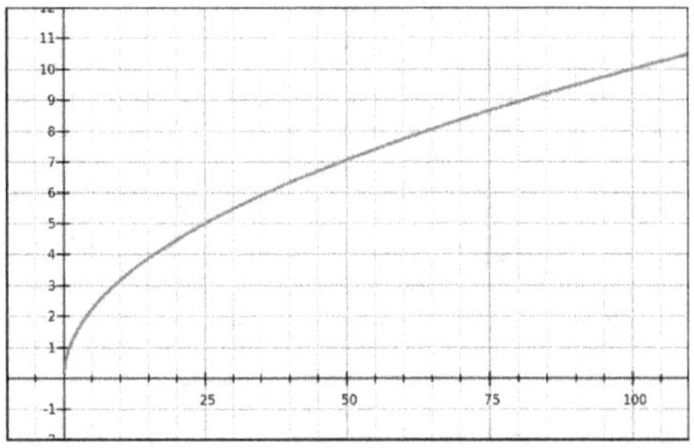
domain		
range		
if $x = 5$, what is y? (approx.)		
if $x = 50$, what is y?		
for what x does the function return a y of 6?		
for what x does the function return a y of 9?		

19) At right are graphs of four functions and one equation which isn't a function. ***Non-functions** return more than one y-value for a given x-value.* Label each using the given names. For each given x-value, use the graph to approximate the y-value for each function and complete the tables. In the next table, state their domain and range.

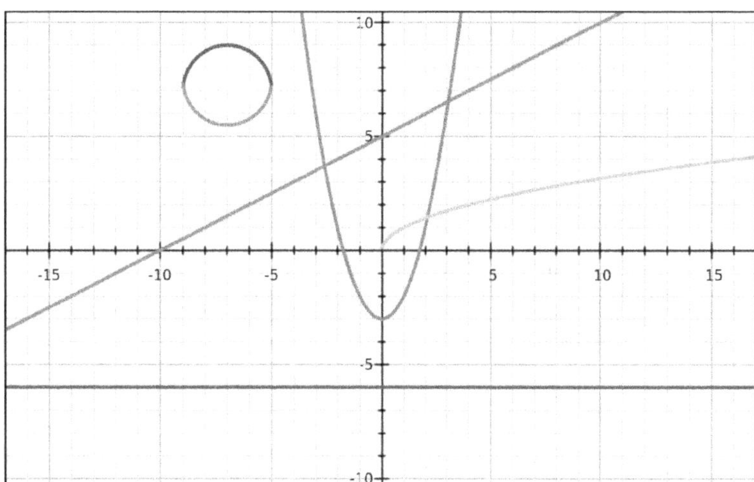

f quadratic

	x	y_1
	0	
	1	
	2	
	3	
	-1	
	-2	
	-3	

g linear

	x	y_2
	-10	
	-5	
	0	
	5	
	10	
	15	
	20	

h constant

	x	y_3
	0	
	100	
	2	
	-8	
	-100	
	$\sqrt{2}$	
	π	

i $\sqrt{\ }$

	x	y_4
	0	
	1	
	4	
	8	
	9	
	10	
	-4	

r notta

	x	y_5
	-9	
	-9	
	-8	
	-8	
	-5	
	-10	
	2	

	function type	domain	range
f	quadratic		
g	linear		
h	constant		
i	square root		
r	not a function		

20) At right are graphs of four functions and one equation which isn't a function. Label each using the given names. For each given x-value, use the graph to approximate the y-value for each function and complete the tables. In the next table, state their type, and the domain and range.

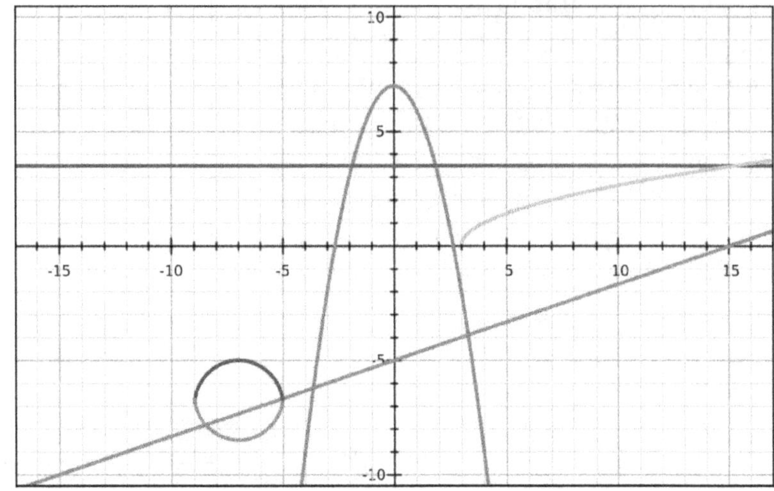

f quadratic

	x	y_1
	0	
	1	
	2	
	3	
	-1	
	-2	
	-3	

g linear

	x	y_2
	-10	
	-5	
	0	
	5	
	10	
	15	
	20	

h constant

	x	y_3
	0	
	100	
	2	
	-8	
	-100	
	$\sqrt{2}$	
	π	

i $\sqrt{}$

	x	y_4
	0	
	1	
	4	
	8	
	9	
	10	
	-4	

r notta

	x	y_5
	-9	
	-9	
	-8	
	-8	
	-5	
	-10	
	2	

	function type	domain	range
f			
g			
h			
i			
r			

21) We often give functions names like f and g, and then state the equations separately as $y_1 = 2x + 7$ and $y_2 = 3x - 9$. But we can also combine the name of the function with the equation by using **function notation**. If we have a function f whose equation is $y = 2x + 7$, we mean that f as a function of x is $2x + 7$. The notation for that is as follows:

	how we write it	how we read it
function	$f(x) = 2x + 7$	"f of x is $2x + 7$"
substitution example	$f(3) = 2(3) + 7$	"f of 3 is $2(3) + 7$"
example continued	$f(3) = 13$	"f of 3 is 13"

Complete the table:

function	how we say that	$f(4)$	domain of f
$f(x) = 2x + 7$			
$f(x) = \frac{1}{2}x - 3$			
$f(x) = 7$			
$f(x) = \sqrt{x}$			
$f(x) = x^2$			

22) Complete the table.

function	$f(-4)$	domain	range
$f(x) = \frac{20}{x}$			
$f(x) = \frac{2}{3}x + 5$			
$f(x) = -3$			
$f(x) = \sqrt{x}$			
$f(x) = x^2 + 7$			

23) Create three functions for each specified domain or range. Write each function as $f(x) = \cdots$

Domain is all Real #'s.	Domain is all Real #'s ≥ 0.
Domain is all Real #'s $\neq 0$.	Range is all Real #'s.
Range is all Real #'s ≥ 0.	Range is all Real #'s ≥ 9.
Range is all Real #'s $\neq 0$.	Range is all Real #'s ≤ 4.

24) Summarize how to find the domain and range of a function from a graph, using examples.

The Take-Home Message: The domain of a function is the set of all x values that the function accepts as input. The range is the set of all y values that the function returns as output. You can determine the domain and range by substituting values for x, by noting the function type, and by examining the graph.

LF8 *Linear Models and Systems*

Name(s)

Date Class/Period/Group

1) Suppose you are traveling from San Diego to San Francisco, and then going further north on a road trip into Oregon which will last as long as your money lasts. Your parents give you a couple options. You can drive from San Diego starting tomorrow morning at 8 AM in your Mini at 80 mph. Or you can fly to San Francisco tonight and drive tomorrow morning at 8 AM in your grandma's 1973 Fiat at 50 mph. Even though you'll start driving at the same time, flying up there would give you a 500 mile head start.

2) Complete the table at right. If you drive the Mini for 20 hours from San Diego, how far will you have traveled? Create an equation for the Mini's distance from San Diego as a function of the hours driven:

$y_1 =$ ________________

hours	Mini option: miles travelled	Fiat option: miles travelled
0	0	500
1	80	550
2		
3		
4		
5		
6		
20		

3) If you choose the Fiat option and drive for 20 hours, how far will you have you traveled from San Diego? Create a function for the Fiat:

$y_2 =$ ________________

4) Rewrite your functions from the previous page here:

5) Graph the functions. Label axes, extend graphs to the edge of the graphing window and label the functions.

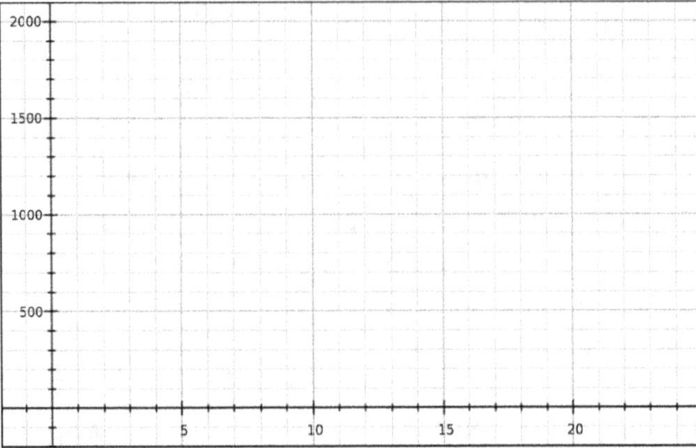

6) Based on the graph, at how many hours will the Mini get you the same distance from San Diego as the Fiat? What distance do both options achieve in this many hours?

7) In order to find exactly how many hours result in the same miles for each option, I want to find the x-value that produces the same y-value in both functions. The following equation will answer that question. Why? Solve the equation, then find the corresponding y-value. Confirm your result with the graph.

$50x + 500 = 80x$

8) How many equations were used to create the equation in #7? ____ Was your answer for #7 a single number or a point? _________ Is that point on both lines? ____ Does that ordered pair make both equations, $y_1 = 45x + 500$ and $y_2 = 80x$ true? _____ Confirm this below.

9) In the previous problem we had two different equations representing two different functions:

$y = 45x + 500$
$y = 80x$

Because we were interested in knowing when these two cars would be the same distance from San Diego, we stopped thinking about the two equations as separate. Instead, we began to think about the distance, y, as the same for for each function for some particular x value. Once you assume that y in one equation is the same as y in another equation (and same for the x's), we are solving a **system** of equations. How many equations do you need for a system? How many variables? Is your **solution** to the system a single number, or an ordered pair, that is, a point? Why?

10) The process of finding an ordered pair (a point) which makes two different equations true is called solving a system of equations. Below is an algorithm you can use to do this when your have two equations as shown below. Complete the steps in the example.

Solving A System of Linear Equations

Example	Steps
$y = 2x - 5$ $y = 4x + 17$	*Make sure you've got two equations in two variables. Make sure you are being asked to solve the system, or find the intersection point, or determine when the functions have the same y-value.*
$2x - 5 = 4x + 17$	*Write an equation which sets the two functions equal to each other in order find the x which returns the same y for both functions. Solve the equation.*
	Substitute the x-value you found into one of the two functions to find the corresponding y-value.
	Optionally, you can confirm your y-value is correct by repeating the last step with the other function.
	Write your solution as an ordered pair (a point). Optionally, you can confirm your solution is correct by graphing the two functions and finding the intersection point visually.

11) Solve the following system of linear equations. Write your solution as an ordered pair.

$y = -2x - 5$
$y = 4x + 7$

12) Solve the following system of linear equations. Write your solution as an ordered pair.

$y = -\frac{1}{2}x - 5$
$y = 3x + 7$

13) Solve the following system of linear equations. Write your solution as an ordered pair.

$y = 1.5x - 5$
$y = 3.5x + 7$

14) You live in Louisville, and your friend lives 200 kilometers closer to the beach than you do. You and he leave at the same time driving towards the coast on the same highway. You drive at 100 km/hr while he drives at only 55 km/hr. Create functions which model your distance and your friend's distance from Louisville after x hours. Label axes and graph.

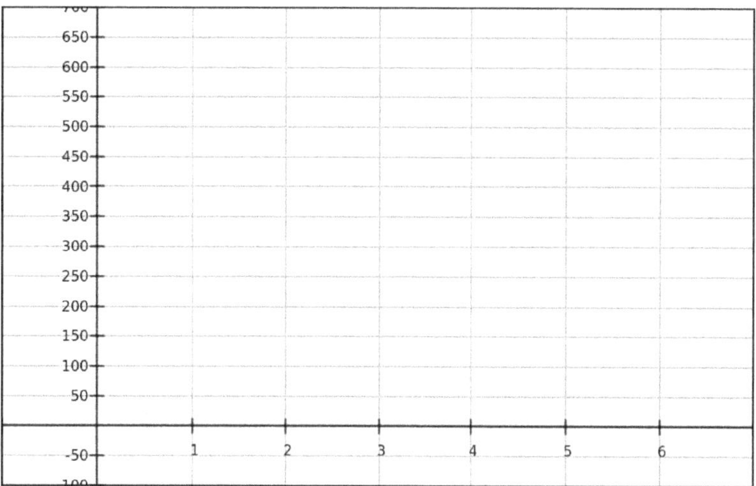

15) Write and solve an equation to determine when you pass your friend. Determine how many kilometers you both have driven by that time. Confirm your result with the graph.

16) One gym costs $\$50$ per month plus an additional one-time fee of $\$160$ to join the gym. Another costs $\$75$ per month but with no one-time fee. Write functions which model the cumulative cost of each gym at x months. Graph. Determine algebraically at which month the gyms have the same cost and what that cost is. Confirm your result with the graph. Which gym would you join, and why?

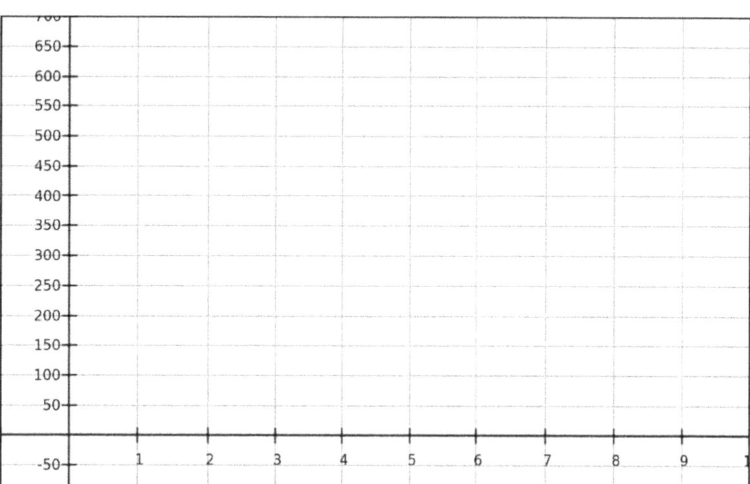

17) Create and solve a linear system. Explain the steps in your solution. Support your explanation with a graph.

18) What is the difference between solving a linear equation in one variable and solving a system of linear equations? Support your explanation with examples.

The Take-Home Message: A system of linear equations is composed of two equations in two variables. To solve the system, we solve for an x-value that returns the same y-value for both functions, then find the corresponding y-value. The solution is an ordered pair which makes both equations true. It is the point of intersection of the two lines.

Courage To Core

"Share the Discovery"

LF9

Forms for Lines and Methods for Systems

Name(s)

Date Class/Period/Group

1) Graph and label each of the following functions on the axes provided. Extend lines to the edges of the graphing window and label. Which of the following methods are you planning on using? __________ Afterwards, circle the ones you actually used.

$y = \frac{1}{2}x$ $y = \frac{1}{2}x + 3$ $y = \frac{1}{2}x - 3$

a) Graphing the y-intercept and one other point that is far away from the y-intercept.

b) Graphing the y-intercept and using slope triangles to get more points.

c) Making a table of values and plotting lots of points.

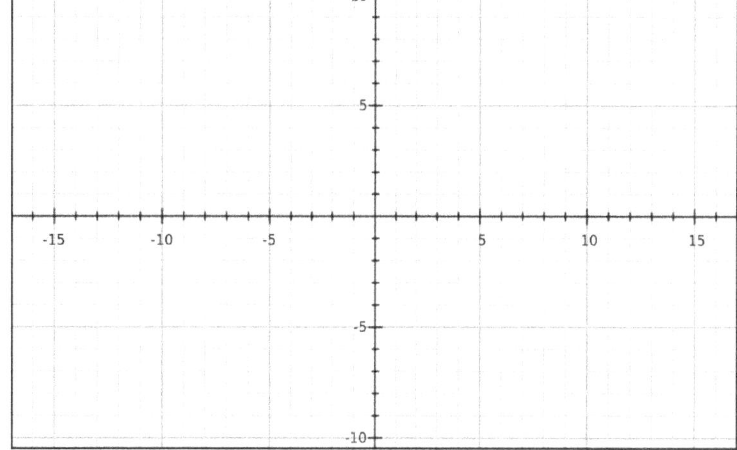

2) What are the similarities and differences between each of the above functions? Explain.

3) What do you think the graph of $y = 2x - 1000$ would look like? Explain, comparing with the functions above. Use a rough sketch to support your explanation.

4) Graph each of the following:

$y = 2x + 2$ $y = -2x + 2$

$y = x + 2$ $y = -x + 2$

$y = \frac{1}{2}x + 2$ $y = -\frac{1}{2}x + 2$

$y = 2$

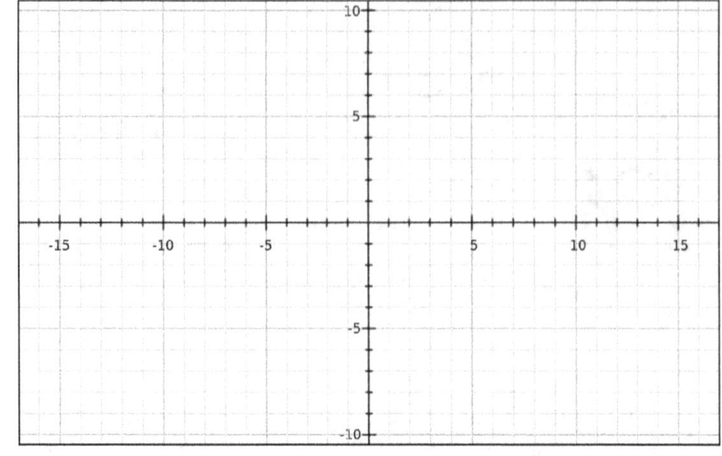

5) What are similarities and differences between these functions? Explain.

6) What do you think the graph of $y = 100x + 2$ would look like? Explain, comparing with the functions above. Use a rough sketch to support your explanation.

7) The following line is in slope-intercept form: $y = 2x + 3$. But you don't have to keep it in that form. With algebra you can change its appearance without changing the fact that it is a linear function with slope of _____ and y-intercept of _____. Below is a series of changes you can make, each one following the previous one. Complete the table to either describe the change or to write the equivalent equation.

equivalent equations	description of algebra
$y = 2x + 3$	original equation
$2y = 4x + 6$	multiplied both sides by 2
$2y - 4x = 6$	
$-4x + 2y = 6$	
	multiplied both sides by -1 (be sure to distribute where appropriate!)
$40x - 20y = -60$	
	multiplied both sides by $\frac{1}{20}$

8) A linear equation like the one in the last line above is said to be in **standard** form. Converting to standard form can be very useful for graphing, as well as for solving systems of linear equations. Here is a table summarizing the two most commonly used forms of a line:

	general form	specific examples
Slope-intercept form	$y = mx + b$	$y = \frac{1}{3}x - 5$ $y = -4x$
Standard form	$ax + by = c$	$2x + 3y = 6$ $\frac{1}{2}x - y = 7$

9) The last example in standard form could be written with integer coefficients. Multiply both sides by **6** below, then simplify. When we write equations in standard form we often prefer integer coefficients.

$\frac{1}{2}x - \frac{2}{3}y = 7$

10) Create a linear function written in standard form. Create a different linear function written in slope-intercept form.

11) Convert $\boldsymbol{y = \frac{1}{3}x + 7}$ into standard form. Convert $\boldsymbol{2x + 3y = 6}$ into slope-intercept form.

12) Standard form gives you an additional method for graphing. Consider the line $\boldsymbol{2x + 3y = 6}$. We know that this is a linear function, and one way to graph a linear function is to plug in x-values and get y-values. We can also substitute y-values and get the corresponding x-values. Below is a table suggesting easy values you can plug in for $\boldsymbol{x}$ and $\boldsymbol{y}$. Complete the table and graph the function.

x	y
$\mathbf{0}$	
	$\mathbf{0}$

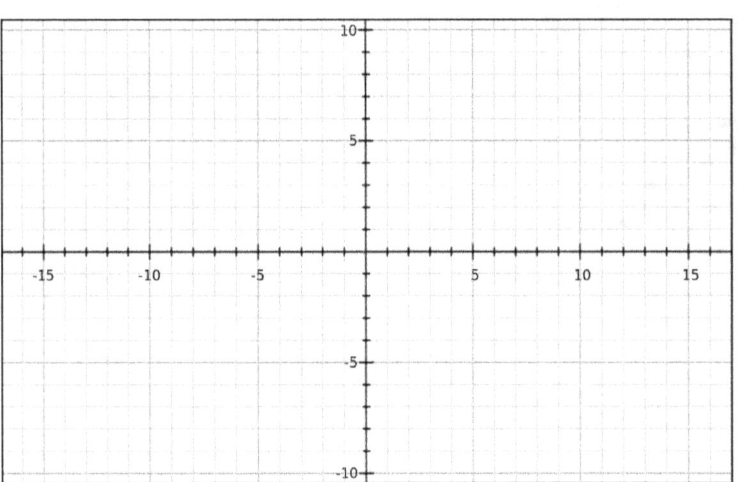

13) When a function is in standard form it can be easy to plug in $\mathbf{0}$ for $\boldsymbol{x}$ and solve for $\boldsymbol{y}$, then plug in $\mathbf{0}$ for $\boldsymbol{y}$ and solve for $\boldsymbol{x}$. What are the advantages and disadvantages of this graphing method?

14) Identify each of the following as in standard form, slope-intercept form or neither.

$\mathbf{2x + 5y = 7}$	$\mathbf{5y = 4x + 8}$	$\mathbf{y = 5x - 9}$
$\mathbf{\frac{1}{2}x + 7 = 8y}$	$\mathbf{\frac{2}{3}x - 7y = \frac{1}{2}}$	$\mathbf{4x - 7 = y}$
$\mathbf{x = 2y - 4}$	$\mathbf{y = 6}$	$\mathbf{3x = y}$

15) Graph the following lines on the axes provided.

$y = 2x + 4$

x	y

$2x - 3y = 6$

x	y
0	
	0

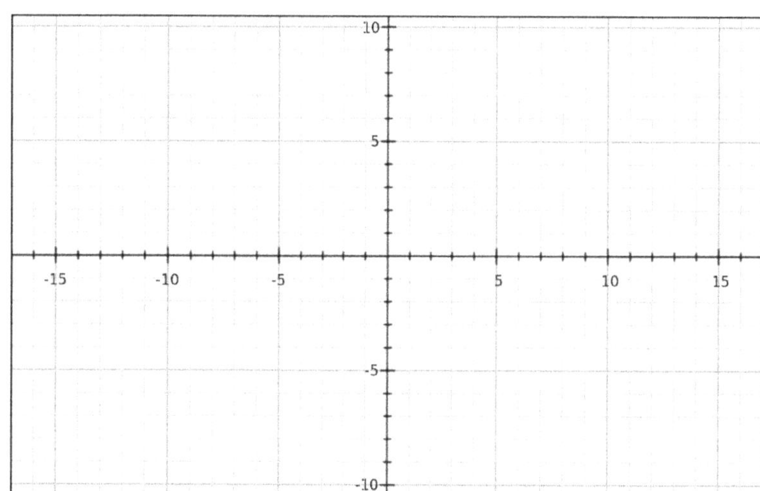

16) Now let's solve the above system of linear equations. One line above is written in standard form, the other in slope-intercept form. At the intersection point, the y-values are the same. This means we can solve the system by substituting "$2x + 4$" from the first equation in place of y in the second equation as shown below. Solve for x, find y, then check your answer with the graph.

1st equation	$y = 2x + 4$
2nd equation before substitution	$2x - 3y = 6$
2nd equation after substitution of $2x + 4$ in place of y	$2x - 3(2x + 4) = 6$

17) Fill in the blanks using the word list below: The substitution method is based on the idea that at the ______________ point of two ___________, the y-values of both functions are the ________. So at that __________, "y" in ______ function is the same as in the ________ function. We solve for ______ to find the ___________ that returns the same _________ in both __________.

one	point	lines	x-value	other
functions	x	y-value	intersection	same

18) Below is the algorithm for using substitution to solve a system of equations. Complete the example by following the steps at right.

Substitution Method to Solve a System

Example	Steps
$-2x + y = 5$ $x + 3y = 1$	*Make sure you've got two equations in two variables. Make sure you are being asked to solve the system, find the intersection point, or determine when the functions have the same y-value.*
$-2x + y = 5$	*Use algebra to re-arrange one of the equations so that either x or y isolated. (In this example, isolate y.)*
	Substitute the expression for y in the above line in place of y in the other equation.
	Solve the resulting equation for the single remaining variable. (In this example, solve for x.)
	Substitute your value for x into either function to find the corresponding y value. Write your solution as an ordered pair.
	Optionally, you can confirm your solution is correct by substituting the ordered pair into the other equation to see that it makes that equation true.
	Optionally, you can confirm your solution is correct by graphing the two functions and finding the intersection point visually.

19) Solve each of the following systems using the substitution method.

$-2x - 5y = -7$ $-2x + y = 4$	$y = 2x - 5$ $4x - 3y = 1$
$x = 3y - 1$ $2x - y = 2$	$y = 3$ $y = 5x - 7$

20) Bill and Dan weigh the same. Mary and Jill weigh the same. Represent these ideas writing some simple equations.

21) Suppose Bill and Mary get on one side of a seesaw together, and Dan and Jill get on the other side of the seesaw together. What will happen? Complete the equation below:

If...	$B = D$
and...	$M = J$
then...	$B + M =$ ______

22) This means that we can add (or subtract) any two equations together as above and get another true equation. We can use this fact to solve systems of equations! Follow the steps shown to complete the example.

Easy Elimination Method to Solve a System

Example	Steps
$x = 11 - 2y$ $3x - 2y = 1$	*Make sure you've got two equations in two variables. Make sure you are being asked to solve the system, find the intersection point, or determine when the functions have the same y-value.*
$x + 2y = 11$ $3x - 2y = 1$	*Re-arrange the equations so that x terms, y terms and constants are vertically aligned. (Standard form works.)*
$4x + 0y = 12$	*In the "easy elimination" method, you'll quickly notice that adding (or subtracting) the two equations will eliminate x's or y's. Add or subtract straight down to accomplish this.*
$4x = 12$ $x = 3$	*Now you can solve for the single variable that remained after elimination.*
	Substitute this into either equation to find y.
	Write solution as an ordered pair.
	Optionally, confirm your solution by substitution in the other equation.
	Optionally, confirm your solution by graphing.

23) Here's another, more complex example using the elimination method. Follow the steps shown to complete the example.

More Complex Elimination Method to Solve a System

Example	Steps
$3x + 5y = 7$ $2x - 7y = 15$	*Make sure you've got two equations in two variables. Make sure you are being asked to solve the system, find the intersection point, or determine when the functions have the same y-value. If needed, re-arrange for vertical alignment of terms.*
$2(3x + 5y = 7)$ $3(2x - 7y = 15)$	*Find the least common multiple between the coefficients of either the x or y terms. In this case it was* 6 *for the coefficients of the x terms. Multiply each equation by what is needed to change those coefficients into* 6.
$6x + 10y = 14$ $6x - 21y = 45$	*Re-write the equivalent equations resulting from the multiplication.*
$0x + 31y = -31$	*Add or subtract straight down in order to eliminate the x terms or y terms. In this case we will subtract to eliminate the x terms.*
$y = -1$	*Solve the resulting equation for the remaining variable.*
	Substitute this into either equation to find x.
	Write solution as an ordered pair.
	Optionally, confirm your solution by substitution in the other equation.
	Optionally, confirm your solution by graphing.

24) Solve the following systems using the elimination method.

$4x - 5y = 8$ $-4x + 2y = 1$	$3x + 4y = 1$ $2x + 3y = -1$
$x + y = 2$ $-3x + 2y = 5$	$\frac{1}{2}x + \frac{1}{4}y = \frac{1}{8}$ $-4x + y = 7$

25) Below are several systems of linear equation. Do not solve them. Just decide which method would be most efficient: substitution or elimination? If you select elimination, state whether you would eliminate x terms or y terms. Explain your choices.

system	method	reasoning
$x = 5y + 4$ $2x + 3y = 9$		
$2x + 5y = 8$ $-2x + 3y = 7$		
$2y = 4x + 8$ $x + 3y = 9$		
$\frac{1}{2}x - 3y = 8$ $x + 4y = 0$		
$\frac{1}{2}y + 2 = x$ $2x + 3y = 1$		

26) Select two systems above and solve them.

27) Create a system which can be solved efficiently using substitution. Solve the system, explaining your steps as you go along.

28) Create a system which can be solved efficiently using elimination. Your system must require multiplying one or both equations. Solve the system, explaining your steps as you go along.

The Take-Home Message: Any linear equation can be manipulated algebraically to create an equivalent equation. At the point of intersection of two lines, the x and y-values are the same, which allows us to use the substitution method to solve the system. Furthermore, any two linear equations are true at their point of intersection, which means we can add them to create another true equation. This gives us the elimination method for solving systems.

LF10 *Parallel and Perpendicular*

Name(s)

Date Class/Period/Group

1) Graph the following lines on the same axes, and label. Extend your graphs to the edges of the graphing window.

$y_1 = 2x - 5$ $y_2 = 2x + 5$ $y_3 = 4x - 10$ $y_4 = -\frac{1}{2}x - 7$

2) Which lines slope positively?

3) Which slope negatively?

4) Which one has the steepest slope?

5) Which ones have the least steep slope?

6) Which lines have the same slope? What is their slope?

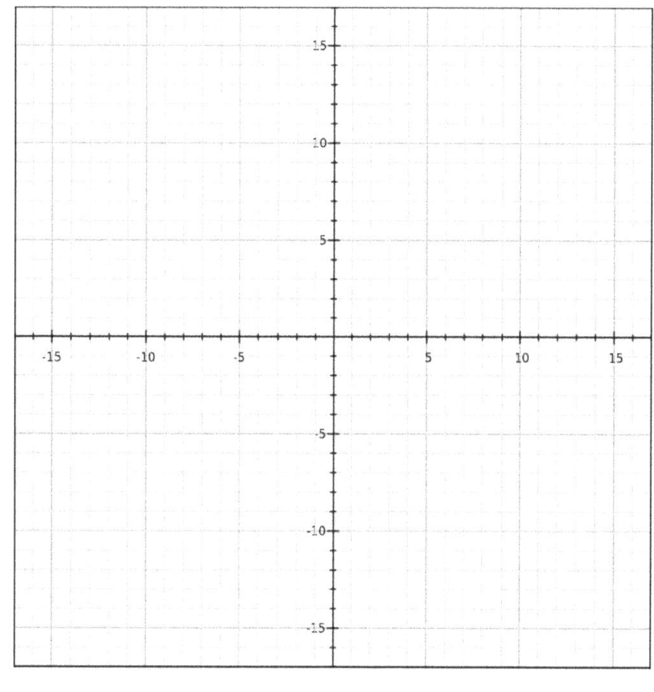

7) The two lines with the same slope are **parallel**. Why? Support your explanation by drawing slope triangles. What does it mean for two lines to be parallel?

8) It is fairly easy to recognize parallel lines and understand how their slopes cause them to be parallel. It can be more challenging to determine if two lines are **perpendicular**. What does it mean for two lines to be perpendicular? Is it sufficient for them to simply intersect? Why or why not?

9) Below are two lines with two points indicated on each. Label the increasing line y_1, and the decreasing line y_2. Do these lines intersect? Do you think they are perpendicular? Why or why not?

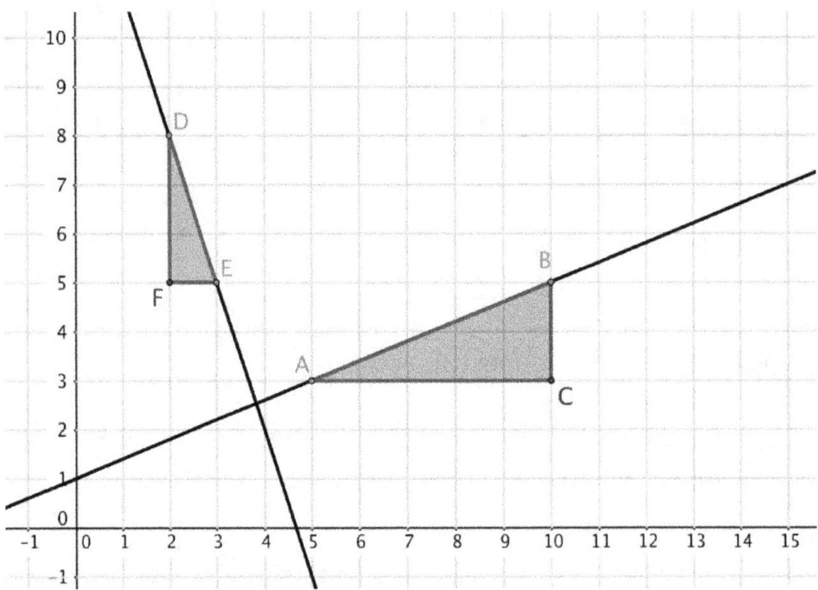

10) One way to see if the lines above are perpendicular or not is to rotate one of them $90°$ and see if they would then be parallel. Find the slope of each line using the given slope triangles. Rotate the slope triangle for y_1 $90°$. Draw the resulting slope triangle, then graph the resulting line and call it y_3. What is the slope of this new line? Is it parallel to y_2? Is it perpendicular to y_1?

11) Below we are given a line and a slope triangle. Label it y_1. We would like to create a line perpendicular to this line. The given slope triangle has been rotated $90°$. Graph a new line through D and C to the edges of the graphing window and label it y_2. Is y_2 perpendicular to y_1? How do you know?

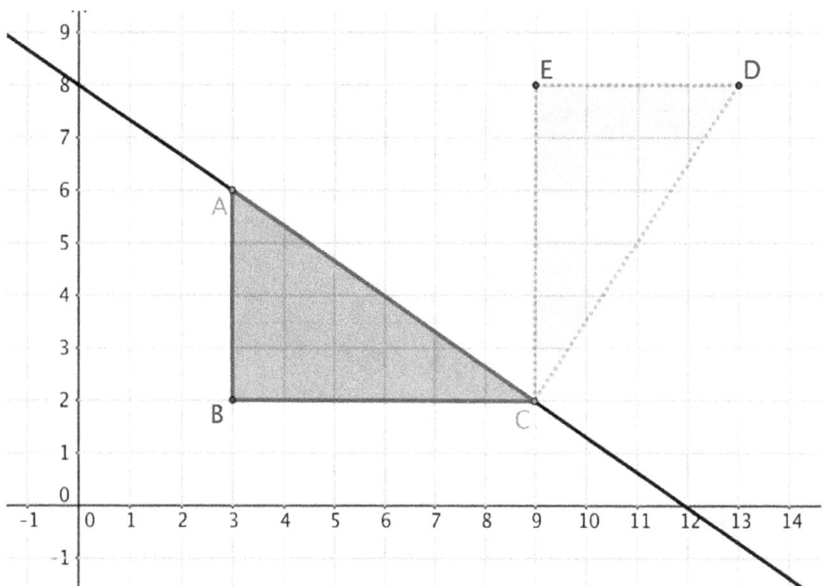

12) What is the slope of y_1? What is the slope of y_2? What is the relationship between the slopes which causes the lines to be perpendicular? Why?

13) Find the equations of both lines above.

14) Here is a summary of the possible relationships between two lines in the coordinate plane:

relationship	graphs	slopes	intersection point
intersecting	The two lines cross.	$\frac{1}{2}$ and 3, for example. Any two lines whose slopes are not the same intersect.	Solve the system and you'll get one solution.
parallel	The two lines do not cross.	$\frac{2}{5}$ and $\frac{4}{10}$ for example. Any two lines whose slopes are the same are parallel, meaning they will not intersect.	Try to solve the system and you'll get no solution.
perpendicular (thus also intersecting)	The two lines intersect at a 90° angle.	$\frac{4}{7}$ and $-\frac{7}{4}$ for example. Rotating the slope triangle 90° for one line gives you a slope triangle for the other. The slopes are **negative reciprocals** of each other.	Solve the system and you'll get one solution, since the two lines do intersect.
coincident	The two lines are the same line.	Two lines we initially think might be different are actually the same. For example: $y = 2x + 3$ and $2y = 4x + 6$. They have the same slope, and the same y-intercept.	Solve the system and you'll get infinite solutions, since two lines that are the same line intersect at all their points.

15) Compare the lines in the first two columns to complete the table. The lines can be intersecting, parallel, perpendicular and intersecting, or coincident.

f	g	relationship
$y = \frac{1}{2}x - 7$	$y = \frac{1}{2}x + 4$	
$y = 3x + 5$	$y = \frac{1}{3}x + 6$	
$y = 3x - 8$	$y = -\frac{1}{3}x + 3$	
$y = -\frac{2}{5}x + 5$	$y = \frac{5}{2}x$	
$y = x$	$y = -x + 100$	
$2x + y = 4$	$y = -2x + 4$	
$3x + 2y = 7$	$6x + 4y = 14$	

16) Graph the following two points, draw a slope triangle, and find the slope of the line: $(5,5)$ and $(0,-5)$. Rotate the slope triangle $90°$ and graph a new line. Find the equation of both the original line and the new line.

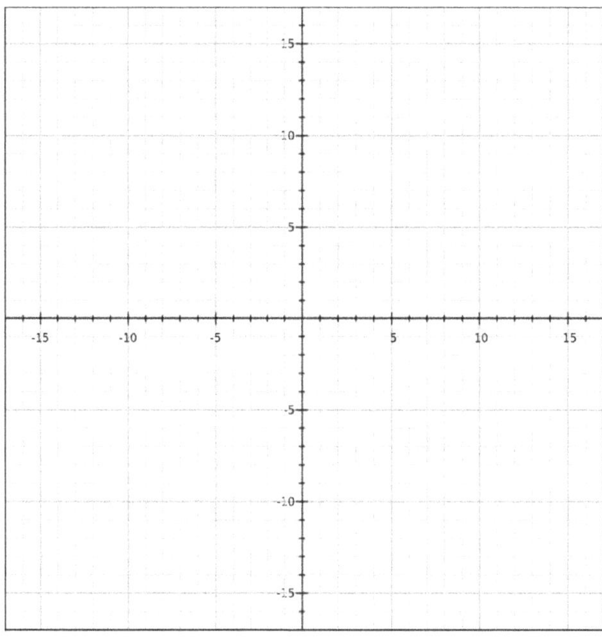

17) Complete the table:

slope of f	slope of any line parallel to f	slope of any line perpendicular to f
2		
-3		
$\frac{2}{7}$		
$-\frac{3}{4}$		
$\sqrt{2}$		
$\frac{\sqrt{3}}{2}$		

18) Given a line f with a slope of 7 and a y-intercept of 3, find the equation for a line parallel to f with a y-intercept of 4.

19) Given a line f with a slope of $\frac{3}{7}$ and a y-intercept of 3, find the equation for a line perpendicular to f with the same y-intercept.

20) Given a line f with a slope of $\frac{3}{7}$, find the equation for a line parallel to f going through the point $(0,5)$.

21) Given a line f with a slope of $\frac{3}{7}$, find the equation for a line parallel to f going through the point $(7,2)$.

22) Given a line f with a slope of $\frac{3}{7}$, find the equation for a line perpendicular to f going through the point $(3,1)$.

23) Find the equation for the line below. Write the equations for and graph one line that is parallel to this line and one which is perpendicular to this line.

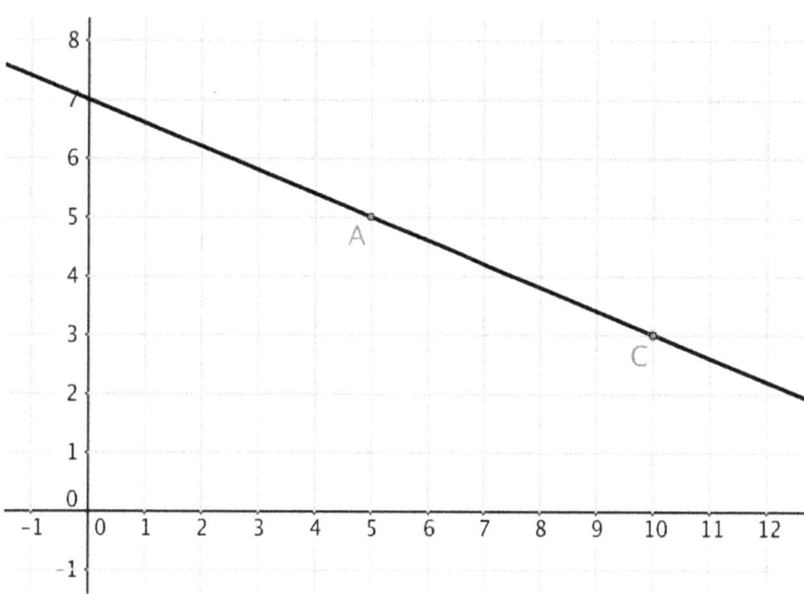

24) Find the equation for the line perpendicular to $\overleftrightarrow{AB}$ going through point C. Graph it.

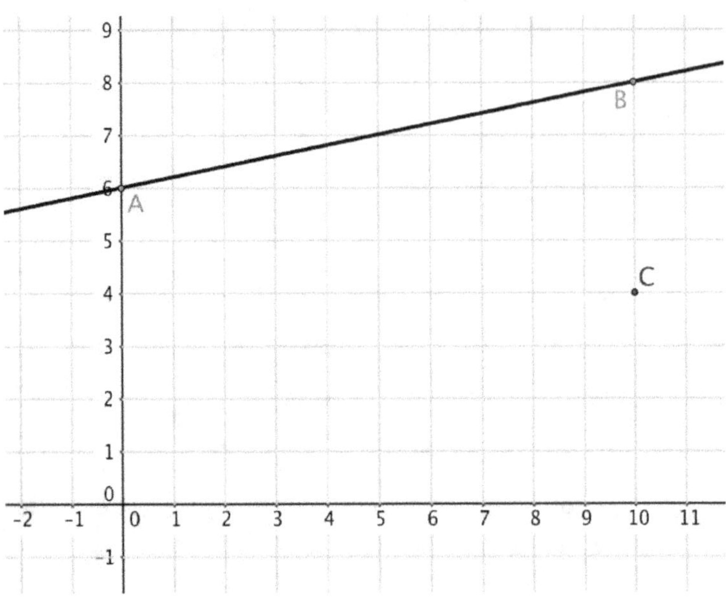

25) Explain why parallel lines have the same slope and why perpendicular lines have slopes who are negative reciprocals of each other. Use examples and sketch graphs to support your explanation.

The Take-Home Message: Parallel lines have the same slope while intersecting lines have different slopes. By rotating a given slope triangle $90°$ we can see that perpendicular lines have slopes that are negative reciprocals of each other. Lines which are coincident have the same slope because they are the same line.

LF11 *Graphing Linear Inequalities*

Name(s)

Date Class/Period/Group

1) Graph the line $y = \frac{1}{2}x + 5$ using the first 3 points in the second column below, and a ruler. The table shows points that are "online," as well as "offline" points which have the same x-values as the online points but different y-values. Graph and label all the offline points as indicated. Complete the table by creating your own online and offline points and answering the questions. The first one is completed for you as an example.

label	point online	label	point offline (but with same x-value)	Is the offline y-value greater than or less than the original?	Is the offline point above or below the line?
A	$(2,6)$	A'	$(2,8)$	greater	above
B	$(-4,3)$	B'	$(-4,0)$		
C	$(0,5)$	C'			
D		D'			
E		E'			
F		F'			
G		G'			
H		H'			

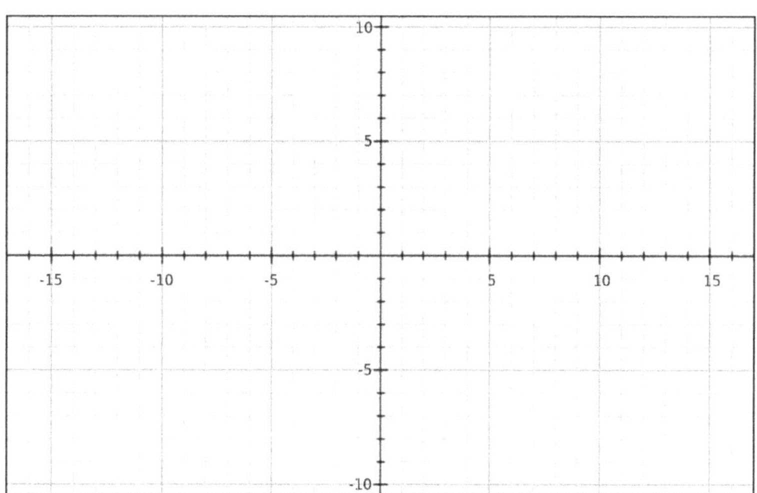

2) Summarize the relationship between the y-value of the offline point and its location above or below the line.

3) The graph at right shows the line $y = \frac{1}{2}x + 5$ along with various online and offline points. Complete the table to describe the relationship between the points and the line, as well as the relationship between the y-values. The first one is completed is an example.

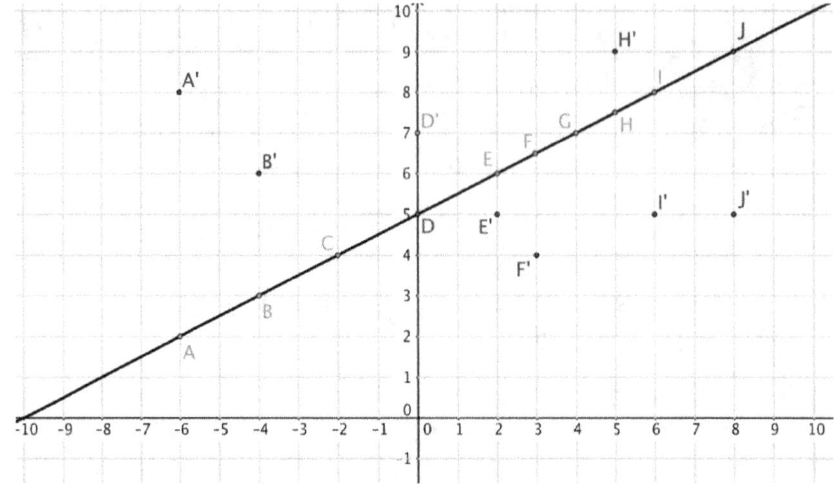

label	point online (x, y)	label	point offline (but with same x-value) (x, y')	$y' > y$ or $y' < y$ or $y' = y$?	Is the offline point above or below the line?
A	$(-6,2)$	A′	$(-6,8)$	$y' > y$	above
B	$(-4,3)$	B′	$(-4,6)$		
C	$(-2,4)$	C′	$(-2,4)$		
D		D′			
E		E′			
F		F′			
G		G′			
H		H′			
I		I′			
J		J′			

4) True or false: Any point on the line $y = \frac{1}{2}x + 5$ make the equation $y = \frac{1}{2}x + 5$ true when substituted for x and y. Explain why or why not, and give examples.

5) True or false: Any point above the line $y = \frac{1}{2}x + 5$ make the **inequality** $y > \frac{1}{2}x + 5$ true when substituted for x and y. Explain why or why not, and give examples.

6) We've observed that points above the line have y-values that are greater than those of the corresponding online points (at the same x-value). Points below the line have y-values below those of the corresponding online points. Online points make the linear equation true, while offline points make an inequality true. Here is a summary of this idea with examples as they relate to $y = \frac{1}{2}x + 5$.

location of points	example	relationship to equation or inequality	example
on the line	$(4,7)$	make the equation true	$7 = \frac{1}{2}(4) + 5$ $7 = 2 + 5$ $7 = 7$ ☺
above the line	$(4,8)$	make the inequality $y > \frac{1}{2}x + 5$ true	$8 > \frac{1}{2}(4) + 5$ $8 > 2 + 5$ $8 > 7$ ☺
below the line	$(4,6)$	make the inequality $y < \frac{1}{2}x + 5$ true	$6 < \frac{1}{2}(4) + 6$ $6 < 8$ ☺

7) Below you have the choice of three statements—one equation and two inequalities. Each of the given points make one of these statements true. Which one? Write each point under the statement it makes true.

Points: $(-1,2)$ $(-3,6)$ $(7,9)$ $(-4,0)$ $(10,-3)$ $(2,2)$ $(0,0)$ $(3,40)$ $(40,3)$

$y = 3x + 4$	$y > 3x + 4$	$y < 3x + 4$

8) When we graph the line $y = 5x - 8$, why do we connect the dots, graph the line and extend the graph to the edges of the graphing window? Why don't we just graph two points and connect them with a segment, or just graph two points and leave it at that?

9) A graph is a visual representation of the points that make an equation or inequality true. That's why we connect the dots when we graph a line—we are trying to visually represent the infinite number of points that make the equation true. So what would you do if you were going to graph the inequality $y > -\frac{1}{3}x + 2$?

10) It's not enough just to graph a few points above the line when graphing an inequality. We need to graph all of them. Below is a graph of $y > -\frac{1}{3}x + 2$. There are two important features to discuss here—one is the shaded part of the graph and the other is the dotted line. How does this effectively represent the graph of $y > -\frac{1}{3}x + 2$?

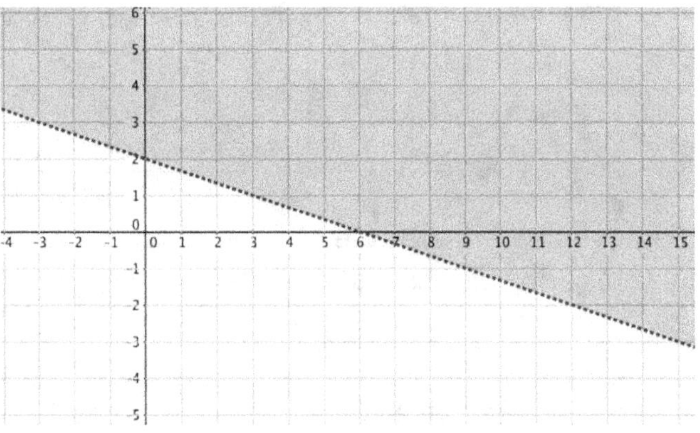

11) Graph the inequality $y < \frac{2}{5}x + 6$ on the axes below.

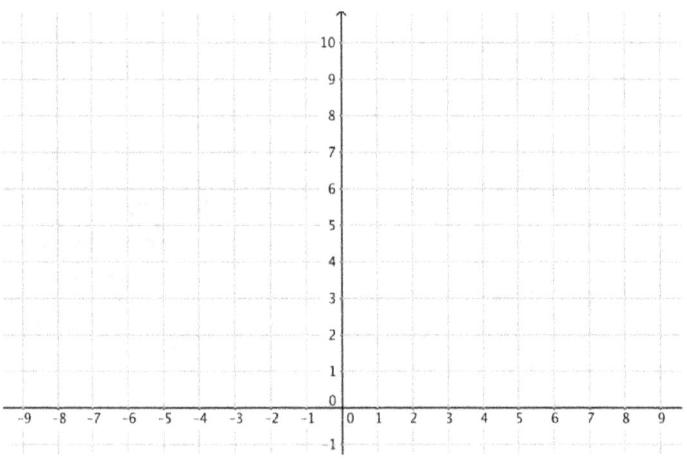

12) What is the difference between $y > \frac{4}{7}x + 5$ and $y \geq \frac{4}{7}x + 5$? How would the graphs be different? Why?

13) If you were graphing each of the following, where you would shade (if anywhere), and would the the line would be dotted or solid? The first two have been completed as examples. Explain your answers.

equation or inequality	shading	line dotted or solid	Why?
$y \geq \frac{4}{7}x + 5$	above	solid	
$y > 8x$	below	dotted	
$y < -10x + 5$			
$y = \frac{1}{3}x + 11$			
$y \leq 0.6x$			
$y = -2x - 54$			

14) Graph $y \geq \frac{4}{7}x + 1$ on the axes provided.

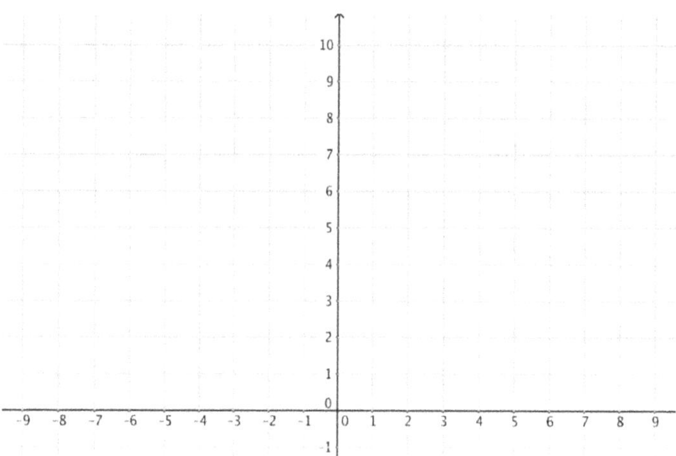

15) Linear equations of course can take many forms, and the same is true for linear inequalities. Below are two commonly used forms for linear equations, with examples. Complete the table by converting these examples into inequalities of your choice.

	general form	specific examples	inequality examples
slope-intercept form	$y = mx + b$	$y = 2x + 3$ $y = \frac{1}{3}x - 7$	
standard form	$ax + by = c$	$2x + 3y = 6$ $\frac{1}{2x} + \frac{2}{3}y = -8$	

16) Let's review first: How do we graph a line in standard form? You have two choices:

Graphing a Line in Standard Form

option 1	example	option 2	example
Convert it to slope intercept form.	$2x - 3y = 12$ $3y = -2x + 12$ $y = \frac{2}{3}x - 4$	Find the y-intercept by plugging in 0 for x.	$2(0) - 3y = 12$ $-3y = 12$ $y = -4$
Graph the line.		Find the x-intercept by plugging in 0 for y.	$2x - 3(0) = 12$ $2x = 12$ $x = 6$
		Graph the line using those two points.	

Practice by graphing the line $4x + 5y = 20$.

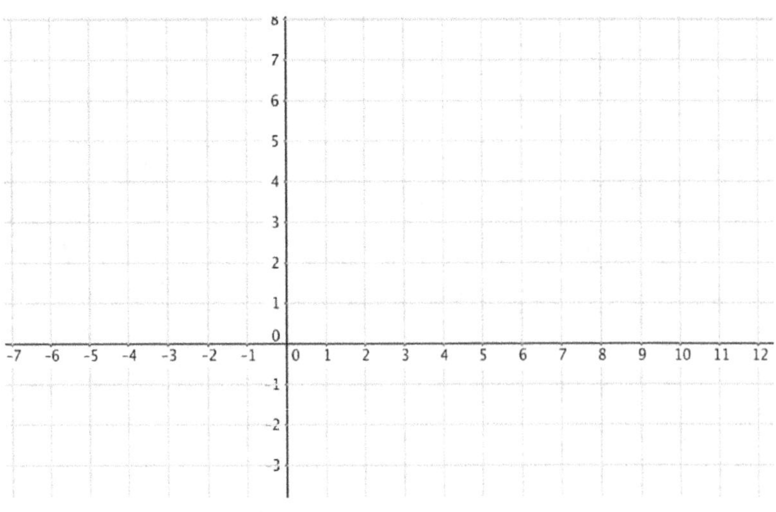

17) To determine where to shade, you have two very similar choices to the ones in #16:

Shading an Inequality in Standard Form

option 1	example	option 2	example
Convert it to slope intercept form. Be careful to flip the sign when you multiply or divide by a negative!	$2x - 3y > 12$ $-3y > -2x + 12$ $y < \frac{2}{3}x - 4$	Substitute any point not on the line for x and y. $(0,0)$ is often the easiest choice!	$2(0) - 3(0) > 12$ $0 > 12$ ☹
Shade appropriately.	In this case we shade below because the sign is $<$.	Decide if that point made the inequality true or false.	$0 > 12$ is false.
		Shade where the point is if true. Shade where it isn't if false.	Shade on the side of the line opposite from where the point $(0,0)$ is.

Practice by graphing the line $4x + 3y \leq 24$.

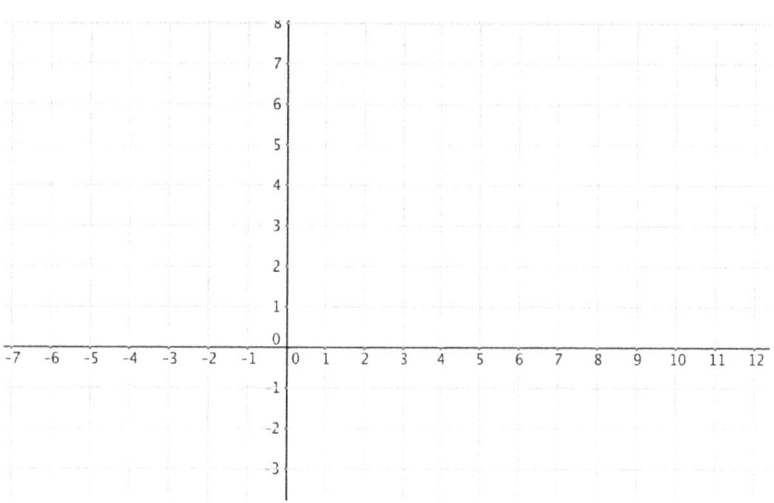

18) Graph the system of inequalities on the axes below. That just means graph them both, double shading any area of overlap.

$y < -\frac{1}{2}x + 5$
$y \geq 2x - 3$

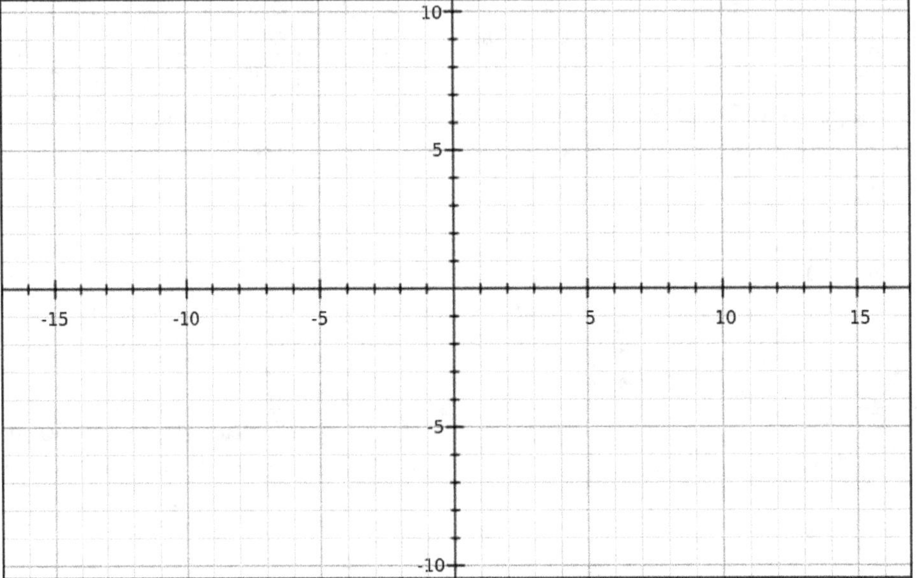

19) Describe in words (and use an arrow pointing to) the region which contains points which make both inequalities true. All these points are the solution to this system of inequalities.

20) Graph the inequality on the axes below.

$2x - 5y > 20$

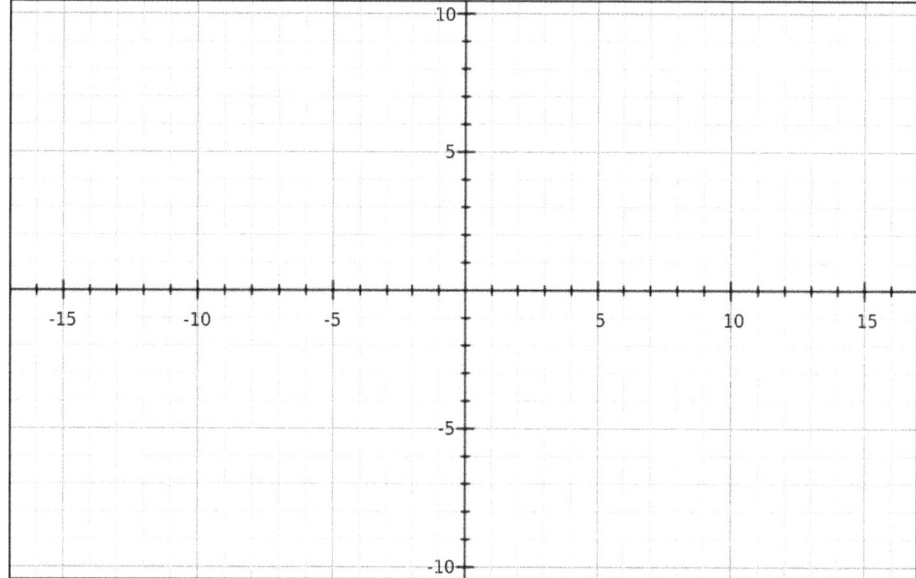

21) Does $(-7, -9)$ make the inequality true or false? How do you know? How is this fact represented in your graph?

22) Graph the system of inequalities on the axes below.

$y > -2x + 4$
$x - 3y \leq 9$

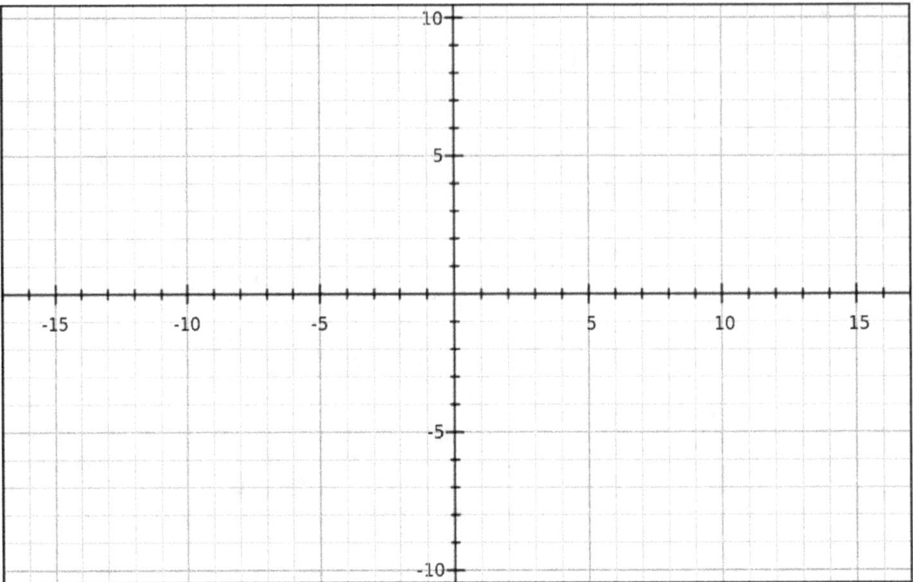

23) Find one point which is part of the solution to the system. Explain how you know that.

24) Explain the differences and similarities between graphing equations and graphing inequalities.

The Take-Home Message: The graph of a linear inequality is a shaded region representing all the points which make an inequality true. We graph the linear equation first then shade above or below using methods appropriate to slope-intercept or standard form.

Courage To Core
Share the Discovery

DS1 Introduction to Statistics

Name

Date Class/Period

1) In some sense we can say that the entire world can be described with numbers. For anything you can imagine, there is a number which summarizes that characteristic. You get a grade on your test, you have a measureable height, the distance from the earth to the moon is a number, the amount of electricity we are using to light this room can be measured. The fact that some things cannot be easily quantified doesn't keep us from assigning numbers to them. List at least 5 measurements you know right now.

2)Describe the most important measurement you can think of in each category. One example is given for you.

what we are trying to assess	what we might measure to assess this
a football player's performance	passes completed
the healthiness of a meal	
the weather	
musical skill	
air quality	
performance in school	
quality of a book	
talent of a hip hop artist	
reliability of an airline	
quality of a chair	

3) We all spend a lot of time either collecting measurements or crunching those numbers with mathematics in order to understand them. We know the number of cars made in Spain each year, the rate of inflation last quarter, the amount of rainfall in Arizona. Actually, we don't know them exactly because we didn't count every car or raindrop, but we were able to collect some **data**, make a mathematically valid guess. Give five examples of data being collected in the world right now.

4) We collect data to understand how the world is working and how it might work in the future. How many cases of Ebola are there in the world right now, and what is the rate of decrease in infection? Is your math grade increasing or decreasing, and how does that relate to the hours you sleep each night? How much rain can we expect in Arizona in February? Will Ford make a lot of cars next year, and will that cause an increase in hiring? The work of collecting and processing data, and using that data to make educated guesses about how the world is working and how it might work in the future--that's the **practice of statistics**. The practice of statistics is the process of coming up with really good estimates. Those estimates, the numbers, those are called **statistics** too. We use the word statistics to describe the entire branch of mathematics involving estimation as well as the actual estimates. Come up with five questions that are interesting to your group which you think that statistics could help answer, and write them below.

5) The goal of statistics is to picture reality as best as possible, then to **quantify** how close to reality that picture is. That's why the weather forecast so rarely states it will rain 100% certainty. They are trying to reflect reality, but they give you a measure of their certainty (or uncertainty) as well. Below are several conclusions or predictions based on statistics. Each statistic is based on a **sample**—it's not the whole picture. For example, when Pepsi claims that 60% of the **population** prefers Pepsi to Coke, they didn't literally ask everyone in the world, they just took a sample. Below are some real statistics from scientific studies. Do you believe them? What else would you want to know before you believed them?

Statistic		Additional Information Desired
Number of hours of sleep for a teenager to feel well rested.	8-10 hrs	
Increase in difficulty for a teen to fall asleep if you add 2 hours of evening screen time.	20%	
Average hours a day spent by teenagers online (including mobile) or watching TV.	7.5 hrs	
Average pulse rate of teenager.	60-90 bpm	
Age at which brain reaches full maturity.	25 years	
Reduction in life span due to smoking.	10 years	
Reduction in life span from sitting 3 hours a day.	2 years	
Grade improvement for teens from increasing exercise to recommended 1 hour per day.	One letter grade	

6) Besides being interesting and even life changing, statistics are something that anyone can create. In order to be statisticians, we first need to collect some data. So first, write your height below, precise to the nearest centimeter. One inch is 2.54 cm. (You may use your calculator.)

Now, write your height in the table on the board. Once everyone has filled in their height, write the data in the table below, in order from smallest to largest, including repeat heights. On the graph, label the x-axis "height", the y-axis "**frequency**," and scale the x-axis appropriately. For each occurrence of a particular height, graph a point. Graph the boys and girls in different colors.

M/F	Height	M/F	Height

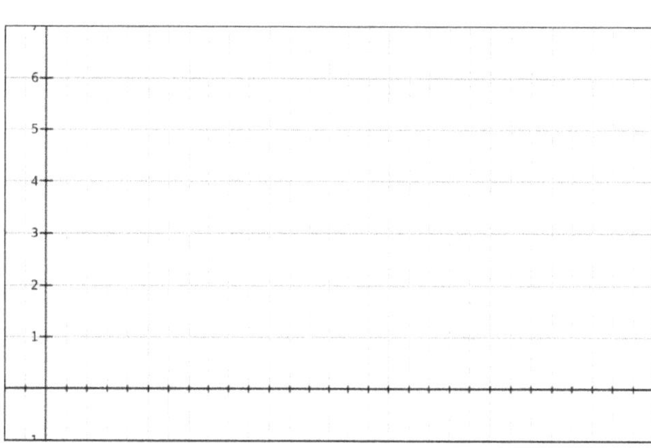

7) Appropriately enough, the above graph is called a **dot plot**. Based on the plot, what would you give as quick approximations for each of the following? Calculate them from the data using your calculator.

	visual approx.	calculated
mean		
median		
mode		

8) Label the mean, median and mode on the x-axis of the graph, and circle and label them in the table.

9) Now find the mean for boys and the mean for girls. Highlight and label in the table, and label on the x-axis of the graph. Write three interesting observations about the data.

10) Looking back on the data from the previous page, what was the **range** of values for height? From ____________ to ______________. If we assign the variable "h" to the height, we could say that h varied between these two values. In other words: ________ $\leq h \leq$ ________ That means that h was between those two values.

11) In your opinion, were there any values for h which seemed like **outliers**? An outlier is a value which is really quite far from the mean compared to other values. If there are some outliers in the data, write them below.

12) Was your height above or below the mean? How far was your personal height from the mean? Are you an outlier?

13) How could you identify the middle 50% of the data? What range of values would represent the middle 50% of the data? Are you in the middle 50%?

14) Write the heights for just your group members below. Find the mean for just your group. How far was your group's mean from the actual mean for our class sample? Now, get the heights from another group, and compute a new average height for the double group. Was this closer to the actual class average? Can you explain why this might be the case?

15) How would this data set change if we did the same collection for a group of 12th graders? Discuss how the mean, median, mode, and range might change.

16) You will only be able to do the first page of this at home—the second page you'll do at school. Collect the following data in your household, and be as accurate as possible.

statistic	value
Number of pets	
Number of people younger than you	
Number of screens (including phones)	
Number of windows	
Number of pieces of fruit	
Number of paintings or photos larger than a sheet of paper on the wall.	
Number of musical instruments	
Favorite animated movie of each inhabitant	
Preferred dessert category of each inhabitant: Ice cream Cake or Pie Chocolate and other candies Pastries and Donuts Churros and other carnival foods Fruit Flans and other pudding-like desserts. Other	
Current time	
Is music playing right now?	
Is someone talking on the phone?	
Is someone interacting with a phone or other computer?	
Are you writing in pen or pencil?	
Are you standing or sitting?	
How many minutes did you exercise today?	
How many hours did you sleep last night?	

17) What was the most interesting question on this survey and why? What additional question should have been on this survey?

18) In your group (back at school), determine at least two questions from the last page that are interesting to you. Think about how you would want to organize all the data from the class for each question. For each question of interest, create a table and a graph of the data. (You can get graph paper for the graph as needed then paste or tape it on this page.) Find the mean, median, mode, middle 50%, and identify outliers. Discuss interesting characteristics of the data.

The Take-Home Message: The practice of statistics is the process of collecting and processing data, and using that data to make educated guesses about how the world is working and how it might work in the future. We take a sample from a population and then produce numbers, also called statistics, like the mean, median, mode and range from the sample.

DS2 *Shape, Center and Spread*

Name(s)

Date Class/Period/Group

1) Your resting heart rate (HR) is the number of beats of your heart per minute (bpm) when you are sitting at rest. Is your HR constant from day to day or hour to hour, or do you think it can vary? Do you think you think that HR's vary from person to person? What do you think a reasonable range of values would be for heart rates for the class population? Give 5 possible reasons for variation in HR from day to day in one individual.

2) Sit so you can see the clock, or ready your watch or phone as a stopwatch. Have your partner place their forearm on the desk, resting with the palm of the hand up. With your forearm stable on the desk, lightly place your index finger and middle finger on the interior of the wrist of your partner, below the heel of the hand. If you press too hard you will not feel a pulse. If either hand moves much you can miss a beat easily. Start the stopwatch, focus on counting without revealing your count. Stop counting at a minute. Write your own HR below. Was it what you expected? Give 5 possible reasons for variation in HR from one individual to the next.

3) Write your group's HR's and mean HR on the board. Complete the table. Label the x-axis "HR." Graph the data as a **bar graph**, with boys in a different color from girls.

M/F	HR	M/F	HR

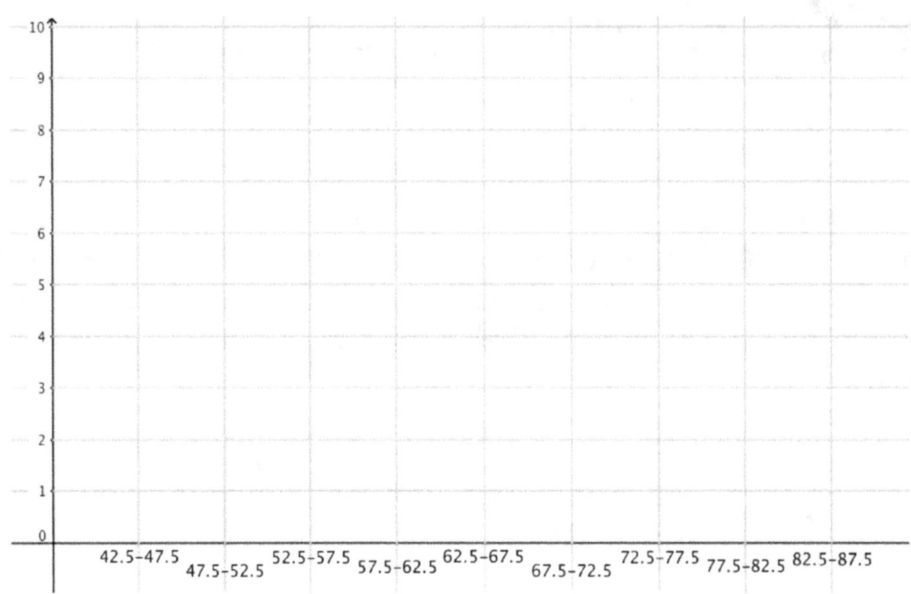

4) Label the mean, median and mode on the x-axis of the graph, and circle and label them in the table.

5) What was the range of values for HR? What range of values represents the middle 50% of the data? Are there any outliers? If so, what are they? Are you in the middle 50%? How far are you from the average?

6) Write the overall mean for heart rate below. Now, calculate the mean of the boys only, then the mean of the girls only. Write three interesting observations about the mean for the boys, the mean for the girls, and the mean overall.

7) Write the HR's for just your group members below. Find the mean for just your group. How far was your group's mean from the actual mean for our class sample? Now, get the HR's from another group, and compute a new average HR for the double group. Was this closer to the actual class average? Can you explain why this might be the case?

8) How would this data set change if we did the same collection for a group of 12th graders? Discuss how the mean, median, mode, and range might change.

9) The practice of statistics is the process of gathering data about a population and using mathematics to produce numbers which help us understand what's going on with that population. The numbers we produce are also called statistics. Some of these statistics measure the **center** of the data, some measure the **shape** of the data and some measure the **spread** of the data. Some statistics like the mean are already familiar to you. Others like outliers, we've only discussed informally. It's now time for us to formalize some definitions of these statistics. For each, decide if the statistic is measuring the shape, center or spread of the data. Then complete the examples based on the following list of heights (cm) of students in a class.

150 160 160 170 175 180 180 180 185 300

statistic	Definition. Does this statistic describe the shape, center or spread?	example
mean	The average. Calculated by adding all the values and dividing by the number of values.	
median	The middle number. Calculated by counting to the center. If there are even number of values, average the two numbers nearest the middle to get the median.	
mode	The most frequently occurring number. There can be more than one of these.	
range	A statement of the smallest to largest values.	
Quartile 1 (Q1)	The median of the first half of the data. Found by finding the median of all the values below the median.	
Quartile 3 (Q3)	The median of the first half of the data. Found by finding the median of all the values above the median.	
Inter-Quartile Range (IQR)	The range of values between Q1 and Q3. Found by subtracting Q1 from Q3.	
outlier	Values far from the median. Found in 4 steps: 1) Calculate the IQR. 2) Calculate 1.5 times the IQR. 3) Calculate Q1 minus that and Q3 plus that. Anything outside those two values is an outlier.	

10) Below are resting heart rates for 20 students aged 16 to 19. Rearrange them from smallest to largest. Complete the chart. Highlight and label important points or sections in the table.

HR	HR (ordered)
70	
63	
67	
82	
72	
26	
54	
59	
68	
76	
81	
95	
53	
64	
68	
59	
71	
74	
87	
77	

statistic	value
mean	
median	
mode	
Q1	
Q3	
IQR	
outliers	

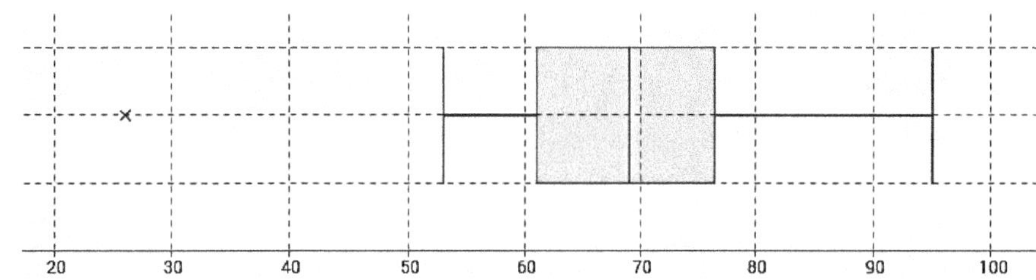

11) The graph above a **box plot** of the data. Along the x-axis, label each of the following: median (M), Q1, Q3, highest value (H) that isn't an outlier, lowest value (L) that isn't an outlier, outlier (O). Along the top, put a bracket showing the IQR. A box plot is a great way to see the shape, center, and spread of the data. For example, between what two values would you expect a student's heart rate to be? Between what two values would you hope to be? Based on the table or graph, what is the chance that their heart rate is between these two values? What value is perhaps a cause for concern or at least curiosity? Is there a value which isn't officially an outlier but might be a cause for concern?

12) Based on the graph or table, what is the chance that a randomly chosen student has a heart rate above Q3?

13) Below is a set of box plots of mid-day temperatures each month in a northern UK city over the past three years. The box plots are oriented vertically along the y-axis. What does the y-axis represent? What does the x-axis represent?

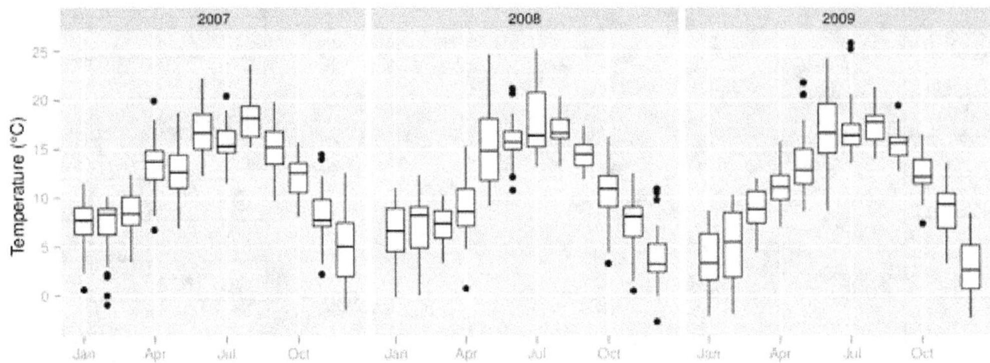

http://www.sr.bham.ac.uk/~ajrs/R/r-gallery.html#timeseries

14) The y-axis specifically represents all the temperatures for each day over the month, so each box plot shows you the shape, center and spread for the temperatures over the month. Complete the table below using the box plots above.

find the...	your answer...
Highest mid-day temperature in July, 2007	
Was that temperature an outlier?	
Median temperature in July, 2007	
Median temperature in July 2008	
Range of values above the median in July 2008	
Range of values below the median in July 2008	
Number of outliers in December 2008	
Range of values in July 2009	
Month, year with greatest range of values	
Month, year with smallest range of values	
Month, year with lowest temperature recorded	
Month, year with highest temp recorded	
Month, year with greatest IQR	
Month, year with smallest IQR	
Two adjacent months with almost the same median	

15) Which month in general seem to have the lowest IQR and/or the lowest range? Why might this be the case?

16) What other patterns do you see in the data? (At least 3 observations.)

17) Below is actual temperature data for Madrid in February and March of 2014. In the second, blank column, order the data from smallest to largest. Complete the table at right. Highlight or label important points in the table. Graph both as box plots on the given graph, one above the other. Label x-axis temperature and scale appropriately.

Feb		Mar	
13		13	
10		17	
8		13	
9		18	
13		18	
17		22	
10		19	
12		21	
8		20	
10		20	
6		20	
8		19	
17		15	
16		18	
10		21	
12		25	
11		24	
13		25	
15		22	
13		23	
12		20	
13		16	
12		14	
12		14	
12		15	
13		14	
13		14	
16		16	
		14	
		15	
		16	

	Feb	Mar
mean		
median		
mode		
Q1		
Q3		
IQR		
outliers		

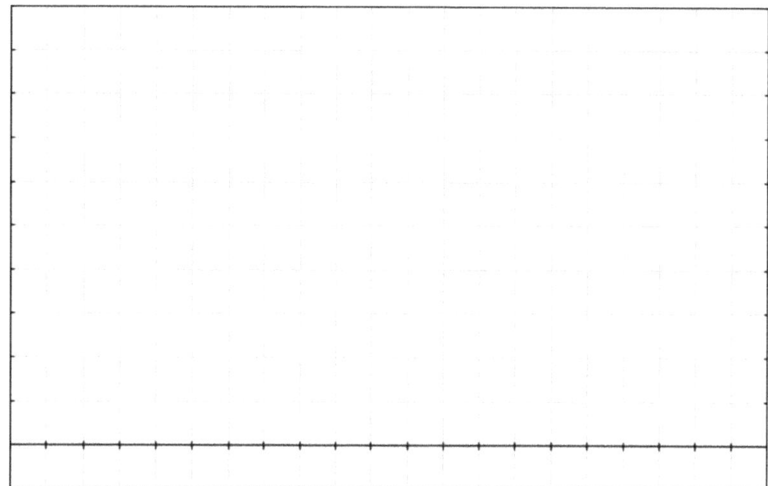

18) What are at least 3 differences and similarities between the two box plots, and what might explain them?

19) The winning time of the 2014 Boston Marathon was $2:08:37$. The table below shows the minutes above 2 hours (rounded down to the nearest half-minute) for the winning times of the Boston marathon from 1983 to the present. Rearrange from least to greatest. Complete the table. Highlight or label important points in the table. Graph as a box plot. Label the x-axis "minutes above 2 hours" and scale appropriately.

Minutes above 2 hours

minutes	minutes
9	
10.5	
14	
7.5	
11.5	
8.5	
9	
8	
11	
8	
9.5	
7	
9	
9	
10.5	
7.5	
9.5	
9.5	
9.5	
9	
10	
10.5	
11.5	
7	
14	
7.5	
8.5	
5.5	
3	
12.5	
10	
8.5	

statistic	value
mean	
median	
mode	
Q1	
Q2	
IQR	
outliers	

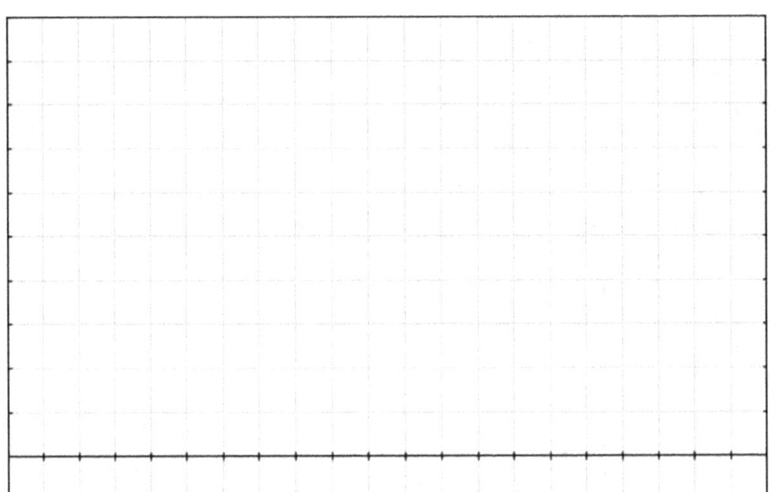

20) Make at least 3 observations of something interesting or surprising about the data and plot, and give possible explanations.

21) Choose 8 numbers at random between 1 and 10. Write them below. Complete the table and graph as a dot plot and as a box plot.

statistic	value
mean	
median	
mode	
Q1	
Q2	
IQR	

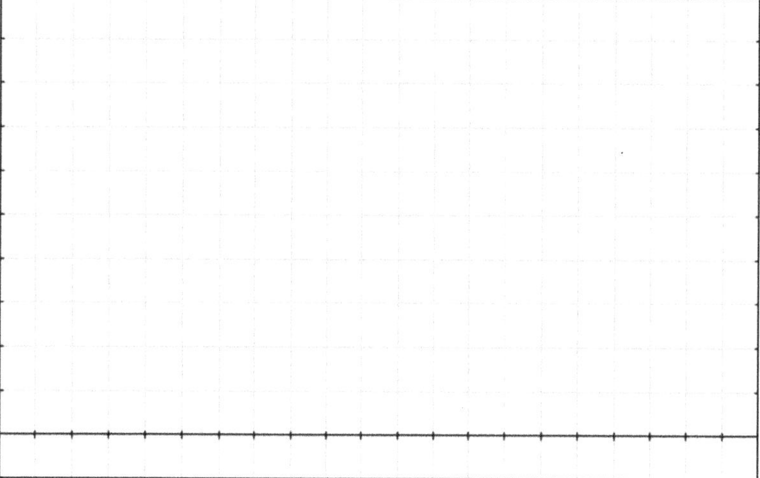

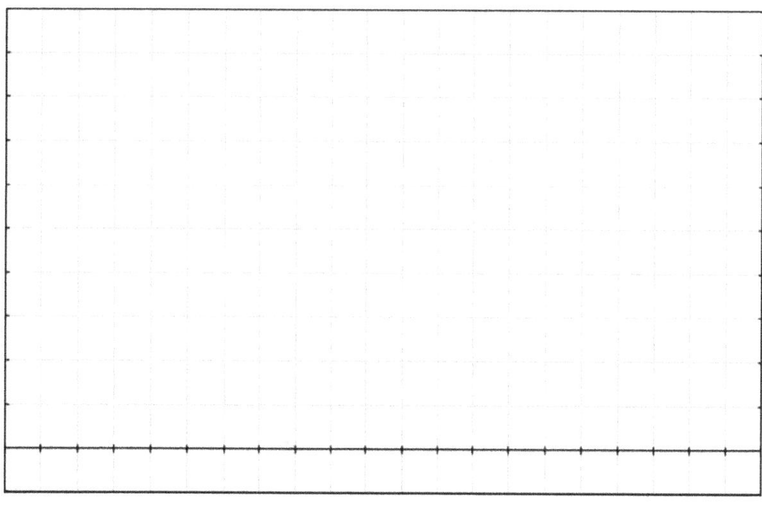

22) Describe the ways that you can represent the shape, center and spread of data. Give examples to support your explanations.

The Take-Home Message: Numerical statistics standing alone or represented graphically tell us about the shape, center and spread of the data. The mean, median, mode, Q1, Q3, IQR and outliers as picture in box and dot plots help us visualize these.

DS3 *The Normal Distribution*

Name(s)

Date Class/Period/Group

1) Over the past year I've taken careful note of the prices of espressos at various cafés I've visited. The prices are listed at right. Graph the data as a bar graph by creating a vertical bar which shows the number of cafes which sold espressos at each price. This means the x-axis is the price and the y-axis is the number of cafes. Label both axes, then graph.

price ($)	frequency (# of cafés)
0.5	1
0.7	3
0.8	4
0.9	7
1.0	10
1.1	8
1.2	4
1.3	2
1.4	1

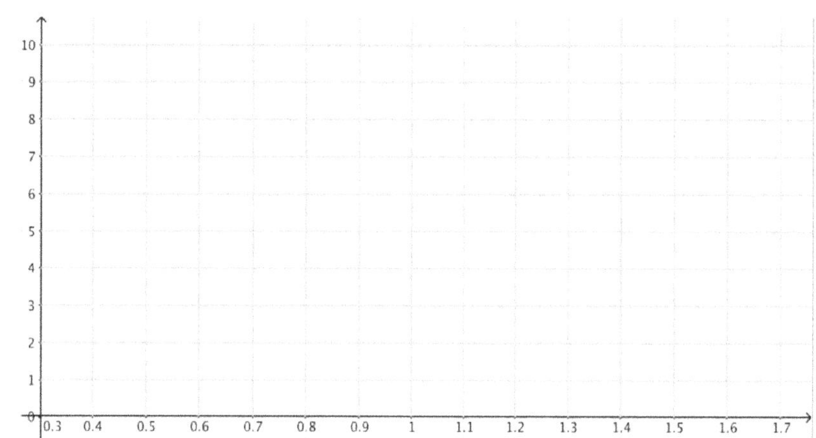

2) Give a quick approximation of the mean based on the graph then use the table to calculate the mean rounded to hundredths. Why was it easy to calculate, and how did you do it? Graph the mean as a vertical line, labeled.

3) Are prices are pretty evenly **distributed** around the mean? Would you say that as you get further from the mean, the number of cafes selling at those prices decreases? What real world explanations can you offer for this?

4) On the last page we made a frequency plot which shows how often each price showed up. The graph is repeated below. Label the x and y-axis appropriately. Explain how the graph shows the frequencies of prices. Give three interesting observations based on the graph and explain your observation. Calculate the mean by adding up all the prices and dividing by the total number of espressos purchased. That calculation looks something like this:

$$Mean = \frac{0.5 + 0.7(3) + 0.8(4) + \cdots}{40}$$

price	freq.	prob.
0.5	1	
0.7	3	
0.8	4	
0.9	7	
1.0	10	
1.1	8	
1.2	4	
1.3	2	
1.4	1	

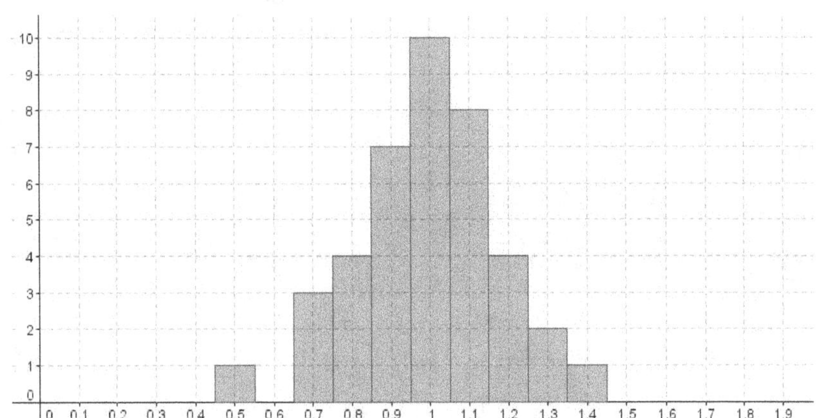

5) Graph the mean as a labeled vertical line. How many cafes did I visit in total? How many of them charge \$1 for an espresso? Based on this distribution, if you randomly choose to visit one of the cafes I visited, what is the probability that the price is \$1? Fill in the probabilities in the table above as reduced fractions or decimals rounded to hundredths.

6) There are a lot of prices close to the mean, and then as you go further from the mean price in either direction, the number of cafes charging those prices declines pretty fast and rather symmetrically. This sort of distribution is very similar to something statisticians call the **normal distribution**. Why were espresso prices distributed this way? Write three other things that might be distributed roughly normally like the espressos.

7) One way to create a normal distribution is to flip coins! Flip a coin 6 times. Each flip is a **trial**. Our **statistical experiment** is flipping the coin 6 times to see how many times we get heads. Conduct this experiment ten times and write the results at right. Then graph a frequency plot. Label the x-axis "number of heads" and the y-axis "frequency." Get the data from an adjacent group and add to the table, then stack these bars on your existing bars in a different color. Calculate the mean number of heads and graph it as a labeled vertical line.

Exp. #	# heads 6 tosses	# heads (2nd group)
1		
2		
3		
4		
5		
6		
7		
8		
9		
10		

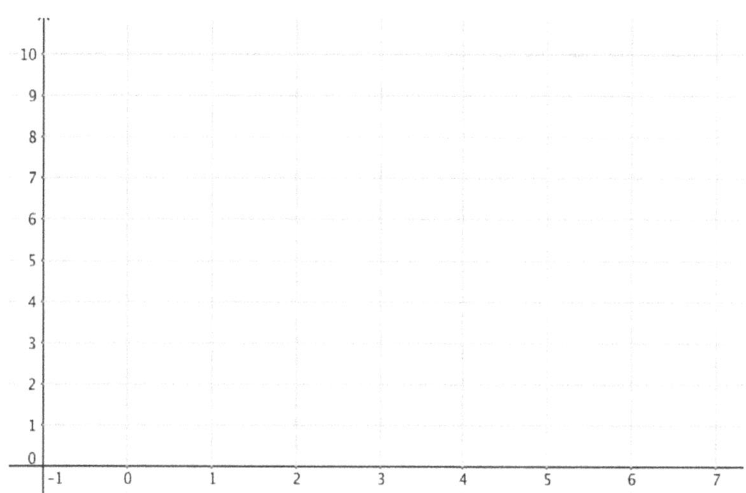

8) Use the table of your experiment above to calculate the probability of each outcome below, as a decimal rounded to hundredths. Bar graph each outcome as x and the probabilities as y. Label x and y-axis. Graph the mean as a labeled vertical line. This graph is called a **probability distribution**. How is this graph similar to and different from the espresso graph? Why?

# heads	Prob.
0	
1	
2	
3	
4	
5	
6	

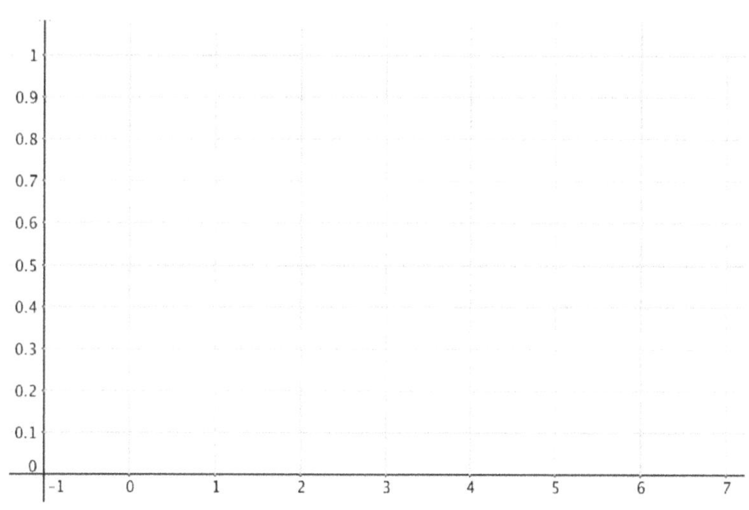

9) Below is a frequency plot of the outcomes for my own experiment of flipping 6 coins twenty times. Label the x-axis "outcome" and the y-axis "frequency." How often did I get an outcome of 3 heads? Based on my experiment, what is the **relative frequency** with which this outcome occurred? Based only on this experiment, what is the probability of this outcome? Was this the same or different from your experimentally derived probability of 3 heads in #8? Why?

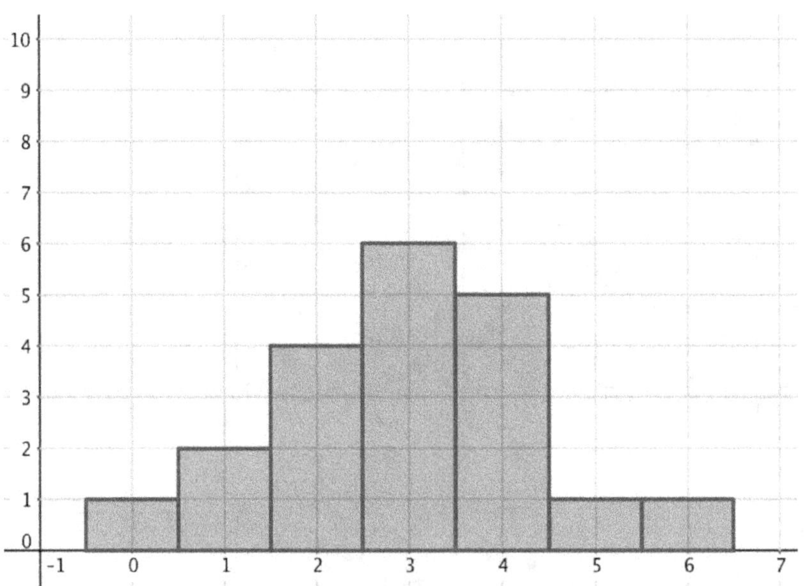

10) Use the table of my experiment above to calculate the probability of each outcome below, as a decimal rounded to hundredths. Graph each outcome as x and the probabilities as y. Label x-axis "outcome" and y-axis "probability." Why is this graph called a probability distribution? Calculate the mean number of heads, and graph it as a labeled vertical line.

# heads	Prob.
0	
1	
2	
3	
4	
5	
6	

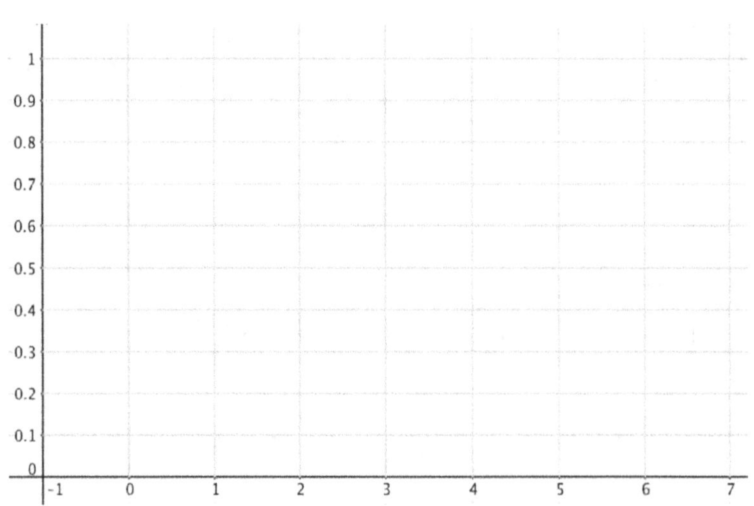

11) My coin flipping experiment on the previous page returned a somewhat normal looking distribution, with a peak around the mean and then symmetrical tails on both sides. Did your experiment produce a roughly normal distribution? How can you tell?

12) Below is a frequency plot showing the results of another repetition of the coin flipping experiment. Label x and y-axis. How often did 2 heads show up? Based on this experiment, what is the probability of this outcome? Does this graph appear roughly normal? Why or why not? Find the mean and graph it as a labeled vertical line.

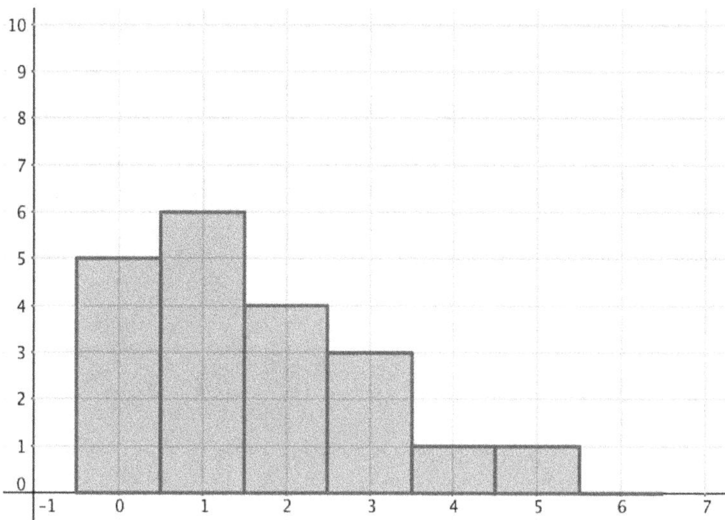

13) Use the table of my experiment above to calculate the probability of each outcome below, as a decimal rounded to hundredths. Graph each outcome as x and the probabilities as y. Label x and y-axis? Calculate the mean number of heads, graph it as a labeled vertical line.

# heads	Prob.
0	
1	
2	
3	
4	
5	
6	

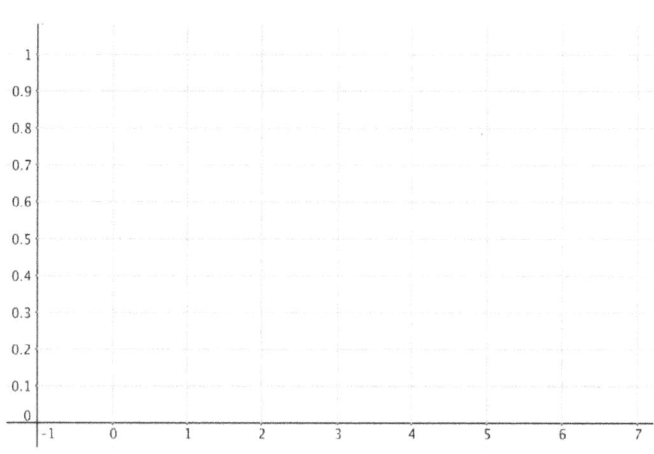

14) On the previous pages you generated three different frequency plots and probability distributions. Why weren't they all the same? What are possible explanations of their differences?

15) To generate each plot on the previous page, we completed 20 experiments where each experiment was to flip a coin 6 times. If we completed 100 experiments, do you think the graph would look more normally distributed? Why or why not?

16) The graph below is the frequency plot for 100 experiments of flipping six coins and counting the number of heads. Label x and y-axis. For how many experiments did I get two heads as a result? Based on this plot, what is the probability of getting two heads? Does this graph look more normally distributed than the plots for 20 experiments? Why or why not?

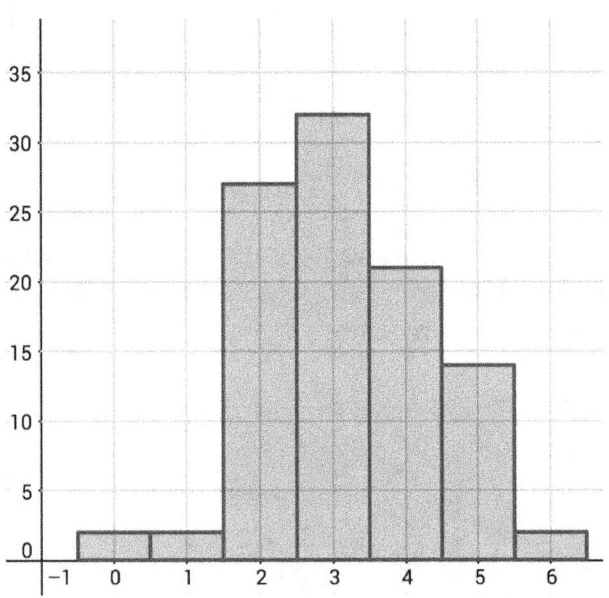

17) The graph below is the frequency plot for 1500 experiments of flipping six coins and counting the number of heads. Label x and y-axis. Based on this plot, how often did I get two heads? What percentage of the time did I get two heads? Based on this plot, what is the probability of getting two heads? Does this graph look more normally distributed than the plots for 20 experiments? Why or why not?

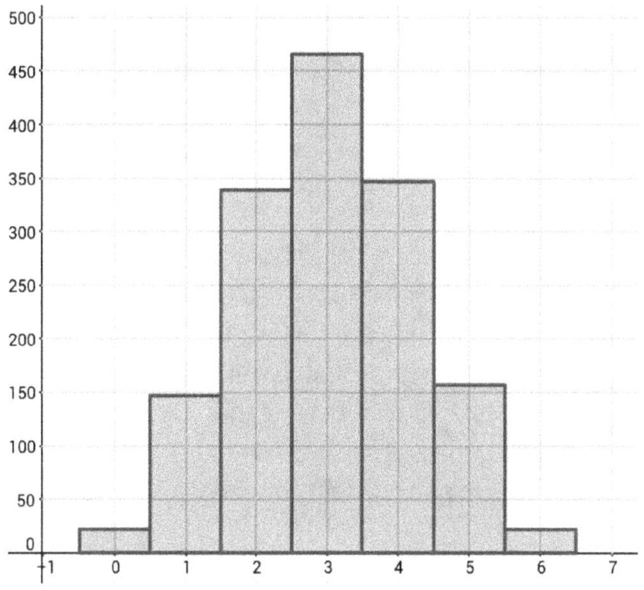

18) If we kept doing this experiment for days and weeks, hundreds, thousands, millions of times with a 50-50 coin, the results would look increasingly normal. And by normal, we mean having the shape of a **normal curve**, a **bell curve**. Imagine completing this experiment of flipping 6 coins an infinite number of times. The graph below is a probability distribution for the results. Label x-axis "outcome" and y-axis "probability." Based on this graph, what percentage of the time will I get the result of 4 heads? What is the probability of getting 4 heads?

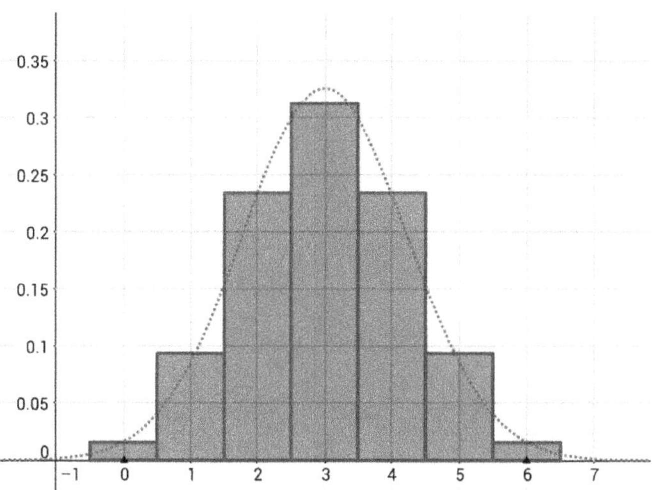

19) What are the differences and similarities between frequency plots and probability distributions? Give examples and sketches as needed to support your explanations.

20) The graph below is a probability distribution for the results for the experiment of flipping 6 coins an infinite number of times. Label x-axis "outcome" and y-axis "probability." What single outcome is the most probable? Some results are further from this mean and some are closer. Is 2 heads a pretty reasonable outcome or a pretty improbable outcome? What about 0 heads? What is the mean outcome? A typical amount of **deviation** from the mean is about 1 head. Some results might deviate more than 1 from the mean and others less than 1, but the typical deviation is 1. We would usually expect to get 3 ± 1 heads, where 1 is the deviation from the mean. So if you flip 6 coins you generally expect to get either 2, 3 or 4 heads.

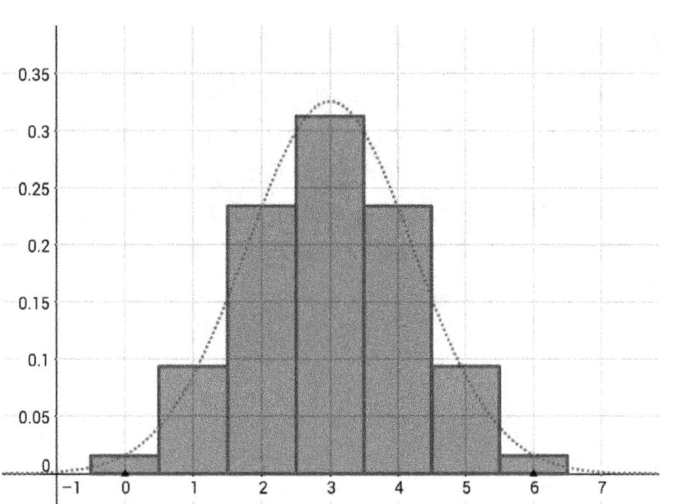

21) What is the probability of getting 3 heads? What is the probability of getting 2 heads? What is the probability of getting 4 heads? What is the probability of getting either 2, 3 or 4 heads? Explain how you know your answer to that last question is correct.

22) Based on your last answer, if we flip a coin 6 times, is it pretty likely that we will get a result of 2,3 or 4 heads? Why or why not?

23) The actual value for the **standard deviation** in the last problem is 1.22, which means that although some results will be further from the mean and others closer, the typical result should be in the range of 3 ± 1.22. That means a typical result will be between what two values? Can we actually get a value of 4.22 heads? Why or why not?

24) Here is the frequency plot for the prices of espressos at various cafes. The mean price was 0.99 cents. Graph and label this as a vertical line on the frequency plot. The standard deviation of the prices of espressos in my sample was 0.18 cents. This means while some prices were further from the mean, and some were closer, the typical deviation from the mean was 0.18 cents. Based on the graph what are the two prices which are exactly one standard deviation from the mean? Graph these prices as vertical lines on the frequency plot. Label them "+1 SD" and "-1 SD." Roughly what percentage of the espressos were within 1 SD of the mean? So what is the approximate probability that an espresso will have a price within 1 SD of the mean?

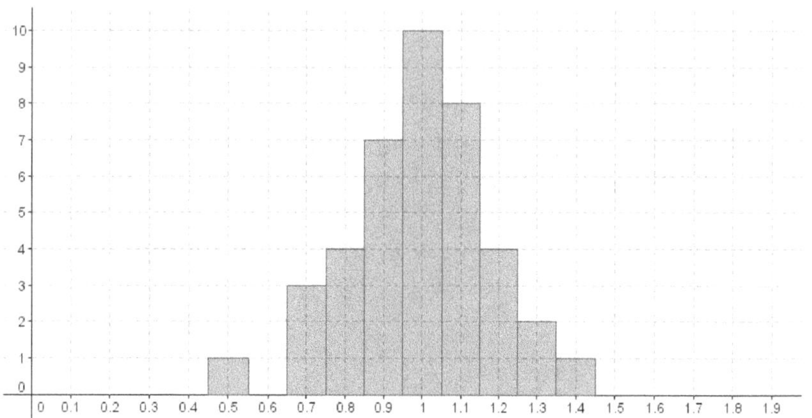

25) In all the above problems we used the roughly normal distribution, the mean, and the standard deviation to describe the shape, center and spread of our data. Sketch the bell curve where requested.

	describes	what it means	how to visualize it
roughly normal distribution	shape	A frequency plot which shows us a mean which occurs more commonly than the other outcomes and symmetrical tails showing frequencies decreasing as we go further from the mean.	A bell curve:
mean	center	The average value. At or near the peak of a roughly normal distribution.	A vertical line in the middle of the curve above.
standard deviation	spread	The typical deviation from the mean. Some values will be more frequent or others less frequent, but this is the typical deviation.	Two vertical lines on the curve above showing the values that are 1 above and 1 below the mean.

26) One of the most popular climbing shoes in the world is called the Anasazi. Below is a bar graph showing the numbers of each size (in European sizes) of this shoe that is available at the climbing store in Tucson, Arizona. Label x-axis "shoe size" and y-axis "number available." Find the mean shoe size of all the Anasazi's at the store. Draw a vertical line on the graph at the mean, and label.

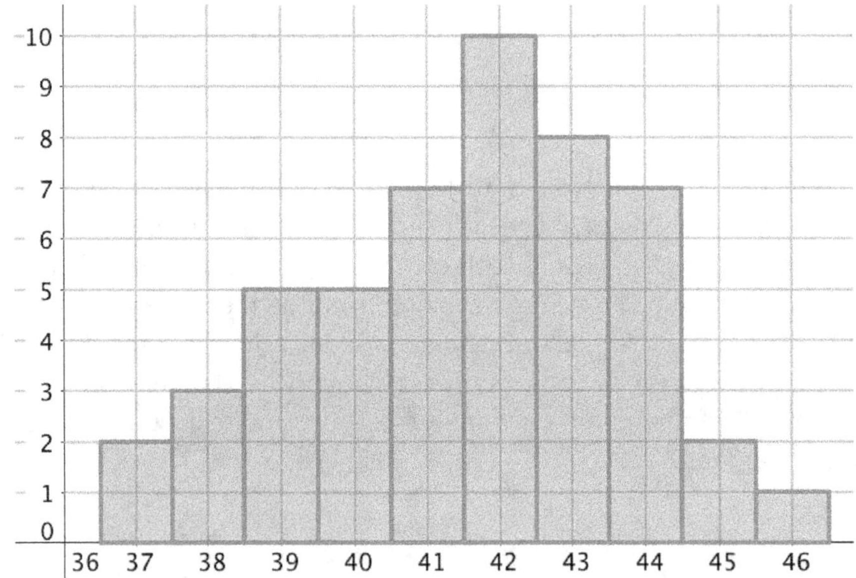

27) The standard deviation for this Anasazi sample is 2.2. What shoe sizes (to tenths place) are one standard deviation from the mean? Put vertical lines at each of these values and label them "+1 SD" and "-1 SD." Approximately what percentage of shoes are within ± 1 standard deviations from the mean?

28) What shoe sizes (to tenths place) are two standard deviations from the mean? Put vertical lines at each of these values and label them "+2 SD" and "-2 SD." Approximately what percentage of shoes are within ± 2 standard deviations from the mean?

29) Does this data appear to be approximately normal? How would the data have to be different in order for it to appear approximately normal? Describe the shape, center and spread of this data by comparing it with the normal distribution, discussing where the mean is relative to the mode and describing the standard deviation.

30) Here is the data from the previous page. Label the x and y-axis, mean and standard deviation, adding the vertical lines as you did previously.

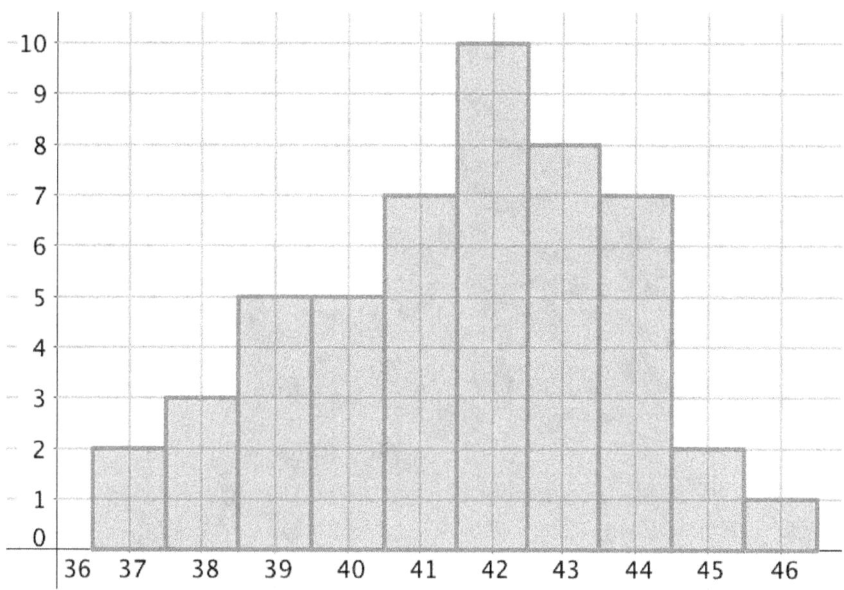

31) Based on the graph above, if an Anasazi climbing shoe is chosen at random from the store, what is the chance that it's a size 42?

32) Based on the graph above, if an Anasazi climbing shoe is chosen from the store at random, what is the chance that it's a size 38 or 39?

33) Based on the graph above, if an Anasazi climbing shoe is chosen at random from the store, what is the chance that it's not a size 44 or 45?

34) Below is a bar graph showing the numbers of each size of the Anasazi climbing shoe that is available at another climbing store in Joshua Tree, California. Label x-axis "shoe size" and y-axis "number available." Find the mean shoe size of all the Anasazi's at the store. Draw a vertical line on the graph at the mean, and label.

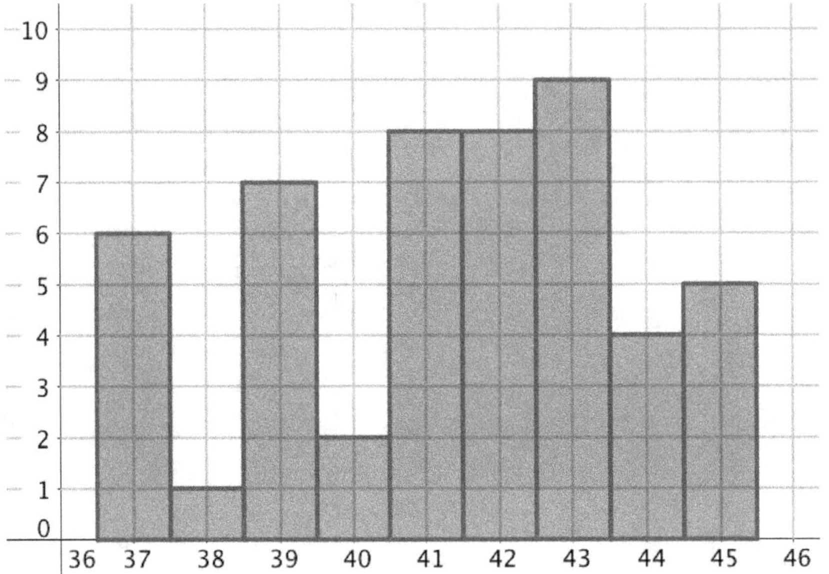

35) The standard deviation for the data above is 2.41. Label and graph ± 1 SD as vertical lines. Roughly what percentage of shoes were between ± 1 SD of the mean? Describe the shape, center and spread of the data in terms of whether the distribution seems normal or not, the mean and its relationship to the mode, and the standard deviation.

The Take-Home Message: Frequency plots and probability distributions show us how data is distributed and the likelihood of different outcomes based on the given distribution. The shape, center and spread of the data can be described by referring to the normal distribution, the mean and the standard deviation. Distributions can approach normal as the number of experiments increases.

DS4 Sampling

Name(s)

Date Class/Period/Group

1) What ratio (part out of total) of candies in a typical bag of m and m's is green? Do you think that the ratio is exactly the same for every bag? Do you think that the factory keeps the ratios the same or varies them? Explain your answers below.

2) What ratio of skittles in the world are green, is there any way you could give me an exact answer? Why or why not?

3) Even though we can't say with absolute certainty what the ratio of green skittles is in the world, we can give a pretty good **estimate**, and say how confident we are in the estimate. One important part of the practice of statistics is using **samples** to generate **statistics** which are pretty good estimates of the actual **parameters** of the entire **population**. Use the information in the table to give your own example of each term below.

term	definition	precision level	example
sample	A preferably random selection of data from the population.	X	Identifying whether candies are green or not in a random selection of 20 candies.
statistic	A summary number that describes the sample.	An estimate. Unlikely to be perfectly precise or perfectly accurate.	The ratio of green candies in the sample.
population	All the data that could possibly be gathered.	X	Identifying all the candies in the world as green or not.
parameter	A summary number that describes the population.	perfectly precise and accurate.	The ratio of green candies in the population.

4) Write the available colors of candies (of a given brand) in the space below. Take a sample of 10 candies. Record the count of each color below, then the ratio of that color to all 10 candies. Each student should take a different sample of 10 candies. (There is space for four students under each color.)

Colors						
Number						
Ratio						

5) Based on your sample, which color seems the most common? Did everyone in your group get the same results? Why or why not? Do you think that increasing the **sample size** from 10 to 20 candies would give you a more precise estimate of the ratios for the entire population? Why or why not?

6) We are going to focus only at the ratio of green candies. Everyone in your group should write their ratio of green candies from #4 on the board. Graph as a dot plot. Label the x-axis "ratio green candies" and the y-axis "frequency of this ratio."

7) Based on the dot plot, what would you guess is the actual ratio of green candies in the world? Why?

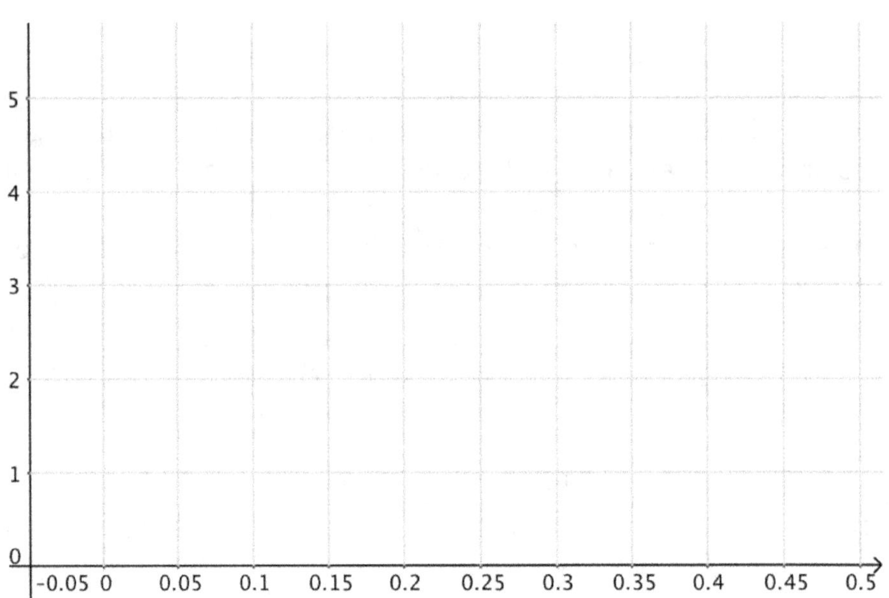

8) Find the mean ratio of candies in all the samples. Graph this as a vertical line and label it.

9) Remember that the standard deviation is a measure of the typical deviation from the mean. Some deviate more from the mean and some less. What would you guess is the typical deviation from the mean?

10) While the standard deviation can be calculated by hand, it is much easier to do so with a graphing calculator. If you have a TI-84+ graphing calculator, follow the steps below to find the mean, standard deviation and other useful statistics. Write the results below rounded to 2 significant figures.

How to Find Mean and Standard Deviation with a Graphing Calculator

- Press STAT then select "Edit..."
- Type values into L_1. Repeated values must be repeated in the list.
- Press STAT then arrow right to see the CALC menu. Select "1-Var Stats."
- Confirm that the list is L_1. Arrow down to select "Calculate."
- The mean is $\overline{x}$. The standard deviation is σx.

11) What ratios are one standard deviation away from the mean? Draw vertical lines on the graph at these locations and label them "+1 SD" and "-1 SD." What percentage of ratios were between these two values?

12) Complete the table summarizing the key statistics of this frequency plot.

Mean	Standard Deviation	+1 SD	-1 SD

13) Do you think that the mean we calculated by using all the sample ratios is pretty close to the mean ratio of green candies in the entire population? Why or why not? The standard deviation is a measure of the precision of our sample ratios. Do you think our sample ratios are precise? Why or why not? What could we do to make them more precise?

14) Take a sample of 20 candies. Do not eat them until after the exercise! Record the count of each color below:

Colors						
Number						
Ratio						

15) We are going to look again at the ratio of green candies specifically. Everyone in your group should write their ratio of green candies from #14 on the board. Graph as a dot plot. Label the x-axis "ratio green candies" and the y-axis "frequency of this ratio."

16) Based on this dot plot, what would you guess is the actual ratio of green candies in the world? Is this pretty close to the mean ratio from the last experiment?

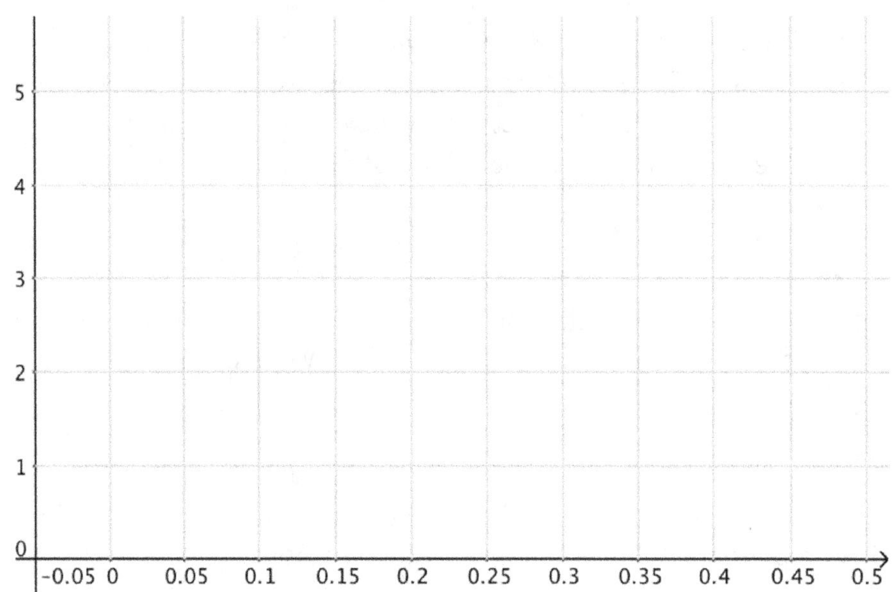

17) Compare the graph of this distribution with the previous one. How is the spread of this distribution different from the last one? Why?

18) Find the mean ratio of green candies in all the samples of size 20. Graph this as a vertical line and label it.

19) What would you guess the standard deviation is this time? Is this more or less than the standard deviation you got using samples of size 10? Why? Use your calculator to find the standard deviation. What ratios are exactly one standard deviation away from the mean? Draw vertical lines on the graph at these locations and label as before. Is the standard deviation smaller or bigger than in the previous experiment? Why?

20) What percentage of ratios of green candies are within ± 1 SD of the mean? Is this close to the percentage you found for the previous distribution? Why is this the case?

21) As we took samples from the population, we found that the number of green candies clustered around a certain mean. In fact, all these samples together arrange themselves in a normal distribution, and the mean of the samples is a good approximation of the actual ratio of green candies in the world! Not only that, as we increased the sample size from 10 candies to 20, we noticed that our samples gave us values which were more tightly clustered around the mean. That means we have greater confidence in our estimate for the actual ratio of green candies in the population. How do you think the distribution would change if we took samples of size 100?

22) The graphs below show the results when other students took candy samples of size 10 and found the ratio of green candies to the total in each sample. (Their candy might be a different brand than yours.) The first graph shows the results for 20 students and the second for 50 students. Answer the following questions for each. The mean and ± 1 SD are shown as vertical lines—label them. What percentage of the ratios are between the values for ± 1 SD? Is the mean close to the mode for the data? Does the distribution appear roughly normal, that is, peaking at the mean and decreasing symmetrically on both sides? Why or why not?

<u>20 Students</u>

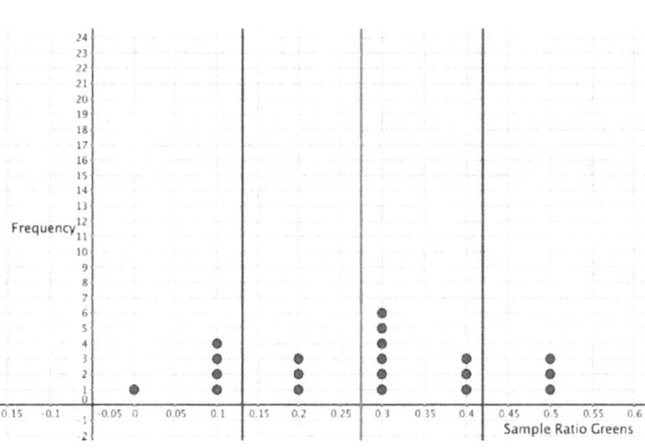

<u>50 Students</u>

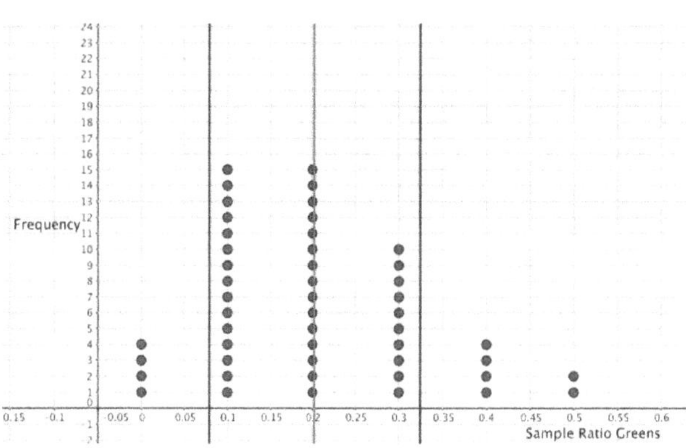

23) Describe differences and similarities between the two plots, and explain them.

24) Let's review some previously learned concepts before moving forward. In all the above problems we used the roughly normal distribution, the mean, and the standard deviation to describe the shape, center and spread of our data. Sketch the bell curve where requested.

	describes	what it means	how to visualize it
roughly normal distribution	shape	A frequency plot which shows us a mean which occurs more commonly than the other outcomes and symmetrical tails showing frequencies decreasing as we go further from the mean.	A bell curve:
mean	center	The average value. At or near the peak of a roughly normal distribution.	A vertical line in the middle of the curve above.
standard deviation	spread	The typical deviation from the mean. Some values will be more frequent or others less frequent, but this is the typical deviation.	Two vertical lines on the curve above showing the values that are 1 above and 1 below the mean.

25) The graphs at right and below show the results when 100 students took samples of size 20, then 50, then 100 and found the ratio of green candies in each sample. The mean and ± 1 SD are shown as vertical lines—label them. Do the distributions appear roughly normal, that is, peaking at the mean and decreasing symmetrically on both sides? Why or why not? Describe the differences and similarities between the plots and explain them, discussing the mean and standard deviation.

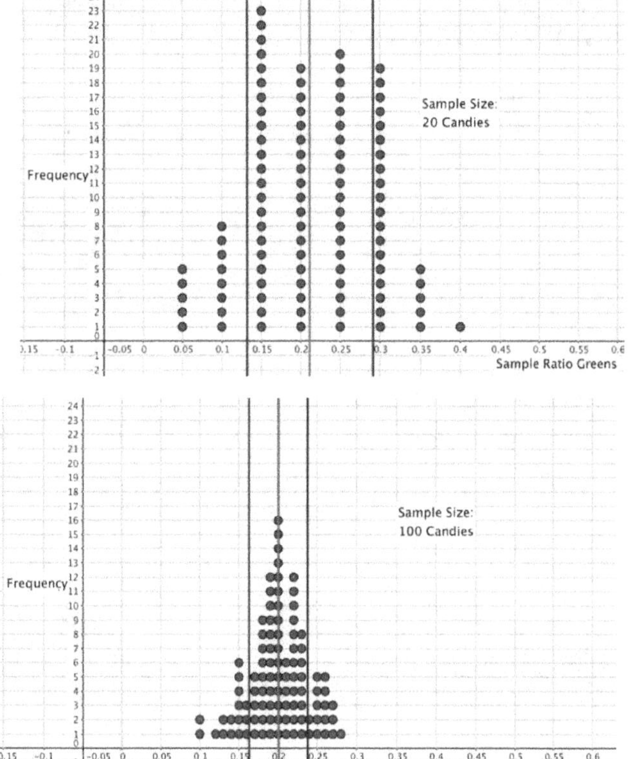

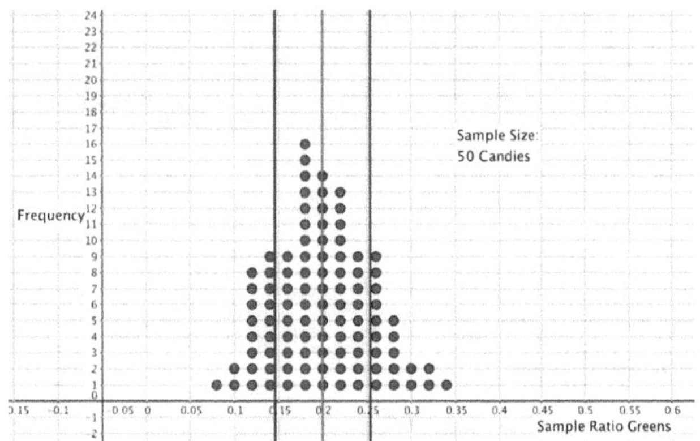

26) Look back at the graphs on the last page. As the sample size increased, what happened to the distribution of the sample means? Specifically, what happened to the standard deviation? Why do you think this happened?

27) Find the percentage of values between ± 1 SD of the mean for each of the plots on the last page. Were the percentages pretty close? And yet we saw that the standard deviation decreased because the sample size increased. That decrease allows us to estimate the ratio of green candies in the population with greater precision. Which sample size gave us the highest level of precision and why? Are there practical limits to the the sample size, or to the number of samples of candies we can take? Describe them.

28) This graph shows the results when 10000 students took candy samples of size 300 and found the ratio of green candies to the total in each sample. Answer the following questions for the plot. The mean and ± 1 SD are shown as vertical lines—label them. Compare and contrast this graph with previous dot plots, discussing the mean and standard deviation.

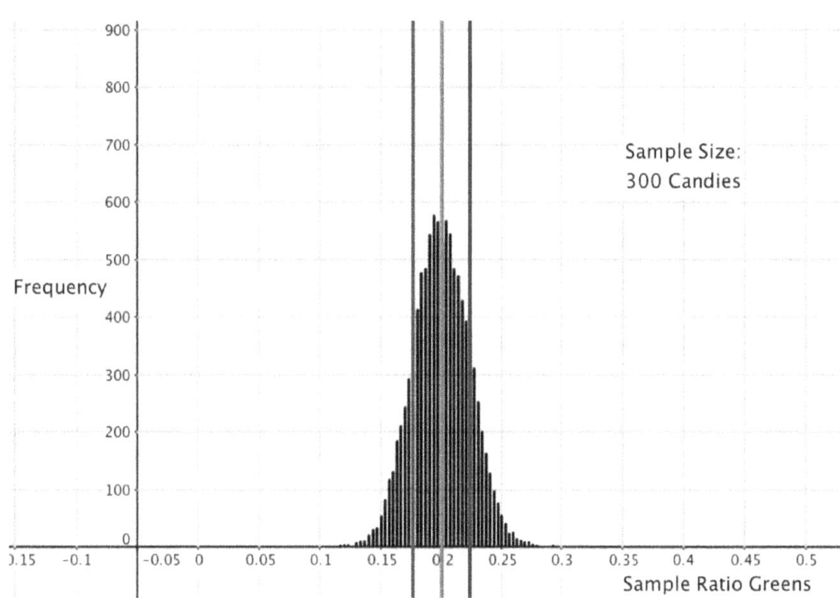

29) Based on the last plot, what do you think the ratio of green candies in the actual population is? The mean above is 0.2 and the standard deviation is 0.02. As you observed earlier, approximately 68% of the sample ratios will be between ± 1 SD of the mean. That means a single sample ratio has a 68% chance of being within 1 SD of the actual ratio of greens in the population. Why?

30) Below is a summary of the impact of changing the sample size or the number of samples on a distribution of sample ratios. Draw before and after sketches below which show the way a dot plot of ratios will change as you increase sample size, and the way it will change when you increase the number of samples. Look back at previous plots for examples if needed. Give verbal descriptions supporting your sketches.

	What happens when you increase sample size?	What happens when you increase the number of samples?
standard deviation	The standard deviation gets smaller. Our 68% confidence in our estimate for the parameter for the population is within a smaller range of values.	It has no impact on the standard deviation, but it does make the plot of the data from the samples look more obviously bell shaped. (See #5 for an example.)
sketch (before)		
sketch (after)		
description		

31) The mean height of NBA players is 2.06 meters, with a standard deviation of 0.08 meters. (That's a mean height of about 6'9" with a standard deviation of 3".) The heights of NBA players are approximately normally distributed, which means a dot plot of all the heights of all the players might look something like the graph at right. Label the x-axis "height in meters" and the y-axis "frequency." Label mean and ± 1 SD.

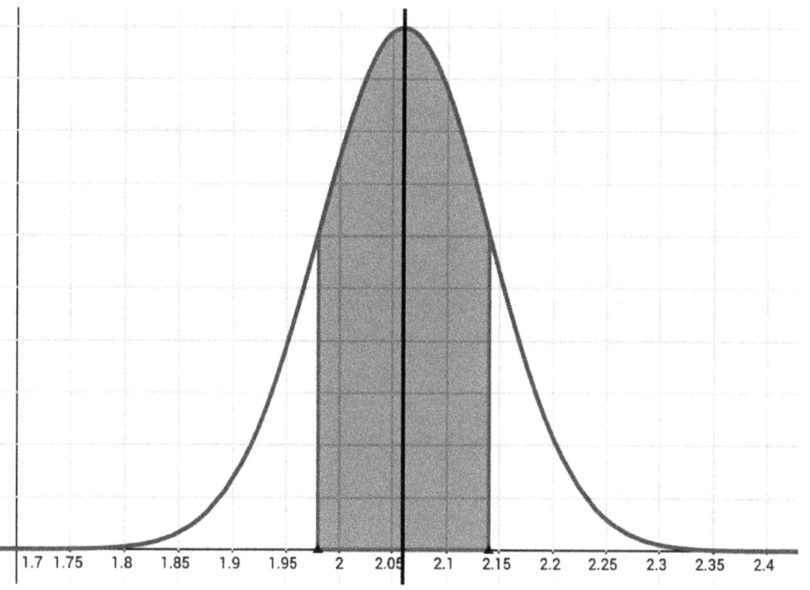

32) Look at the graph in #28. There we graphed the results for 10000 samples as a dot plot, and there were so many dots that the graph almost looked like a shaded space. For our graph of NBA player heights, we imagine an infinite number of players, even though that isn't realistic. But the point is for us to have a useful model, and our model is a perfect normal distribution. According to the normal distribution above, what is the most common height of players? According to the model, which is more common: a height of 2 meters of a height of 1.9 meters? How does the graph show you this? Does this make sense in terms of what you know about NBA players? Why?

33) What is are the values for $+1$ SD and -1 SD? If we pick a player at random, what is the chance that he would be within ± 1 SD? Does the shaded area above appear to be about this percentage of the area under the normal curve? What is the total area under this curve?

34) Define each of the following. Give an example or sketch to support your definition.

term	definition and example or sketch
mean	
standard deviation	
± 1 SD	
sample	
statistic	
population	
parameter	
increasing sample size	
increasing number of samples	
estimating parameter of a population with 68% confidence	

35) A scientist took a sample of 45 cans of a soda and measured the grams of sugar in each. Below is the data in a table showing the frequency. Graph as a dot plot, labelling the x-axis "grams of sugar" and and y-axis "frequency." Use your calculator to find the mean and standard deviation, rounding to 3 significant figures. Graph the mean and ± 1 SD as vertical lines, and label.

value	frequency
34	1
34.5	1
35	3
35.5	6
36	8
36.5	7
37	6
37.5	5
38	4
38.5	2
39	2

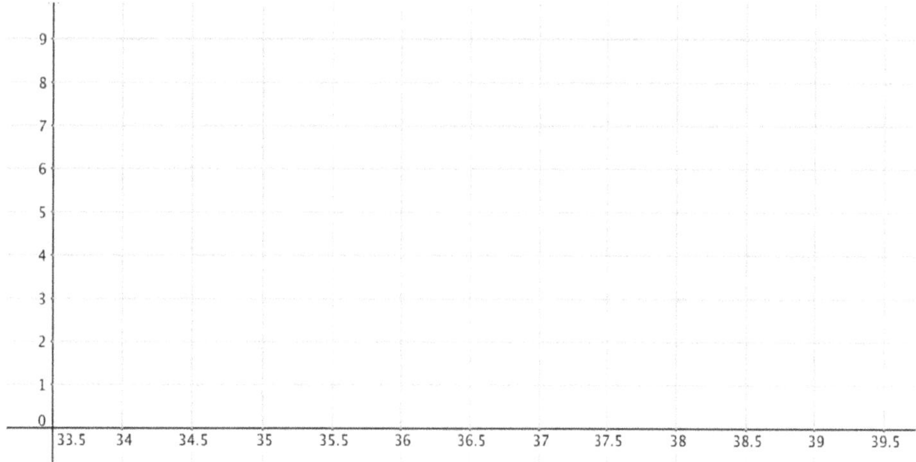

36) Even though we only have one sample above, we have a sample mean and a sample standard deviation. And we know that one sample mean has a 68% chance of being within 1 SD of the actual mean. That works in reverse as well—the actual mean has a 68% chance of being within 1 SD of the sample mean. Therefore, you can say that you are 68% confident that the actual mean grams of sugar in the population is between which two values? Why?

37) Find, invent or create a data set like the one on the previous page. Perform the same graphing and analysis as you did in the last problem.

The Take-Home Message: A greater sample size produces a tighter distribution with a smaller standard deviation. This allows us to estimate population parameters with greater confidence. We are always about 68% confident that the actual parameter is within ± 1 SD of the mean.

Courage To Core

Share the Discovery

DS5 Linear Regression

Name(s)

Date Class/Period/Group

1) Up to now we've just looked a distributions of single random variables, like the distribution of heights of NBA players or means of samples of candies. But we aren't just interested in knowing the shape, center and spread of single random variables. We also want to know how those variables influence other variables. Sure, NBA players have an average height of 2.06 meters with a standard deviation of 0.08 meters. But how does that influence the scoring capacity of those players? Do you think that as height increases, scoring increases? Why or why not? Do you think that other variables besides height also influence scoring? Give examples.

2) Do you think that all the variability in scoring by players can be explained by height, or that only some of the variability can be explained by height? Does anyone in your group think that height influences none of the variability in scoring? Why or why not?

3) Do you think that there is a strong positive **correlation** between height and scoring? (A correlation is a relationship, and if the relationship is strong and positive then we would see that scoring very clearly increases in line with height.)

4) Define each of the following and give an example.

Term	Definition and Example
Correlation	
Strong Positive	
Strong Negative	
Weak Positive	
Weak Negative	
No correlation	

5) At right is a graph of the points scored every 40 minutes (y-axis) vs. height of NBA player in inches. Each point represents one player. Label the x and y axis. Strong positive, weak positive, no correlation? Why?

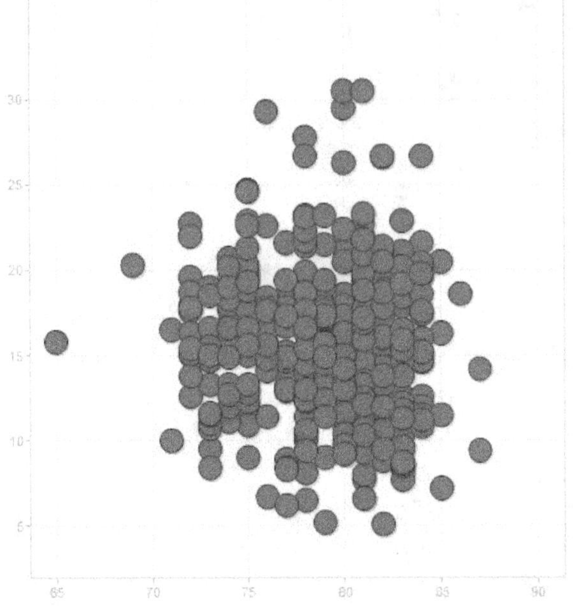

http://hoopdata.com/motioncharts.aspx

6) What is a rebound in basketball? At right is a graph of rebounds vs height. Label the axes. Do you think there is a strong positive or weak positive correlation between rebounds and height? What might explain this? Do you think the relationship between rebounds and height is linear or quadratic?

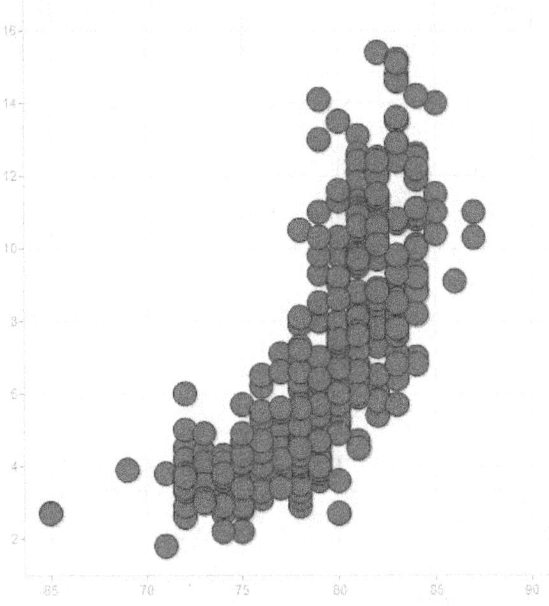

7) If a player is 75 inches tall, and you had to pick just one number, what would you expect his number of rebounds to be? What range of values would you expect them to be within? Roughly what is the average number of rebounds for a player who is 845 inches tall? If you magically increase your height from 75 inches to 85 inches, roughly how much do the rebounds increase? Draw a slope triangle on the graph representing this. Roughly what is the rate of change of rebounds per inch of increase in height?

8) We saw on the last page that there was a positive correlation between rebounds and height. Do you think there are other factors which influence the number of rebounds for a player? Give examples. Do you think that increasing height causes an increase in rebounds? Why or why not?

9) At right is a chart showing corn production in the US and global coffee production. Graph the data on the axes below right, with corn on the x-axis and coffee on the y-axis. Label and scale the axes. Describe the correlation, if any, between coffee and corn production.

Year	US corn (billions of bushels)	world coffee (millions of bags)
1995	7.4	83
1996	9.2	102
1997	9.2	100
1998	9.8	107
1999	9.4	130
2000	9.9	111
2001	9.5	108
2002	9.0	121
2003	10.1	106
2004	11.8	115

10) Do you think the correlation between production of these two products demonstrates that one causes the other, in other words, the **correlation implies causality**? Why or why not? What are some other variables which may have caused production of both products to increase? Is it possible that both increases were caused by some other variable? Give examples.

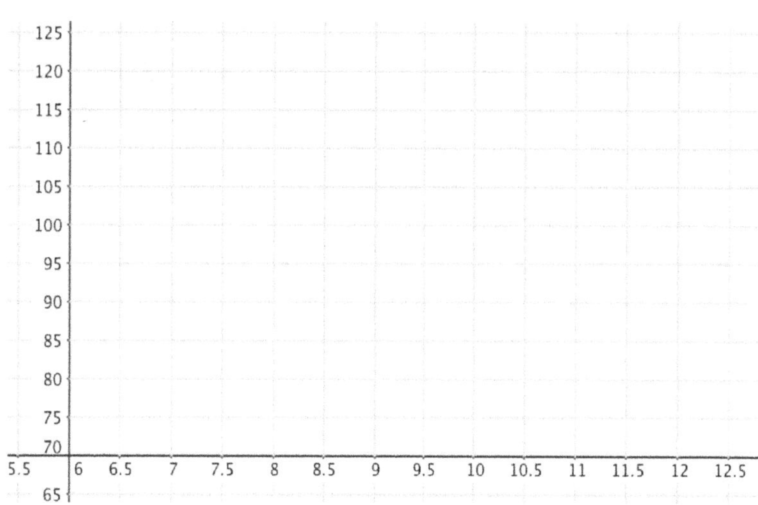

11) Following is a table showing the mass of a sample of different cars and their corresponding fuel efficiency in miles per gallon. Graph the data on the axes provided. Scale and label the axes appropriately. Describe the correlation, if any.

Type of Car	kg	mpg
Mercedes Benz SL550	2018	16
Chevrolet Cobalt	1276	25
Pontiac G5 GT	1286	26
Acura TSX	1549	25
Mitsubishi Lancer	1392	24
Volkswagen Rabbit	1348	24
Chevrolet Malibu	1577	25
Mazda 6	1628	23
Toyota Camry	1498	25
KIA Rondo	1543	22
Nissan Titan 2WD	2389	14
Cadillac SRX	1913	18
GMC Envoy	1962	16
Nissan Armada	2564	14

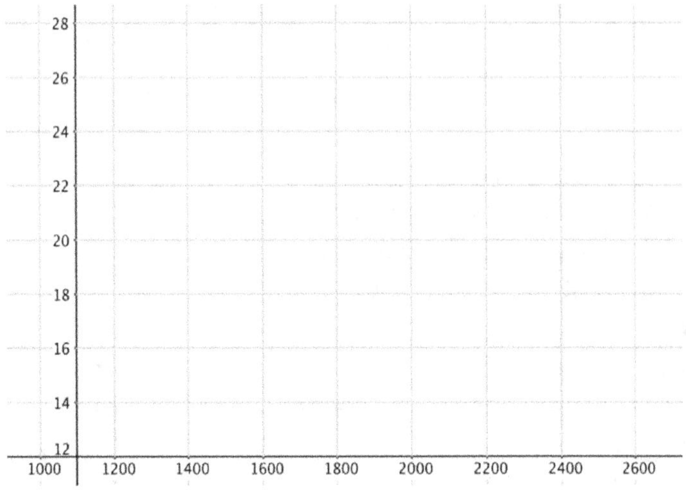

12) Based on the graph, is there is a correlation between mass and fuel efficiency? Is there causality, that is, in your opinion, does a change in mass cause a change in fuel efficiency? Why or why not?

13) Look at the trend. If a car has a mass 1300 kg and you had to pick one number, what would you guess for its fuel efficiency? Your guess doesn't have to be a point already on the graph. If a car has a mass of 2400 and you had to pick one number, what would you guess for its fuel efficiency? Graph these two points in a different color and label them A and B. Connect these points with a straight line extended to the edges of the graphing window and draw the resulting slope triangle. The **trend line** you've drawn is a visual summary of the relationship between fuel efficiency and mass. Find the equation for this line. Use this line to predict the fuel efficiency for a car with a mass of 1700 kg.

14) In fact, there are a few ways to find the best line to fit data which appears to show a linear relationship between one variable and another. The **line of best fit** is the one which minimizes the amount of variation of the actual data points from the line. You can think of it as a line which finds the middle of the data, sort of like the mean for single variables. When we calculate the standard deviation from this line, it's the smallest it can be. Go back to #10 and graph an approximate line of best fit.

15) Of course, our calculator can find the line of best fit for us. Complete this procedure for the data in #11.

- Press "2nd, 0" select DiagnosticsOn. Press ENTER. (You can skip this step from now on.)
- Press STAT then select "Edit..." to create list L_1 of the x-coordinates.
- Create list L_2 of the y-coordinates.
- Press STAT then arrow-right to CALC.
- Select "LinReg(ax+b)"
- Write down the values for a, b and r rounded to three significant figures.
- Press WINDOW to create a graph window and scale appropriately, like the graph in #11.
- PRESS "2nd, Y="
- Select "On," and make sure L_1 and L_2 are Xlist and Ylist.
- Press "Y="
- Type in "ax+b" using your values for a and b.
- Press GRAPH. You should see all the data points as well as the line of best fit.

16) Write the equation for the line of best fit with slope and y-intercept rounded to three significant figures. On a scale from 0 to 1, how well do you think the line fits the data? Graph this new line dotted or in pen to distinguish it from the line you already graphed. Hopefully they are pretty close!

17) Our calculator has also told us how well the line fits the data. This number is called the **correlation coefficient**, and it helps us understand the level of correlation between the two variables. If there is a perfect positive correlation, the correlation coefficient is 1. If there is a perfect negative correlation, the correlation coefficient is -1. If there is no correlation at all then the correlation coefficient is 0. A correlation coefficient above 0.8 suggests a strong positive correlation and a weaker but not meaningless one is between 0.5 to 0.8. The correlation coefficient in our car data was given by the calculator as r. What was r? Describe the correlation. What explanations can you offer for this?

18) Find data or create some fictitious data which has a correlation. Explain what x and y represent. Make a table for the data which includes at least 10 data points. Graph the data, scaling and labelling axes appropriately.

x	y

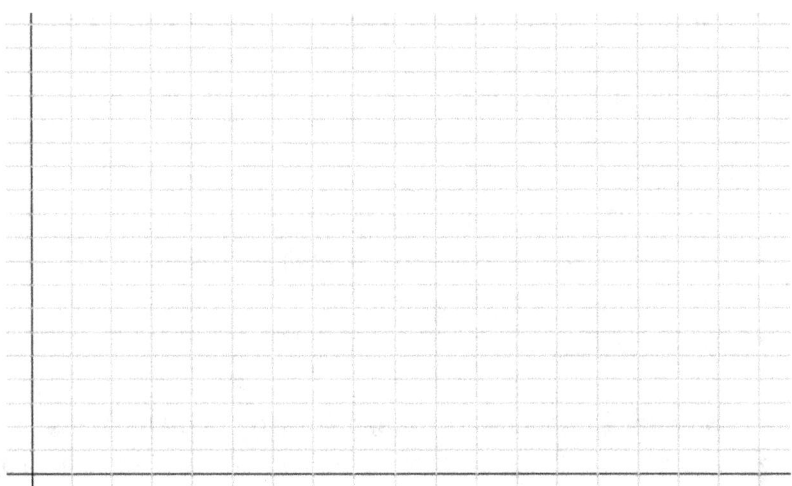

19) Create lists for x and y, and use the linear regression function on your calculator to find the line of best fit. Write the line of best fit and the correlation coefficient below. Graph the line. Describe the meaning of the slope of the line in the context of the data. Use the equation for the line to predict a y value for an x-value not represented in the original data set.

20) Here is one (fictional) surfer's data set relating hours of surfing one summer and her happiness rating from 1 to 5, with 5 being the happiest. Graph the data on the axis provided, labelling and scaling the axes appropriately.

Surfing per week (hrs)	Happiness Index (1-5)
1	1
5	4
3	3
2	2
7	4
10	5
9	5
1	2
12	5
15	5
18	4
10	4
7	4
2	1
6	3
0	1

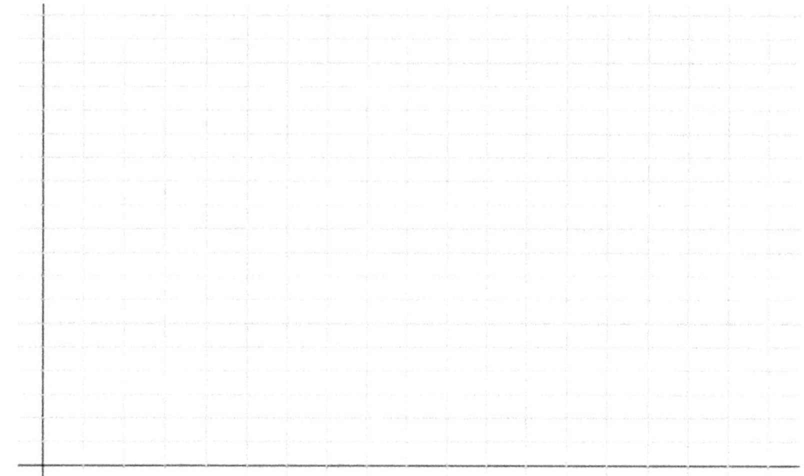

21) Use your calculator to determine the line of best fit. Write down the correlation coefficient. Graph the line above. Use the line to predict a level of happiness for 8 hours of surfing. For every increase of one hour of surfing, how much do we expect our surfer's happiness to increase?

22) The typical adult vocabulary ranges from 20,000 to 35,000 words. Fifteen (fictional) students tracked the number of (fictional) novels they read through high school and college, reported what they had read and then were tested to estimate their vocabulary. The results are in the table below. Perform a complete analysis of the data as you did on the previous page.

Novels Read	Estimated Vocabulary (thousands)
30	23
10	45
57	31
34	39
93	42
104	37
12	25
28	27
17	30
39	34
86	31
28	17
70	25
150	12
19	28

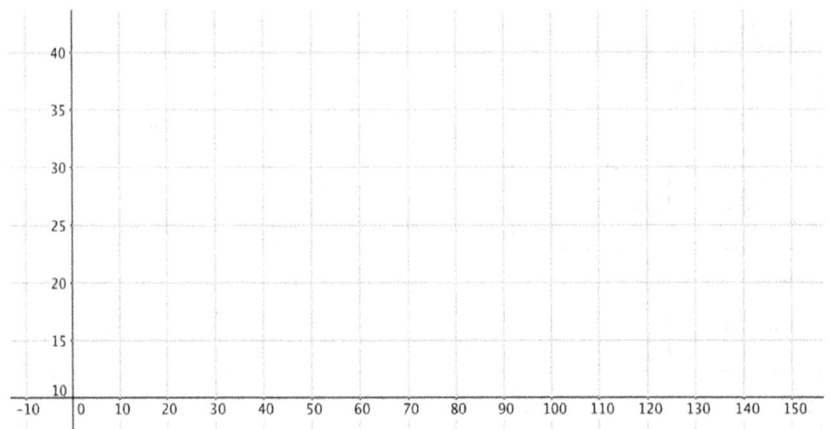

23) What problems might there be with a study like this? What besides novels might influence vocabulary development? Does the test produce a precise number for each person's vocabulary or is this also an estimate? Are there some data points which seem like outliers? Are there some data points which seem suspicious? Why?

24) Find data or create some fictitious data which has a correlation. Explain what x and y represent. Make a table for the data which includes at least 10 data points. Graph the data, scaling and labelling axes appropriately.

x	y

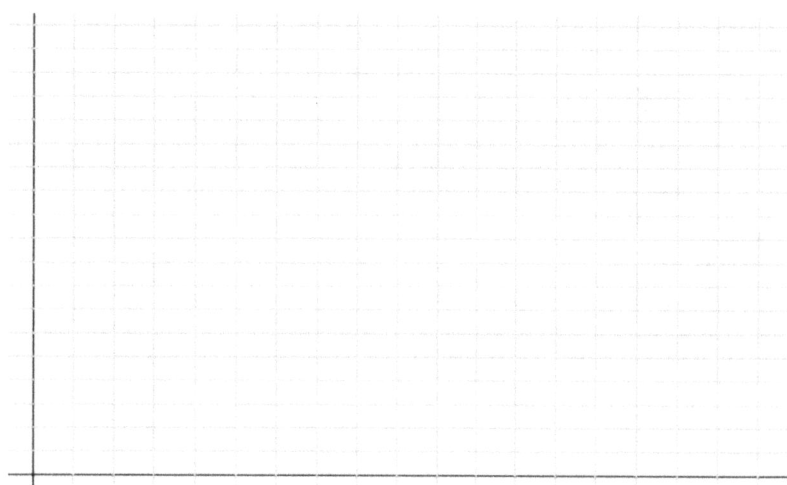

25) Create lists for x and y, and use the linear regression function on your calculator to find the line of best fit. Write the line of best fit and the correlation coefficient below. Graph the line. Describe the meaning of the slope of the line in the context of the data. Use the equation for the line to predict a y value for an x-value not represented in the original data set.

26) Define each of the following terms and give examples or sketches to support your definitions.

term	definition and sketch or example
single variable statistics	
two variable statistics	
correlation	
causality	
trend line (line of best fit)	
linear regression	
correlation coefficient	

The Take-Home Message: A correlation between two variables mean that changes in one correspond to changes in the other. This can be seen in the graph. Causality means that the change in one causes the change in the other—this can't be known just from the graph. We can use linear regression to estimate the line of best fit and find the correlation coefficient, thus summarizing the relationship between the two variables.

Courage To Core
Share the Discovery

EQ1 *Exponential Functions*

Name(s)

Date Class/Period/Group

1) Linear functions can be used to model many situations. Write the equations for the linear functions which model each of the following situations. Explain what x and y mean for each one.

Situation	Linear Model
A genie takes the number of candies you give him, multiplies them by 2 and adds 3.	
A gym membership costs 20 euros to start and then 15 euros a month after that.	
A wingsuit flier descends 2 meters vertically for every 5 meters of horizontal movement. He jumps from a height of 500 meters.	
A car is racing at 150 km per hour with a 100 meter head start over its competitors.	
The cost of a special variety of apple in 2000 (year 0) was 10 cents. Now, in the year 2015, the same apple costs 50 cents.	
A scatterplot shows that as morning rainfall in cm increases, more students are late to school.	

2) But of course, there are many situations which are modeled by non-linear functions. Many physics equations express relationships which are not linear. First, use words to describe the relationship which the equation is representing. Then classify each function as linear, quadratic, rational, square root, or exponential. Hint: When we say, for example, "y is a function of x in the equation $y = mx + b$," we are saying that m and b are known constants, and y is the output for the function while x is the input.

Equation	Meaning	Example Equation	Type of function
$D = \frac{m}{V}$		D as a function of V: $D = \frac{6}{V}$	
$E = mc^2$		E as a function of m:	
$F = ma$		F as a function of m:	
$C = \pi d$		C as a function of d:	
$s = \frac{1}{2}at^2 + v_i t$		s as a function of t:	
$y = \sqrt{x} + d$		y as a function of x	
$y = 2^x + e$		y as a function of x:	
$y = a(3)^x + 5$		y as a function of x:	

3) Suppose you have 5 candies. You run across a genie who tells you he will double the candies for you, not just once, but many times. In fact, if you give him 3 magic pebbles, he will double the candies 3 times! You are able to find exactly 3 magic pebbles. you give the genie the magic pebbles and your candies, and he gives you how many candies? ________ Remember, he doesn't multiply your candies by 3, he doubles them 3 times! Are you sure about your answer? ______

4) The table below shows different amounts of candies, different genies, and different amounts of magic pebbles. Write the numerical expression for the amount of candies you will have, then write the simplified value. The first one is done for you as an example.

Candies	Genie	Magic Pebbles	Expression	Value
5	doubler (times 2)	3	$5(2)^3$	40
10	tripler (times 3)	2		
7	times 4	1		
100	times $\frac{1}{2}$	4		
12	times 1.5	3		
8	times 10	4		
1	times 6	3		
1900	times 1	8		
900	times $\frac{1}{3}$	5		
114	times 2.5	x		
130	times 15	x		

5) Just like the general form for a line is $y = mx + b$, the above function has a general form: $y = a(b)^x$. Describe in words what a, b, x and y all mean.

6) Consider the following real world situation: A scientist has a petri dish with a single bacteria cell in it. Every day, this species of bacteria reproduces, that is, divides. In the following blank, put a guess as to how many bacteria will be in the dish after 8 days. __________ Complete the table and graph. Write an equation for function for the number of bacteria as a function of time. Use your function to find the number of bacteria after 16 days.

x	y
0	1
1	2

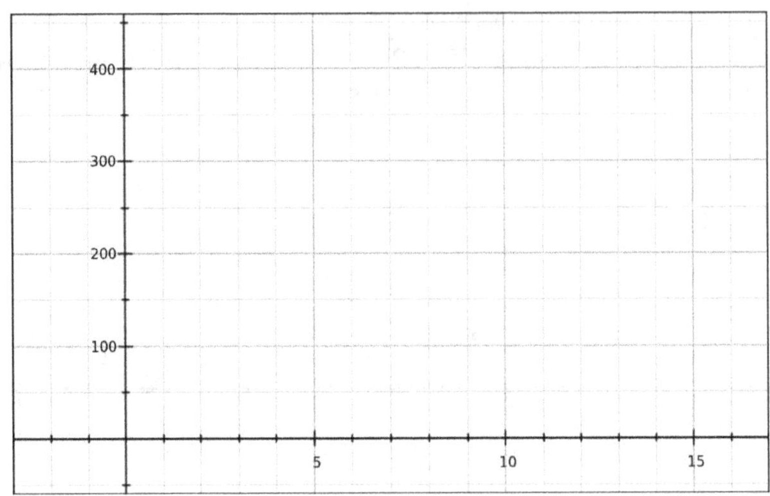

7) What if you started with 45 bacteria instead of 1? Make a table showing the resulting bacteria population for days 0 through 5. How is this graph similar to, and different from, the one you made above? Write an equation for a function for the number of bacteria as a function of time. Use your function to find the number of bacteria after 16 days.

x	y
0	45
1	90

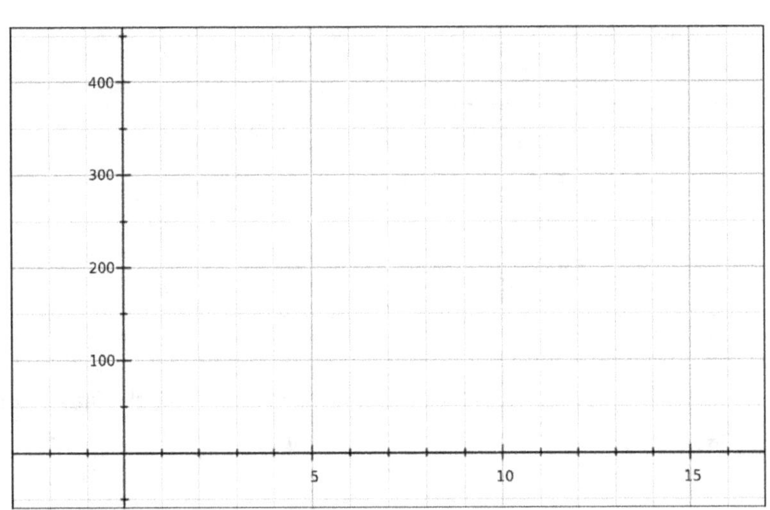

8) One week, our researcher went on vacation and forgot to put in the food necessary to keep the bacteria alive. The petri dish had 1278 bacteria cells in it. Every day, half of them died. Complete a table, and create an equation for the function for the bacteria after x days. When the researcher returned (after 7 days), how many cells were still in the dish? Graph this data. How is this graph similar to, and different from the ones you made above? Use your function to find the number of bacteria after 16 days.

x	y
0	1278

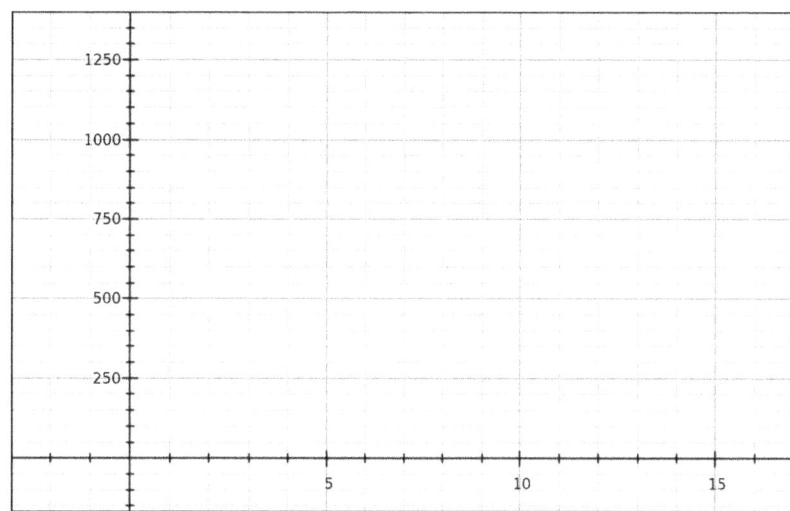

9) Supposing you put $\$100$ in a bank account that earns 5% interest every year. Make a chart that shows how much money you have for years 0 through 10. Make a graph of this data. How is it similar to, and different from the various bacteria graphs? Write an equation for the function and determine how much money you would have after 60 years.

x	y
0	100
1	105
2	110.25

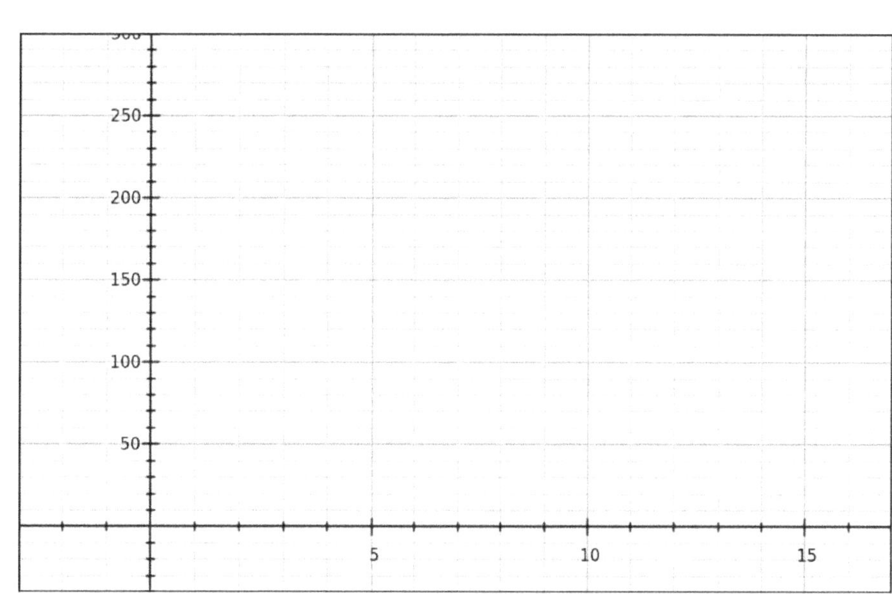

10) Create equations for functions which model each of the following scenarios.

Scenario	Equation
A genie takes your candies, multiplies them by 3 and adds 20.	
A doubling genie has 5 candies and doubles them based on the number of magic pebbles you give him.	
A car is racing at 140 km/hr with a 200 km head start.	
A gym costs 45 euros per month with an initial cost of 25 euros.	
60 bacteria double every day.	
100 bacteria triple every day.	
3000 bacteria increase by 14% every day.	
2450 bacteria die off by $\frac{1}{2}$ every day.	
1200 bacteria decrease by 70% every day.	
1500 dollars are put in a bank account and the value increases by 8% every year.	
A genie takes the square root of the number of candies you give him then adds 500.	
A genie squares the number of candies you give him then subtracts 12.	

11) Graph the following two functions and use the graph to determine when they cross.

$y = 10x$

x	y

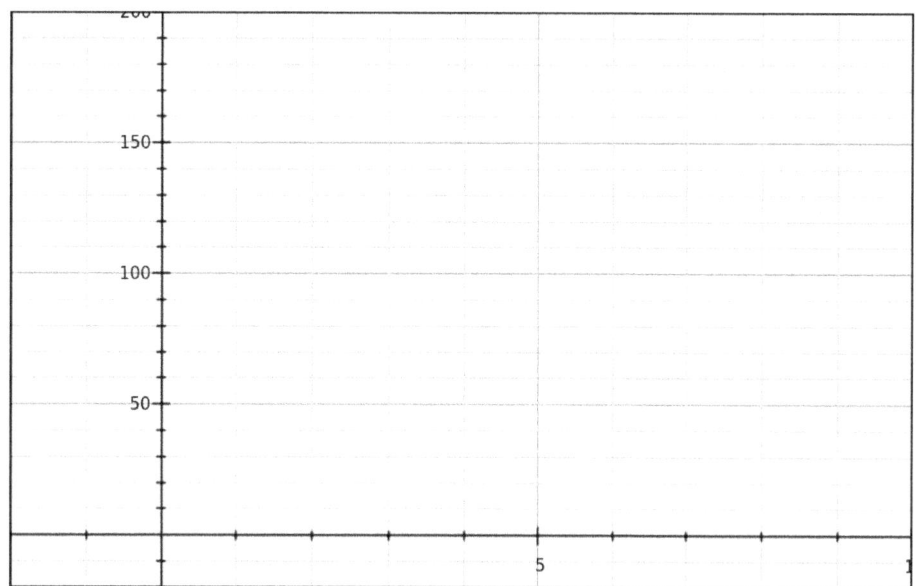

$y = 2^x$

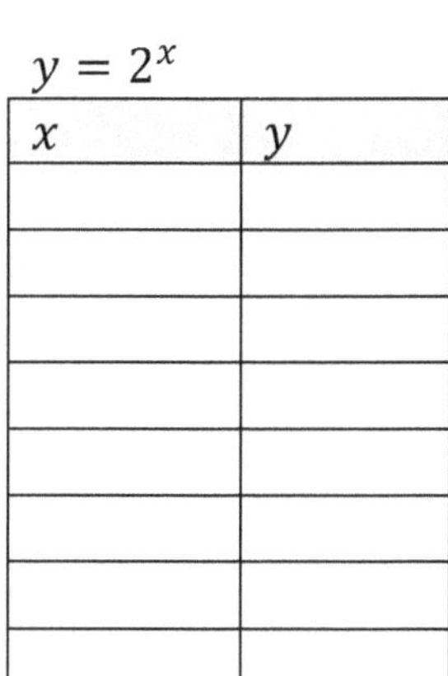

x	y

12) What is the y-intercept for the linear function above? Does the slope of the linear function stay constant, or increase? What is the slope?

13) What is the y-intercept for the exponential function above? Does the slope of the exponential function stay constant or increase?

14) Genie Gene offers to multiply your candies by two, seven times. Another Genie George offers to add two to your candies, seven times. Which genie do you prefer and why? Which genie is exponential? How do you know?

15) Complete the table for the following function, and graph it: $y = 2^x$.

x	y
-4	
-3	
-2	
-1	
0	
1	
2	
3	

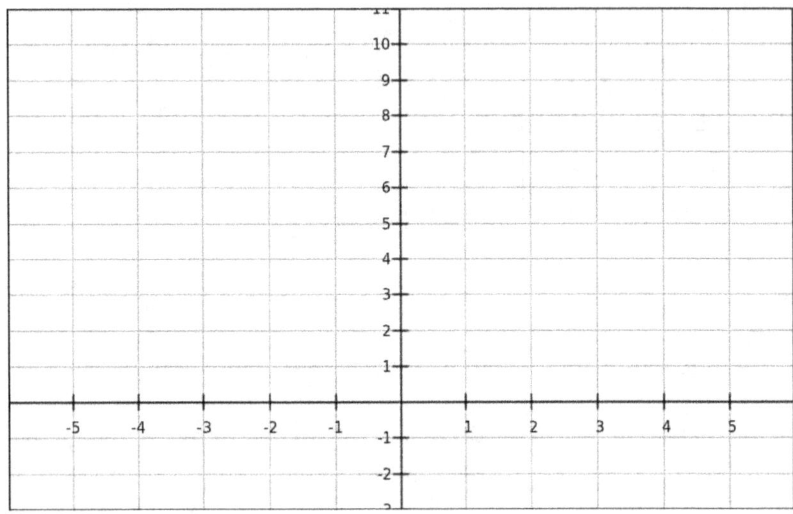

16) Can this function produce a y-value of 0? Why or why not?

17) Complete the table for the following function, and graph it: $y = 2^x - 2$.

x	y
-4	
-3	
-2	
-1	
0	
1	
2	
3	

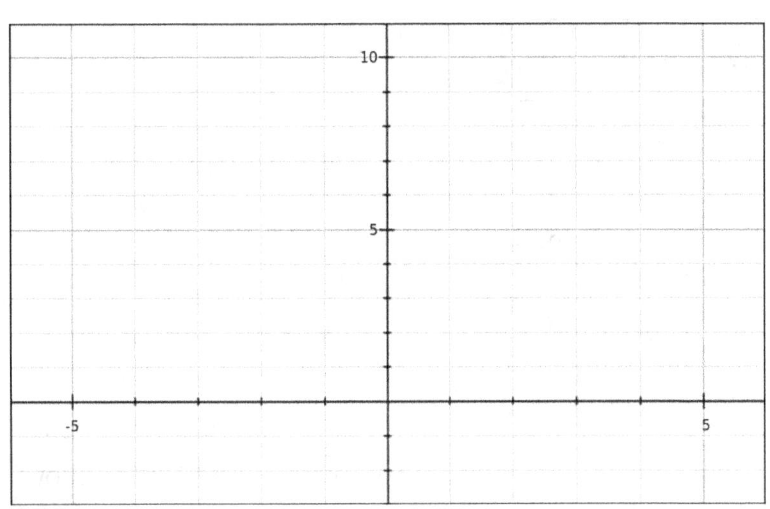

18) Describe 3 differences and similarities between this graph and the previous one.

The Take-Home Message: Linear functions have the form $\boldsymbol{y = mx + b}$ while exponential functions have the form $\boldsymbol{y = a(b)^x}$. For linear functions, the input is multiplied by a number (the slope), and the y-intercept $\boldsymbol{b}$ is the initial value returned for $\boldsymbol{x = 0}$. For exponential functions, the input counts the number of times an initial value $\boldsymbol{a}$ is multiplied by a constant $\boldsymbol{b}$.

Courage To Core

Share the Discovery

EQ2 Comparing Linear, Exponential and Quadratic Functions

Name(s)

Date Class/Period/Group

1) Consider the following situations. Your family wants to set up a bank account for you. There are two options. For the first option, which Uncle Larry prefers, you start with $300 and they add $100 every year. For the second option, which Aunt Elizabeth is keen on, you start with $300 and they multiply your previous year's money by 1.1 every year. Which one is a better option, and why? Use the table to help you. You may use a calculator for Elizabeth's plan.

Year	Larry's Plan ($)	Year	Elizabeth's Plan ($)
0	300	0	300
		1	330
		2	363
x		x	
30		30	

2) In the 2nd to last row you should have generated an expression which represents the amount of money you'll have after x years. Write them here as equations:

$y_L =$

$y_E =$

3) What kind of function is Larry's plan? How do you know? What kind of function is Elizabeth's plan? How do you know?

4) Which one is better after just 3 years? Why? Which one is better after 30 years? Why?

5) Your Uncle Q-Bert has an entirely different plan. His plan is that he will start with $300, and that is multiplied times the square of the number of years. This means that after 1 year you still have just $300. Complete the table. Write an equation for the function for Q-Bert's plan at right.

Year	Q-Bert's Plan ($)
1	300
2	
x	
30	

6) Of all three plans, which one is the best after 5 years? _________ Why?

7) Of all three plans, which one is the best after 30 years? _________ Which one is the worst? _________
Why?

8) Thanks to life extending technology, we all expect to live for 500 years. Which of the three plans is better at the 100 year mark? _______ Why?

9) Below are the functions for the three different plans above. Write the type for each function. Discuss the meaning of each number and variable in each function, and the differences and similarities between the three functions.

Name	Function	Type	Meaning of numbers and variables
Larry	$y = 100x + 300$		
Elizabeth	$y = 300(1.1)^x$		
Q-Bert	$y = 300x^2$		

10) Compare each of the numbers first by guessing which one you think is bigger. Put a check by the biggest and put an X by the smallest. No calculator here, you have to guess! You can do some mental calculation but no writing. (Note that $4(3)^2 = 4(9) = 36$.)

$11(8)^2$	$11 + 8(2)$	$11(2)^8$
$5(1.1)^{10}$	$5 + 1.1(10)$	$5(10)^2$
$100 + 1.05(42)$	$100(1.05)^{42}$	$100(42)^2$
$12(0.5)^{16}$	$12(16)^2$	$12 + 0.5(16)$
$100\left(\frac{1}{4}\right)^7$	$100 + \frac{1}{4}(7)$	$100\left(\frac{1}{4}\right)^2$

11) Now, go back and use arithmetic by hand (and your calculator) to determine the biggest and the smallest. Which results were surprising and why?

12) In each row below are three functions. Think of each of them like ATM's for each of three of your bank accounts. You input the number of years since the account was started for x and it returns the amount of money. You get to go to each ATM only once in your life! The linear ATM is called Larry, the exponential one is called Elizabeth, the quadratic one is called Q-Bert, and the constant one is called Carl. Give the correct name for each ATM and answer the following questions for each row: Which one do you want to go to right now? Which one would you go to in the medium term (say, 5 years)? Which one do you want to go to in the long term (say 30 years)? Write "now," "medium term" or "long term" for each one. Explain your choices in terms of what the function does.

$y = 100 + 2x$	$y = 100(2)^x$	$y = 100x^2$
$y = 20(5)^x$	$y = 20 + 5x$	$y = 20(5)^2$
$y = 600\left(\frac{1}{2}\right)^2$	$y = 600 + \frac{1}{2}x$	$y = 600\left(\frac{1}{2}\right)^x$

13) The slope-intercept form for a line is $f(x) = mx + b$. An example is given below. Create four more examples. Roughly sketch the graphs and compare and contrast them.

$h(x) = \frac{1}{3}x - 5$				

14) Standard form for a quadratic function is $g(x) = ax^2 + bx + c$. Write 5 examples of your own below. Compute $g(3)$ for each.

15) Standard form for an exponential function is $h(x) = a(b)^x$. Write 5 examples below. Compute $h(3)$ for each.

16) How can you tell if a function is linear, exponential or quadratic?

17) Below are three tables for three functions. Determine which one is linear, which is exponential and which is quadratic. Explain how you know. Then write the equation for each function. Explain how you found it.

x	$f(x)$	x	$g(x)$	x	$h(x)$
0	15	0	15	0	0
1	17	1	30	1	15
2	19	2	60	2	60
3		3	120	3	135
4		4		4	240
5		5		5	
x	$f(x) =$	x	$g(x) =$	x	$h(x) =$
30		30		30	

18) Below are three tables for three functions. Determine which one is linear, which is exponential and which is quadratic. Explain how you know. Then write the equation for each function. Explain how you found it.

x	$f(x)$	x	$g(x)$	x	$h(x)$
0	10	0	0	0	10
1	30	1	10	1	
2	90	2	40	2	16
3		3	90	3	19
4		4		4	
5		5		5	24
x	$f(x) =$	x	$g(x) =$	x	$h(x) =$
30		30		30	

19) Identify the functions below as linear, quadratic or exponential. Explain how you know. Find the equation for each function. Explain how you found it.

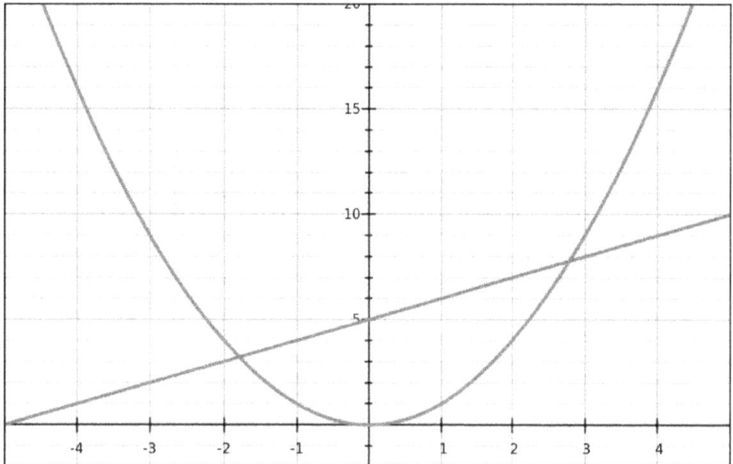

20) Identify the functions below as linear, quadratic or exponential. Explain how you know. Find the equation for each function. Explain how you found it.

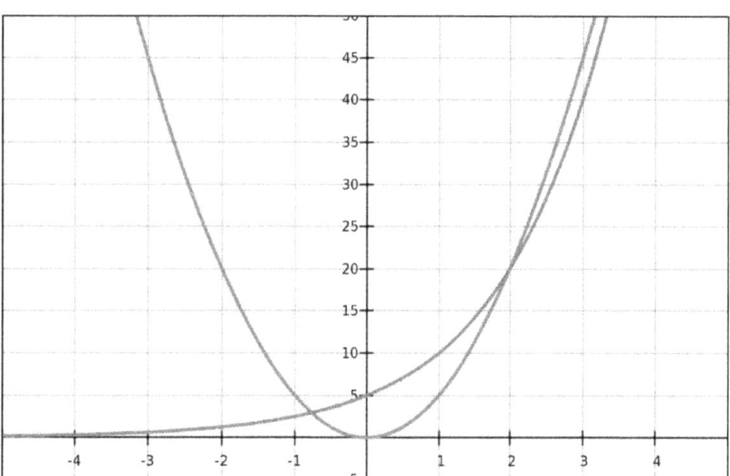

21) Graph each pair of functions on the axes provided. Use a table to assist you. Remember to plug in negative numbers for x in addition to 0 and positive numbers. State if each is exponential, quadratic, or linear and how you know. Explain how the graph confirms your identification of each.

$f(x) = x^2$ $\quad$ $g(x) = 2x$

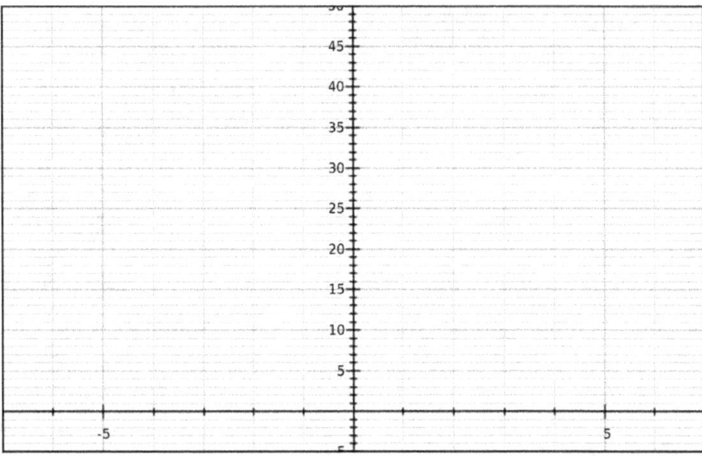

$f(x) = 4(2)^x$ $\quad$ $g(x) = \frac{1}{3}x + 20$

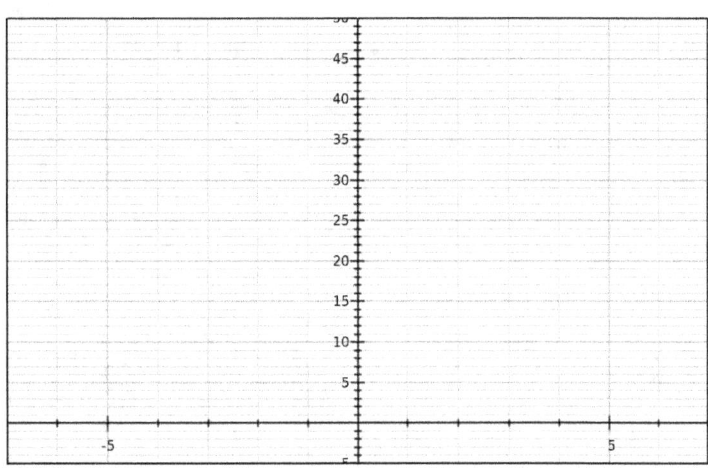

22) Graph each pair of functions on the axes provided. State if each is exponential, quadratic, or linear and how you know. Explain how the graph confirms your identification of each.

$f(x) = \frac{1}{2}x^2$ $g(x) = -2x + 40$

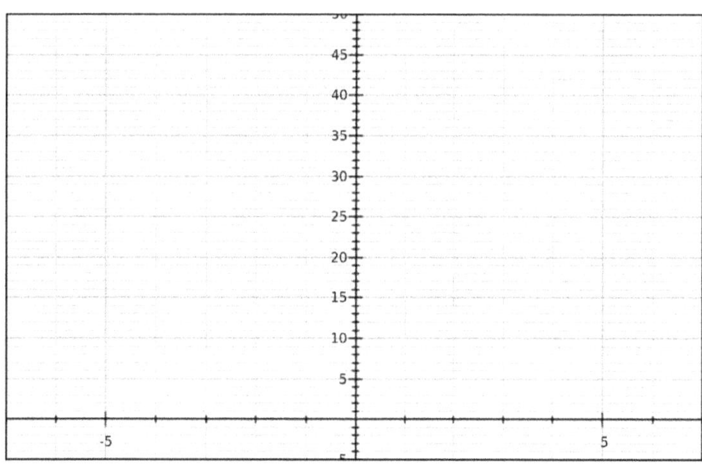

$f(x) = 10(1.5)^x$ $g(x) = -x^2 + 50$

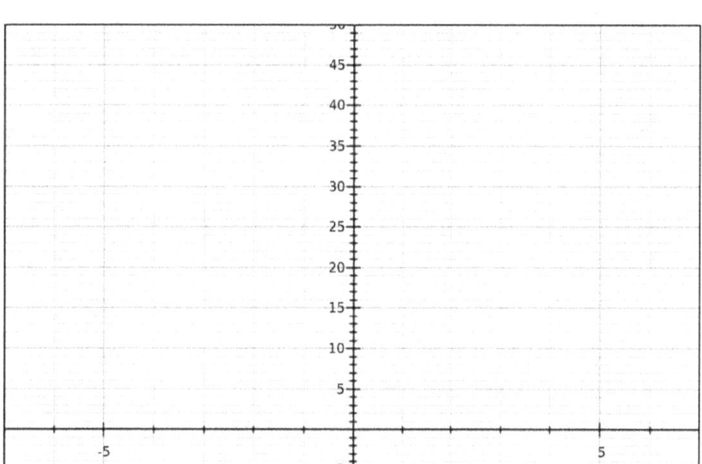

23) Identify the functions below as linear, quadratic or exponential. Explain how you know. Find the equation for each function. Explain how you found it.

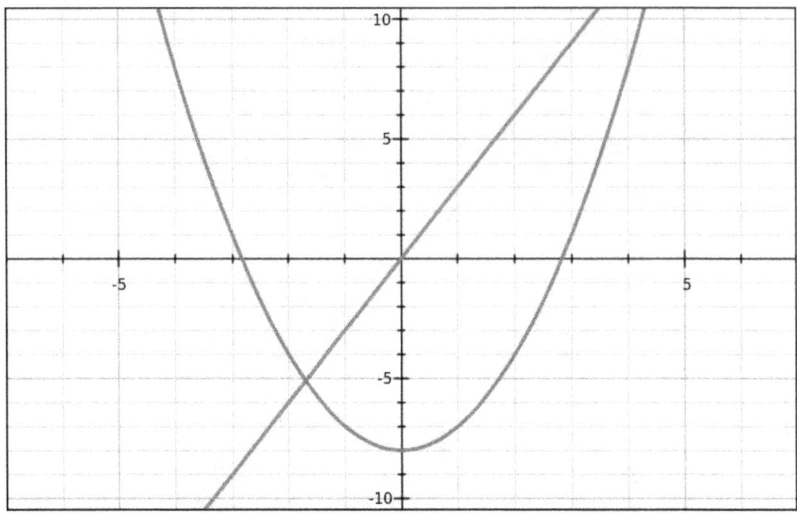

24) Identify the functions below as linear, quadratic or exponential. Explain how you know. Find the equation for each function. Explain how you found it.

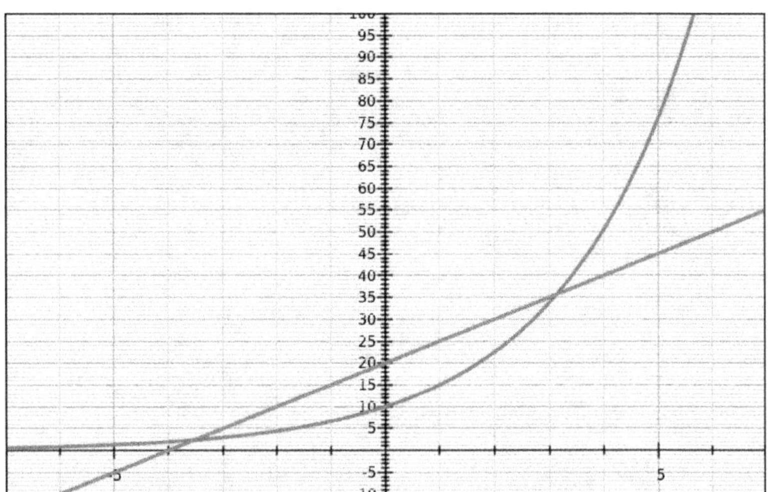

25) Below is a table showing information for just three functions, scrambled in each row. Put an "L" next to the linear function, an "E" next to the exponential function and a "Q" next to the quadratic function in each row.

type	linear	exponential	quadratic
function	$g(x) = 5(2)^x$	$h(x) = x^2 - 20$	$f(x) = 2x + 10$
graph	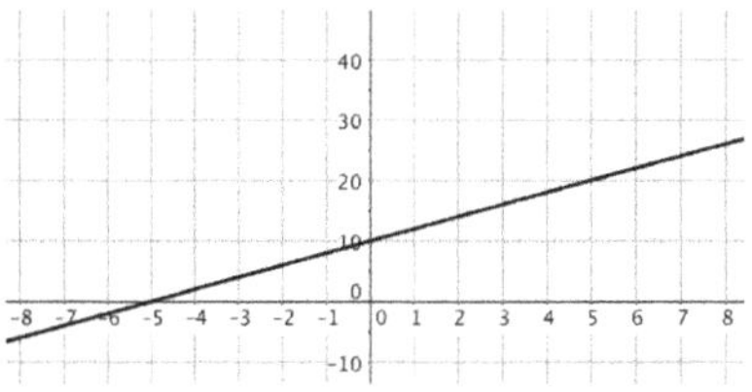	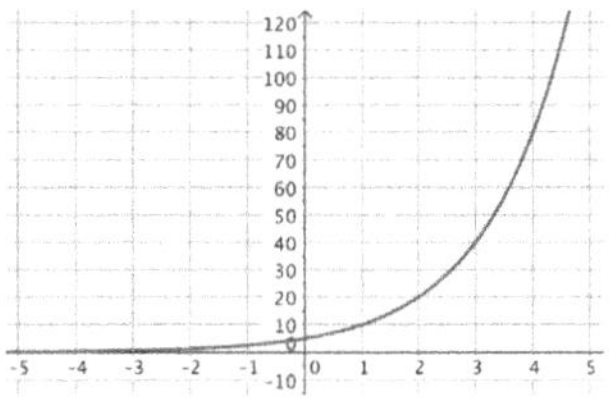	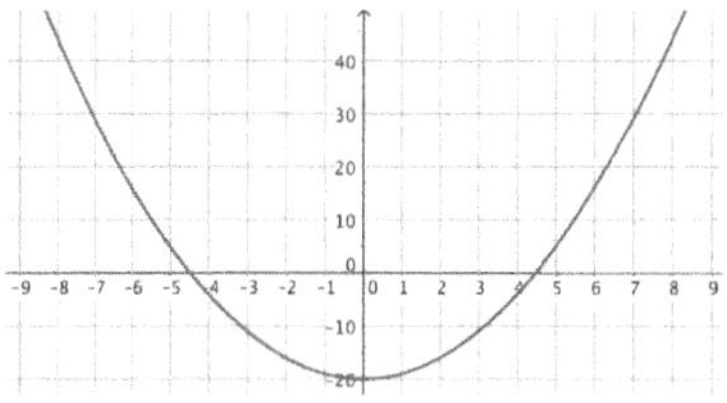
table	x | y -5 | 5 0 | -20 5 | 5 10 | 80	x | y -3 | 4 0 | 10 3 | 16 4 | 18	x | y -1 | 2.5 0 | 5 1 | 10 2 | 20
form	$y = a(b)^x$	$y = mx + b$	$y = ax^2 + bx + c$
in words	Constants a, b, and c determine a polynomial consisting of a required quadratic term, and optional linear and constant terms.	An initial value a is multiplied by a constant b, x number of times.	An initial value b has m added to it, x number of times.
example	Your brother puts $10 in a piggy bank and adds $2 every week.	The 5 starfish in a tank double every month.	A genie will square the number of candies you bring him, but keeps 20 of them before returning you the rest.

26) How can you recognize a linear function from a graph? From a table? From its equation? Give examples.

27) How can you recognize an exponential function from a graph? From a table? From its equation? Give examples.

28) How can you recognize a quadratic function from a graph? From a table? From its equation? Give examples.

29) What are some differences and similarities between these three types of functions? Give examples.

The Take-Home Message: Linear functions have graphs that are straight, and increase at a constant rate called the slope from an initial value called the y-intercept. Increasing exponential functions have graphs which curve steeper as x increases, because x counts the number of times an initial value is multiplied by a constant. Quadratic functions with a positive leading coefficient have a required quadratic term and increase symmetrically to the left and right of a lowest value.

EQ3 *Vertical and Horizontal Shifts*

Name(s)

Date Class/Period/Group

1) Your Uncle Edward wants to create a savings account for you. The account starts with $\$600$ and increases by an impressive 12% every year. Complete the table for y_1 and create an equation for a function for the amount of money in the account after x years. Label the x-axis "years" and the y-axis "money," then graph the function y_1 with a smooth curve.

x	y_1	y_2
0	600	
1		
2		
3		
4		
6		

$y_1 =$ ____________

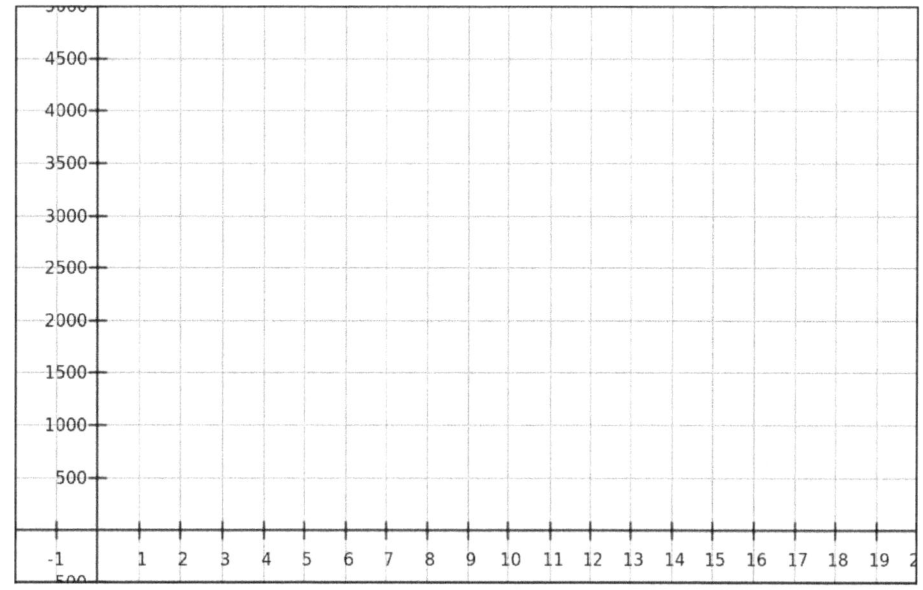

2) The graph shows you a picture of how much money you would receive if you withdraw the money x years from now. Now imagine that your uncle has a secret plan that on the day you withdraw your money he will simply hand you an extra $\$1000$. Wow, that's generous! So now, the amount you will receive will be $\$1000$ more than the amount in the account in the year of withdrawal. For example, even at year 0, if you take the money, you'll get $\$600 + \$1000 = \$1600$. Put that as y_2 for $x = 0$. Then complete the remaining values for y_2 and graph. Write the equation for the function below. How is y_2 similar to and different from y_1?

$y_2 =$ ____________

3) Are the functions below linear, exponential or quadratic? Before graphing them, make a guess about the characteristics of each, and discuss how their differences and similarities. Then graph them below on the same axes.

Function	Characteristics	Differences and Similarities
$f(x) = x^2$		
$i(x) = x^2 + 7$		

x	$f(x)$	$i(x)$
0		
1		
2		
3		
4		
5		
6		
-1		
-2		

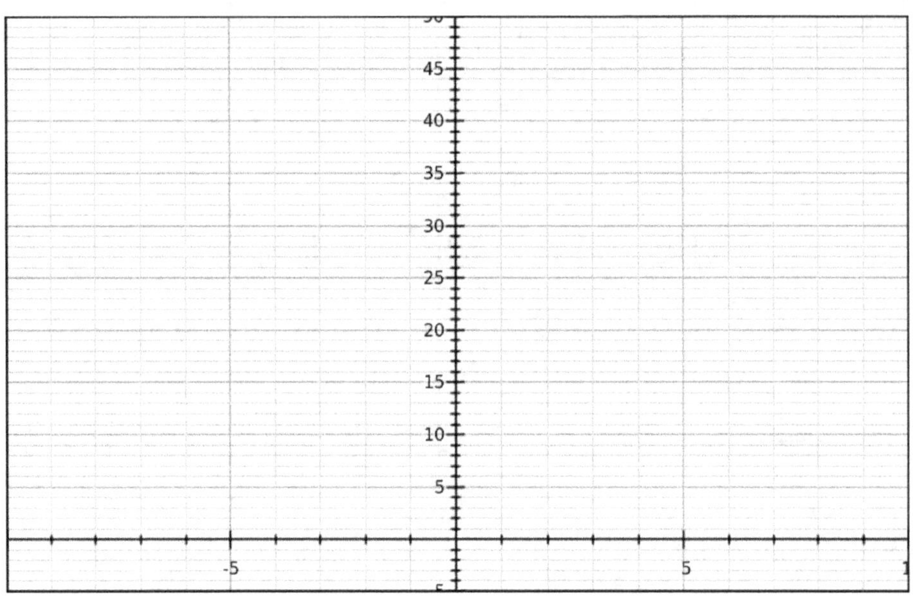

4) Were your guesses in #3 correct? What are additional characteristics of each graph which you may not have anticipated? What are some differences or similarities which you may not have realized until you graphed them?

5) You can think of the second function, $i(x)$, above as a **transformation** of $f(x)$. Specifically, it is a **vertical shift** of 7 units. How is y_2 on the previous page a transformation of y_1?

6) On the last two pages we performed a vertical shift on an exponential function, then on a quadratic function. Any type of function can be shifted vertically. In each column below are three functions represented in three different ways. Shift each vertically by -5 units then write or graph the result.

<table>
<tr><th></th><th>f</th><th>g</th><th>h</th></tr>
<tr><td>Pre-image function, that is, the original function:</td><td>$f(x) = x^3$</td><td><table><tr><td>x</td><td>y</td></tr><tr><td>2</td><td>7</td></tr><tr><td>5</td><td>1</td></tr></table></td><td> 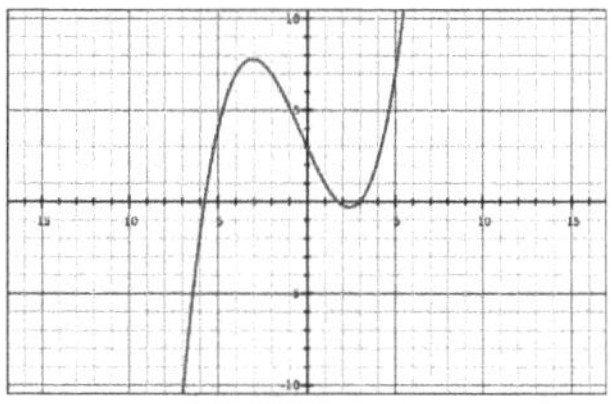</td></tr>
<tr><td>Image function, that is, the new function resulting from transformation:</td><td>$i(x) =$</td><td><table><tr><td>x</td><td>y</td></tr><tr><td>2</td><td></td></tr><tr><td>5</td><td></td></tr></table></td><td> 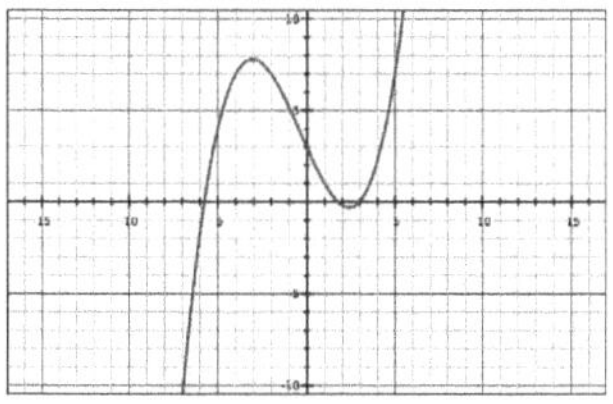</td></tr>
</table>

7) If $f(x) = x^6$, write the equation of the function which results when we shift $f(x)$ vertically by 12 units.

8) It's helpful to practice verbalizing transformations. Describe each transformation in words below. The first is given for you.

given $f(x) = \sqrt{x}$...	how we say it
$i(x) = 7 + \sqrt{x}$	When $f(x)$ is shifted vertically by $+7$ units, the result is $i(x)$.
$j(x) = -9 + \sqrt{x}$	

9) Describe the following transformations in words.

given $f(x) = x^7$...	how we say it
$i(x) = x^7 + 100$	
$j(x) = x^7 - 17$	

10) Describe the transformation which has been applied to each function below.

<table>
<tr><th></th><th>f</th><th>g</th><th>h</th></tr>
<tr><td>Pre-image</td><td>$f(x) = \sqrt{x}$</td><td><table><tr><th>x</th><th>y</th></tr><tr><td>4</td><td>−7</td></tr><tr><td>−9</td><td>13</td></tr><tr><td>12</td><td>−2</td></tr></table></td><td></td></tr>
<tr><td>Image</td><td>$i(x) = 9 + \sqrt{x}$</td><td><table><tr><th>x</th><th>y</th></tr><tr><td>4</td><td>−8</td></tr><tr><td>−9</td><td>12</td></tr><tr><td>12</td><td>−3</td></tr></table></td><td></td></tr>
<tr><td>Description</td><td></td><td></td><td></td></tr>
</table>

11) We can also perform horizontal shifts of functions. Graph the function below. Try to transform the function by performing a horizontal shift of −4 units. If you can't write the equation for the image function, don't worry—just try to perform the transformation on the graph and table, then try to write the equation for the function afterwards.

x	y
3	9

$f(x) = x^2$

x	y
−1	9

$i(x) =$ __________

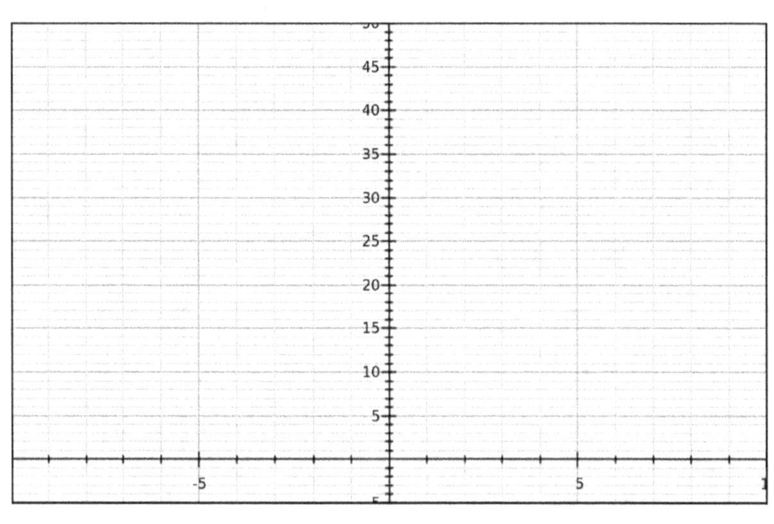

12) The function below has been transformed twice. Graph the function and the two resulting transformations. Were these the results you expected? Why or why not? Can you explain the results?

$f(x) = x^2$

x	y

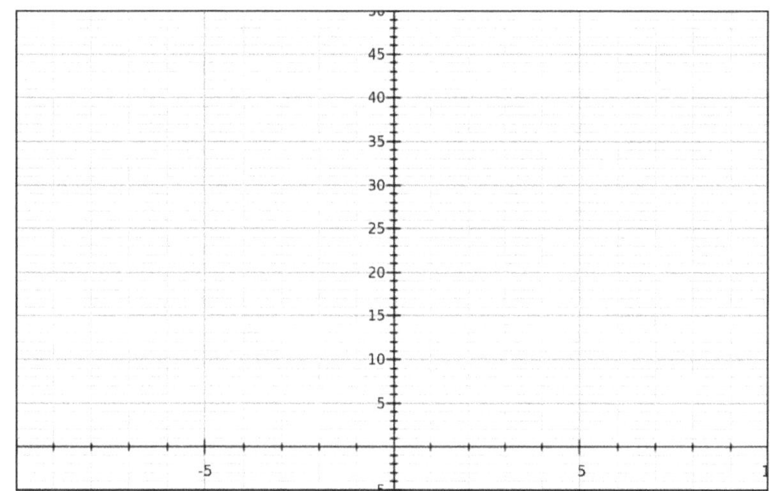

$i(x) = x^2 + 8$

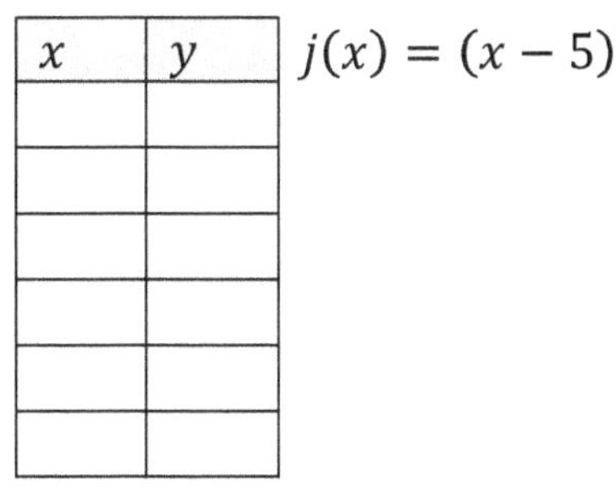

x	y

$j(x) = (x - 5)^2$

x	y

13) The second transformation, $j(x) = (x - 5)^2$, is a horizontal shift of 5 to the right. This is a bit counter-intuitive—you might expect it to shift the function to the left. One way to understand this is to think about the **vertex** of each quadratic above. Remember that the vertex is the point with the lowest y-value. Give the vertex for each function:

	$f(x) = x^2$	$i(x) = x^2 + 8$	$j(x) = (x - 5)^2$
vertex			

14) Why is the vertex for the last function $(5,0)$ and not $(-5,0)$? Discuss and explain the vertices of the first two **parabolas** as well.

15) Perform a horizontal shift of -8 units on each function.

	f	g	h
Pre-image	$f(x) = x^3$	x: 2, 5; y: 7, 1	
Image	$i(x) =$	x: (blank), (blank); y: 7, 1	

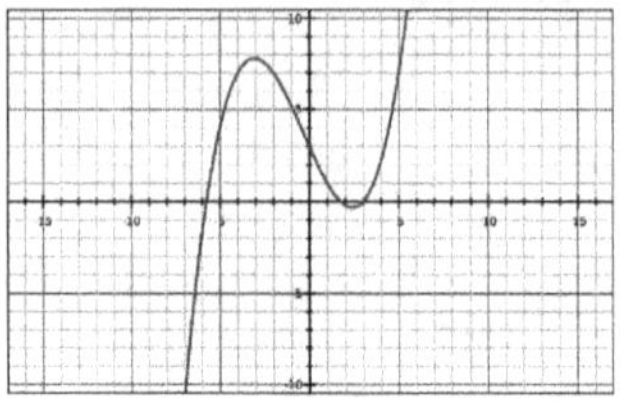

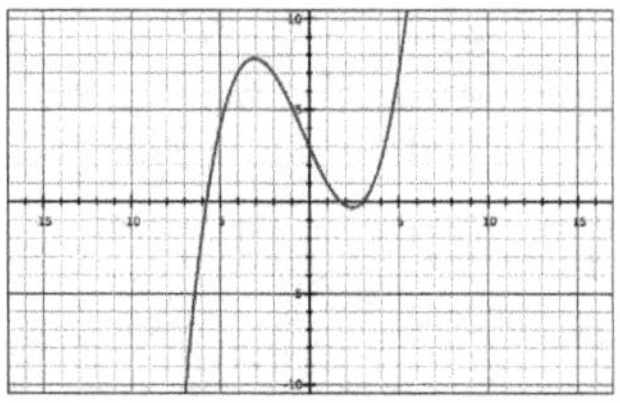

16) Describe the transformation which has been applied to each function below.

	f	g	h
Pre-image	$f(x) = \sqrt{x}$	x: 4, -7, 12; y: 1, 4, -2	
Image	$i(x) = \sqrt{x - 7}$	x: 2, -9, 10; y: 1, 4, -2	
Description			

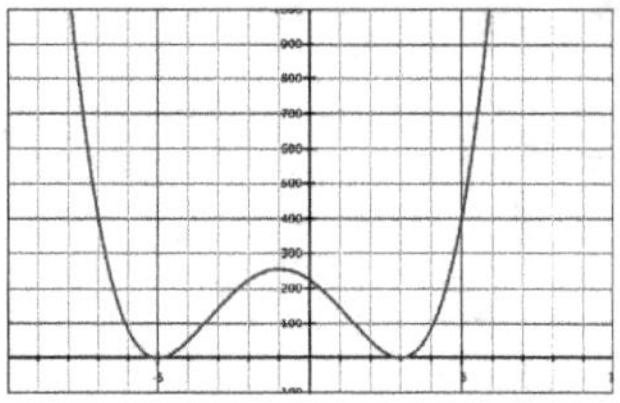

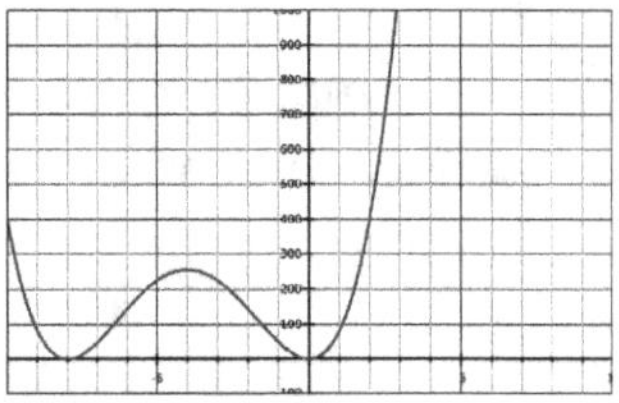

17) Perform the indicated transformation on each function and write the equation for the resulting function. The first one is given for you as an example.

pre-image	transformation	image
$f(x) = \sqrt[3]{x}$	vertical shift 5 units	$i(x) = 5 + \sqrt[3]{x}$
$g(x) = x^2$	horizontal shift -10 units	
$h(x) = 2^x$	vertical shift -700 units	
$p(x) = 3x$	horizontal shift 12 units	
$q(x) = \frac{5}{x}$	vertical shift -19 units	
$g(x) = x^2$	vertical shift 4 units and horizontal shift -17 units	

18) Write the equation for a function $g(x)$ below. Create a vertical and horizontal shift of your choosing. Write the equations for the image functions as indicated in the table. Complete the short table of values for each.

function	horizontal transformation	vertical transformation
$g(x) =$ ________________	horizontal shift _____ units	vertical shift ______ units

	pre-image	after horizontal shift only	after vertical shift only	after both
equation				
table	x | y	x | y	x | y	x | y

19) The function below has been transformed twice. Graph the function and the two resulting transformations. Give the vertex for each parabola.

$f(x) = x^2$

x	y

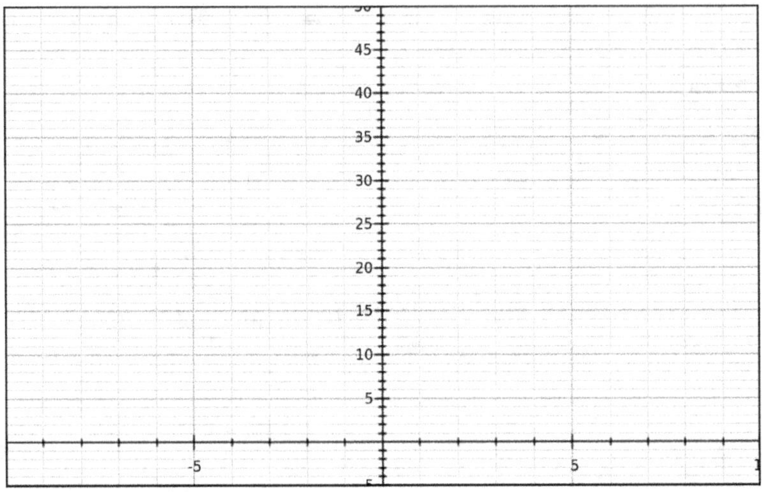

$i(x) = x^2 + 5$

x	y

$j(x) = (x - 3)^2 + 6$

x	y

20) How can the vertex of each of the above parabolas be determined directly from the equation without graphing or using a table? Describe in detail and explain why this method works.

21) We can represent a transformation of a function even if we don't have an equation, table or graph for it. In order to do this, we imagine a function, $f(x)$, and we write its "equation" as follows:

$y = f(x)$

Not much of an equation, right? But we can still represent various transformations of this function, and that's the point. This method of writing an "equation" for a function gives us a way to represent transformations quickly. Complete the tables using the example in the first row.

function	after horizontal shift 1 only	after vertical shift −8 only	after both
$y_1 = f(x)$	$y_2 = f(x-1)$	$y_3 = f(x) - 8$	$y_4 = f(x-1) + 8$
$y_1 = g(x)$			

function	after horizontal shift 12 only	after vertical shift −4 only	after both
$y_1 = h(x)$			
$y_1 = p(x)$			

22) What transformation is represented by each of the following?

$y = f(x) + 5$	$y = f(x-3)$	$y = f(x) - 6$	$y = f(x+1) - 7$

23) Fill in the blanks: We can see that the __________ shifts above directly affect the input, x, and __________ shifts affect the output, $f(x)$. We always apply __________ shifts to the x in the function, and __________ shifts are applied "afterwards," that is, to the y values produced by the function. Let's practice again with a simple function before moving on to more complex functions.

function	$f(x+5)$	$f(x)+9$	$f(x+5)+9$
$y = f(x) = x^3$	$y =$	$y =$	$y =$
$y = f(x) = 2x + 7$	$y =$	$y =$	$y =$
description of transformation			

24) Apply the transformations to each of the functions below. The first one is given as an example. Note that these could be simplified, but we will keep them un-simplified for now.

function	$f(x+5)$	$f(x)+9$	$f(x+5)+9$
$y = f(x) = x^2 + 3x$	$y = (x+5)^2 + 3(x+5)$	$y = x^2 + 3x + 9$	$y = (x+5)^2 + 3(x+5) + 9$
$y = f(x) = 7x^3$	$y =$	$y =$	$y =$
$y = f(x) = (x+4)^2$	$y =$	$y =$	$y =$
$y = f(x) = 8\sqrt{x}$	$y =$	$y =$	$y =$
$y = f(x) = -2x + 1$	$y =$	$y =$	$y =$
$y = f(x) = 2(3)^x$	$y =$	$y =$	$y =$

25) Write the equation for a function $p(x)$ below. Create a vertical and horizontal shift of your choosing. Write the equations for the image functions as indicated in the table. Graph.

pre-image	$p(x) =$
after horizontal shift ___ units	
after vertical shift ___ units	
after both	

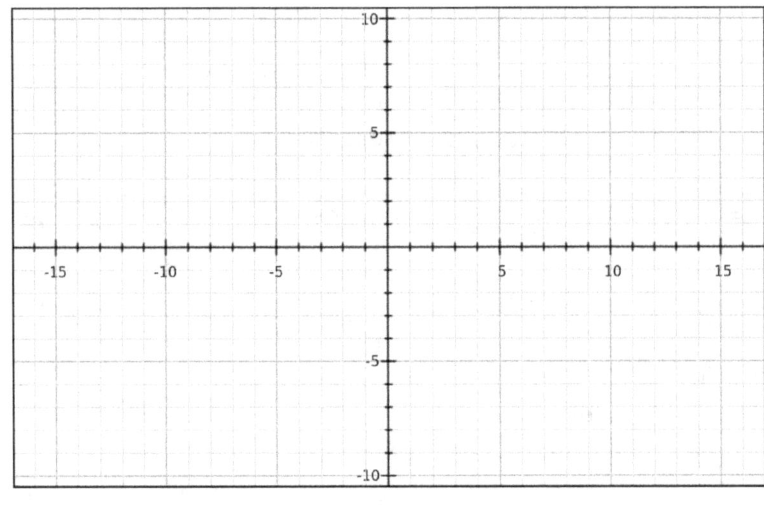

The Take-home Message: Vertical and horizontal shifts can be performed on any function given its equation, table or graph. Vertical shifts are applied to the y-values while horizontal shifts are applied to x-values.

Courage To Core – Share the Discovery –

EQ4 Arithmetic and Geometric Sequences and Series

Name

Date Class/Period/Group

1) A scientist has a petri dish with 3 amoebae in it. Every day, this species of amoeba reproduces, that is, divides. In the following blank, put a guess as to how many amoebae will be in the dish after 8 days. ___________ Complete the table, label axes and graph. Write an equation for the number of bacteria as a function of time. Use your function (and your calculator) to find the number of bacteria after 16 days. Extend your graph to the edges of the graphing window.

x	y
0	3
1	

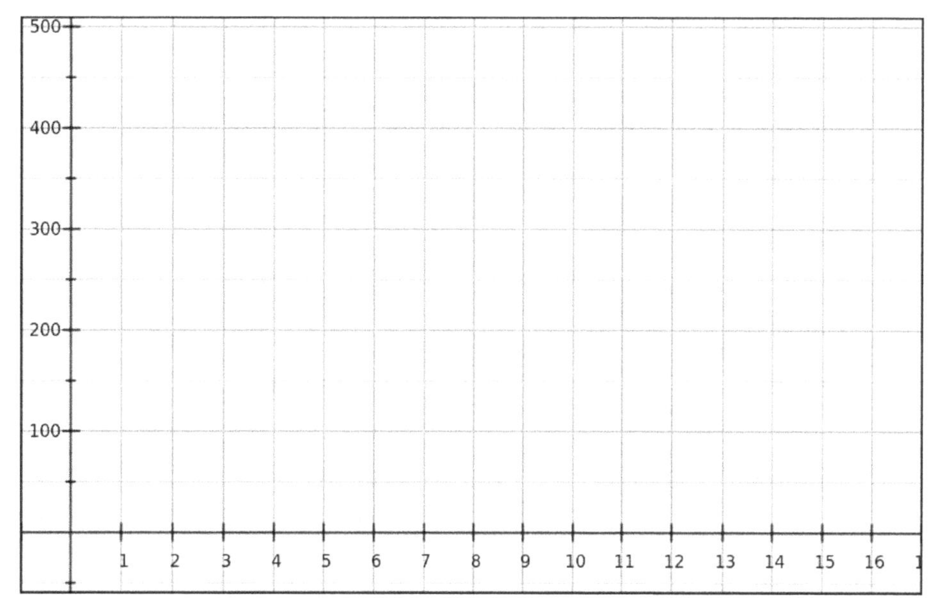

2) A second scientist has a petri dish which has 100 amoebae in it. Each day, when 20 amoebae have divided, she adds a chemical which stops the division process. This means that each day there are 20 more amoebae in the dish. Create an equation for the function for the number of bacteria in this second dish over time, and graph it on the axes above, extending the graph to the edges of the graphing window. When does the number of bacteria in the first dish surpass that in the second dish? How many bacteria does each dish have on this day?

3) What two kinds of functions are represented in #1 and #2? Compare and contrast their equations, tables and graphs.

4) Below are examples of linear and exponential scenarios. Write linear or exponential next to the description of each. Describe the input and output for each function, then write the equation for each. The first is given as an example.

Scenario	meaning of x	meaning of y	Equation
A magic genie takes the candies you have in your pocket, multiplies them by 4 and adds 15, then returns them to you.	the number of candies in your pocket	the number of candies the genie returns	$y = 4x + 15$
57 bacteria in a petri dish double every day.			
A car is racing at 180 km/hr with an 80 km head start.			
A gym costs 35 euros per month with an initial cost of 17 euros.			
60 bacteria triple every day.			
$\$3000$ in a bank account increases by 14% every year.			
2450 bacteria die off by half every day.			
Your friend is much smaller than you, and eats half the calories that you eat for lunch, plus a 200 calorie soda.			
1800 bacteria decrease by 60% every day.			
1500 dollars are put in a bank account and the value increases by $\$400$ every year.			

5) Slope intercept form for a line is $y = mx + b$. Standard form for an exponential function is $y = a(b)^x$. What does b represent for the linear function? Why? What does a represent for the exponential function? Why? How are these similar and different?

6) Two high school basketball players are competing to see who can improve the most over the 20 game season. James starts by scoring 10 points in the pre-season game we will actually call game 0. He improves to 12 points in the first real game, game 1. In fact, every game he scores 2 more points than in his previous game. Write an equation for the function for James' points in each game. Label axes and graph.

x	y
0	10
1	12

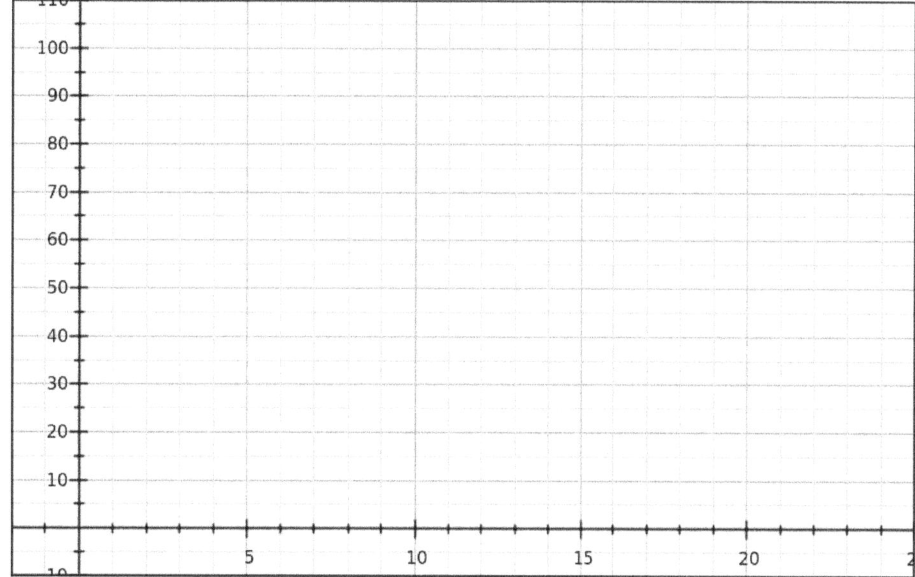

x	y
0	1
1	2

7) Steph scores 1 point in the pre-season game, and improves to 2 in the first real game. In the 2nd game he scores 4 points, and in the 3rd he scores 8. In fact, in every game he scores twice as many as he scored in the previous game. Write an equation for the function for Steph's points in each game. Graph on the axes above. In what game does Steph score more points than James?

8) Look at the tables for the two functions. As x increases by 1, y increases very predictably in both cases, taking the points from the previous game and performing one arithmetic operation on that value. What operation is performed for each function to change the previous game's points into the points for the current game?

9) In order to expand on the scenarios described on the previous page we need to review horizontal shifts. A horizontal shift is a transformations of a function focused on the input of the function. Below are two exponential functions. Complete the table for each and graph.

$f(x) = 2^x$

x	y
0	
1	
2	
3	

$i(x) = 2^{x-3}$

x	y
3	
4	
5	
6	
7	
8	
9	
10	
0	
1	
2	

10) Label the first points from each table point A and point B on the graph. Point B has an x-value of 3. Why did I select that value to begin? How does point A relate to point B? Label the 5th point in each table C and D, and draw an arrow from C to D. How does this show you that the second function is a transformation of the first? Discuss the table, the graph and the equation.

11) Perform the indicated transformation on each function and write the equation for the resulting function. The first two are given as examples. Note that while it is sometimes possible to simplify, it's not necessary here.

pre-image	transformation	image
$f(x) = 4(3)^x$	horizontal shift 7 units	$i(x) = 4(3)^{x-7}$
$g(x) = 12x - 8$	horizontal shift -9 units	$j(x) = 12(x+9) - 8$
$h(x) = 1.5^x$	horizontal shift -8 units	
$p(x) = 3x + 5$	horizontal shift 12 units	
$q(x) = 5(0.9)^x$	horizontal shift -19 units	
$g(x) = \frac{1}{5}x$	horizontal shift 4 units	

12) Horizontal shifts are a bit counter-intuitive—if you want to shift a function 4 units, you change x in the function to $x - 4$. One way to understand this is to think about the easiest values to compute for linear and exponential functions. While we can plug in whatever we want for x, we often start with $x = 0$. Below is a linear function. Graph it, and label the y-intercept point A. Transform the function by shifting it 4 units to the right, and graph the result. As this was a horizontal shift, where did point A move to? Label this points B. Draw an arrow from A to B. Why is it that if we want to shift a function 4 units, we change x in the function to $x - 4$?

x	y
0	
1	
2	
3	

$h(x) = \frac{1}{2}x + 3$

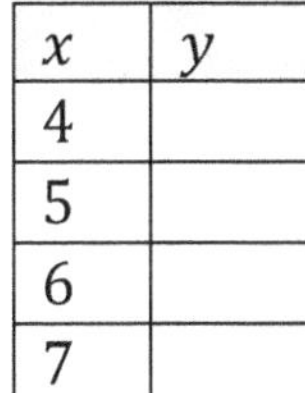

x	y
4	
5	
6	
7	

$i(x) =$ __________

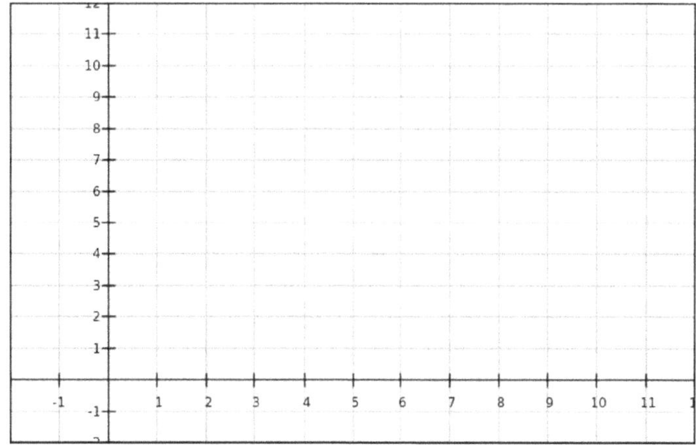

13) During their freshman year, Tim and Kevin are competing to see who can improve the most over their 20 game basketball season. In the pre-season game Tim scores 24 points, and this increases by 3 every game thereafter. Kevin scores 1 point in the pre-season and doubles his point total every game. Find the functions, label axes and graph both. Label the y-intercepts A for Tim and B for Kevin.

x	y
0	

$t(x) =$ ______

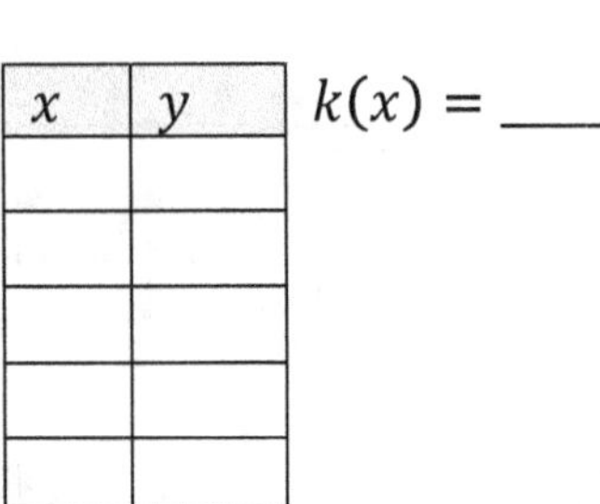

x	y

$k(x) =$ ______

14) In which game does Kevin score more than Tim?

15) In the second season (also 20 games), the scoring unfolds exactly as above, except that due to an injury, Kevin doesn't start playing until the 14th regular season game! Kevin still scores 1 point in his first game, but his first game is now the 14th game of the season. That means his graph gets horizontally shifted 14 units. Write his new function below and graph it on the axes above. Where did point B move to? Label this point C and draw an arrow from B to C.

x	y
14	

$i(x) =$ ______

16) Tim's graph is the same for both seasons while Kevin's is shifted. Who scores more points in the 20th game in the second season?

17) In the third season for Tim and Kevin, the scoring is exactly as in the first season, except there is no pre-season game, so they both start scoring in game 1. (This is a horizontal shift, for both.) Write the functions as transformations of $t(x)$ and $k(x)$ from the first season on the previous page (no simplification), label axes, and graph. Do not extend the graph before $x = 1$ or after $x = 20$. Why not?

x	y
1	

$T(x) =$

x	y

$K(x) =$

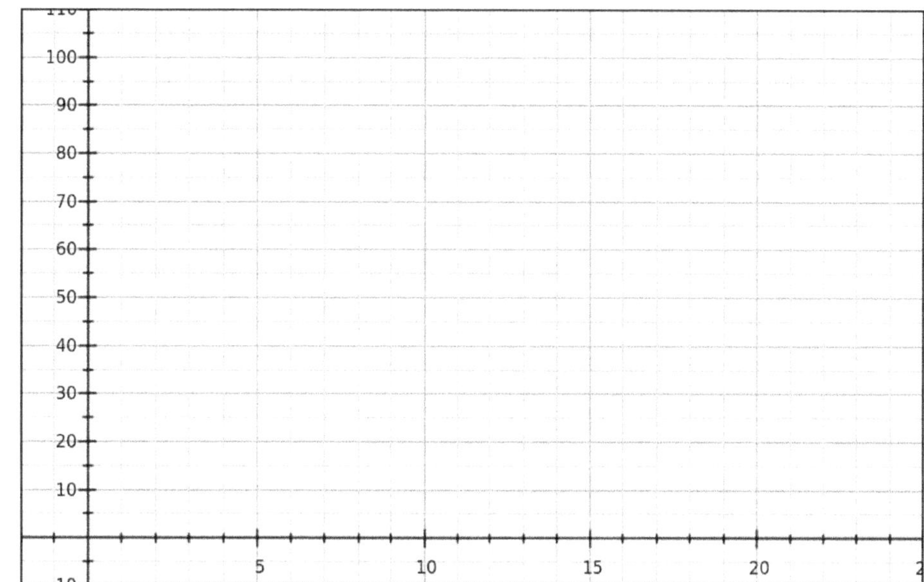

18) Suppose Steph, James, Tim or Kevin score according to a linear or exponential function like those above, but with different slopes, different y-intercepts, etc. What questions would you want to ask to determine who would score the most by the 20th game?

19) One of the most important elements which determines who scores more is when they start playing in the season. If Steph doesn't start playing until the 8th game, the games before that don't show up on his graph. By the way, does Steph actually score any points between the 8th and 9th game? So does it really make sense to connect the points on these graphs? What could we do instead?

20) So far in this mission we have modeled scenarios with linear and exponential functions. We shifted those functions horizontally, realistically starting the models at game 1 instead of game 0. Finally, we observed that it doesn't make sense to connect the points on the graphs, because no points scored between games. Linear and exponential functions that have been modified like this are called **sequences.** Specifically, a linear function so modified is called an **arithmetic sequence**, while an exponential function so modified is called a **geometric sequence.** Can you think of reasons for these names?

21) Both types of sequences have a domain which is **restricted** to the natural numbers. We also write equations for sequences differently from how we usually write functions. The following table summarizes the notation and meaning for both types of sequences. Complete the sections that are not complete.

<table>
<tr><th>type</th><th>arithmetic sequence</th><th>geometric sequence</th></tr>
<tr><td>example scenario</td><td>Steph scores 3 points in his first game, and his points increase by 7 each game after that.</td><td>James scores 6 points in his first game, and his points double each game after that.</td></tr>
<tr><td>table for this example</td><td><table><tr><th>x</th><th>y</th></tr><tr><td>1</td><td></td></tr><tr><td>2</td><td></td></tr></table></td><td><table><tr><th>x</th><th>y</th></tr><tr><td>1</td><td></td></tr><tr><td>2</td><td></td></tr></table></td></tr>
<tr><td>function for this example</td><td>If he had scored 3 points in game 0:
$S(x) = mx + b$
$S(x) = 7x + 3$
Now horizontally shift that 1 unit, so that he scores 6 points in game 1, not game 0:
$S(x) =$ ______________</td><td>If he had scored 6 points in game 0:
$J(x) = a(b)^x$
$J(x) = 6(2)^x$
Now horizontally shift that 1 unit, so that he scores 6 points in game 1, not game 0:
$J(x) =$ ______________</td></tr>
<tr><td>sequence notation</td><td>$a_n = a_1 + (n-1)d$</td><td>$a_n = a_1(r)^{n-1}$</td></tr>
<tr><td>meaning of variables in sequence notation</td><td>Notice that the formula for a_n is the same as for $S(x)$ above, just slightly re-arranged with d in place of m and a_1 in place of b.
a_n is read "a sub n" and represents the nth term (or output) in the sequence.
a_1 is "a sub 1," the first term in the sequence.
d is the common difference (or slope). It's the amount that is added to a term to get the next one.
n is the term number (or input). In our example it's the game number, starting with game 1.</td><td>Notice how this is the same as $J(x)$ above, with a_1 in place of a and r in place of b.
a_n is ______________________
a_1 is ______________________
r is the common ratio (or multiplier). It's the amount a term is __________ by to get the next one.
n is ______________________</td></tr>
<tr><td>example in sequence notation</td><td>$a_n =$ ______________</td><td>$a_n =$ ______________</td></tr>
</table>

22) Use sequence notation to write the formula for each of the following sequences. Use your formula to find the value of the 9th term in the sequence. The first two are started as examples.

description	type	formula	9th term
Tim scores 5 points in his first game and double each game thereafter.	geometric	$a_n = 5(2)^{n-1}$	$a_9 =$ ______
Kevin scores 27 points in his first game and his score increases by 2 each game thereafter.	arithmetic	$a_n = 27 + (n-1)(2)$	$a_9 =$ ______
A vending machine sells 4 candy bars the first day it's installed, and its sales increase by 5 candy bars each day thereafter.			
A coffee shop sells 5 coffees on the first day its open, and its sales triple each day thereafter.			Use calculator.
Serena makes $\$7000$ in sales of her biography the first day its on sale. Her sales increase 30% each day after that.			Use calculator.
A gamer scores 1200 points the first day she plays a game. Her score improves 300 points every day thereafter.			
An arithmetic sequence has an initial value of $a_1 = 14$ with a common difference $d = 3$.			
A geometric sequence has an initial value of $a_1 = 15$ and common ratio of $r = 1.6$.			Use calculator.
$\{2,5,8,11,\ldots\}$			
$\{2,6,18,54,162,\ldots\}$			Use calculator.

23) So far you've been given both the initial value and either the common difference or common ratio in order to quickly find the formula for a sequence. Imagine that you only start following the basketball player Anthony in his 9th game. He scores 35 points in that game. In the next game he scores 37 points, and then in the next game he scores 39. What kind of sequence is this? What is the common difference? _______The first three steps to find the initial value a_1 are given below—solve the resulting equation. Then write the formula for the sequence.

n	a_n
9	35
10	37
11	39

Finding the common difference.	Finding the initial value.	Writing the formula for the sequence.
	$a_n = a_1 + (n-1)d$ $a_9 = a_1 + (9-1)2$ $35 = a_1 + (8)2$	

24) On the 5th day after a video game is released, the high score is 25,000 points. Two days later the score is 29,500. You suspect that the score is increasing according to an arithmetic sequence. Find the common difference. (Hint: the common difference is the slope.) Then find the initial value. Then find the formula for the sequence. Use your formula to find the high score on the 20th day.

n	a_n
5	25000
7	29500

Finding the common difference.	Finding the initial value.	Writing the formula for the sequence.
	$a_n = a_1 + (n-1)d$	

25) You take over management of a coffee shop on the 12th day that its open. It earns $450 in revenue that day, then $495 the following day. You have a feeling that you will be able to increase the store's revenue by the same percentage every day. Will this be an arithmetic or geometric sequence? How do you know? The first two steps for finding the common ratio are given below. Solve the resulting equation. The first three steps to finding the initial value is also given. Find the initial value, rounded to hundredths. Then write the formula for the sequence.

n	a_n
12	450
13	495

Finding the common ratio.	Finding the initial value.	Writing the formula for the sequence.
$a_{13} = a_{12}(r)$ $495 = 450(r)$	$a_n = a_1(r)^{n-1}$ $a_{12} = a_1(1.1)^{12-1}$ $450 = a_1(1.1)^{11}$	

26) For an arithmetic sequence, $a_6 = 46$ and $a_9 = 58$. Find the formula for the sequence. Use this to find the 14th term of the sequence.

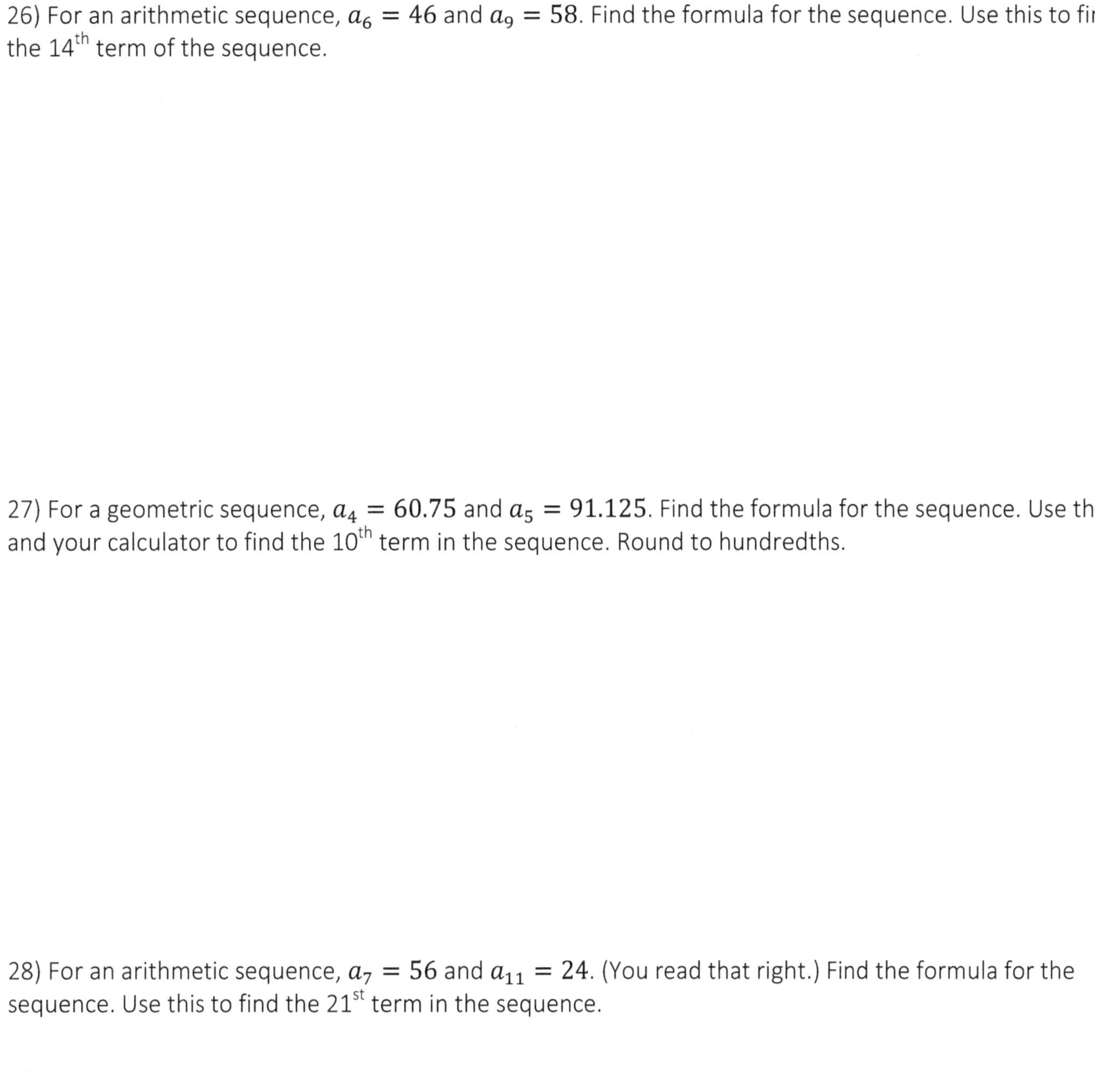

27) For a geometric sequence, $a_4 = 60.75$ and $a_5 = 91.125$. Find the formula for the sequence. Use this and your calculator to find the 10th term in the sequence. Round to hundredths.

28) For an arithmetic sequence, $a_7 = 56$ and $a_{11} = 24$. (You read that right.) Find the formula for the sequence. Use this to find the 21st term in the sequence.

29) Below is a table showing some values of a geometric sequence. The first steps to finding the common ratio are provided. Complete these steps to find the common ratio, then find the initial value, rounded to hundredths. Then write the formula for the sequence.

n	a_n
5	15
6	
7	
8	25.92
9	

Finding the common ratio.	Finding the initial value	Writing the formula for the sequence.
$a_8 = a_5(r)^3$ $25.92 = 15(r)^3$ $r^3 = \frac{25.92}{15}$	$a_n = a_1(r)^{n-1}$	

30) In a geometric sequence, $a_2 = 13.5$ and $a_5 = 45.5625$. Find the formula for the sequence.

31) In #24 on the previous page we wrote the equation $a_{13} = a_{12}(r)$. Why is this equation true? Why is r called the common ratio?

32) In #28 above we wrote the equation $a_8 = a_5(r)^3$ to find the common ratio. Why is this equation true?

33) Supposing Josh starts his 20 game basketball season scoring 3 points in the first game. His score increases 2 points every game after that. First, find the formula for the sequence and use your calculator to determine how many points he scores in the 20th game. Is your answer reasonable? After that, try to determine how many points in total Josh scored in the season.

34) Naturally, there is a systematic way to answer the last question. Let's try to make a simpler problem first. Supposing the season was only five games long. Then Josh's sequence would be:

$\{3,5,7,9,11\}$

We want to find the following value, S, without having to add them up directly:

$3+5+7+9+11=S$

Here is that sum written twice, once forwards and once backwards. Adding straight down gives you twice the desired value:

$$\begin{array}{l} 3+5+7+9+11=S \\ \underline{11+9+7+5+3=S} \\ 14+14+14+14+14=2S \end{array}$$

Wow, there are five 14's there! 14 is the sum of the first and last term, and five is the number of terms. So if we add the first and last term, then multiply that by the number of terms, we will have twice our desired value. In other words:

$(3+11)(5)=2S$ (Solve it!)

In general, if you want to add the first n terms of an arithmetic sequence:

$$(a_1+a_n)\cdot n=2S_n \quad \text{or} \quad S_n=\frac{(a_1+a_n)}{2}\cdot n$$

Which one do think is easier to remember and why?

35) Now use what you learned on the last page to determine the total number of points scored by Josh in his season. Remember, the season was 20 games long, he scored 3 points in the first game, and his increased by 2 points every game after that. You also already found the points he scored in the 20th game on the previous page.

36) Tim scores 5 points in the first game of the season, and his points increase by 3 every game after that. The season is 20 games long. How many points does he score in the season?

37) Steph scores 1 point in his first game, and he doubles his points in each game after that. Use the formula below to determine the sum of the first 20 terms of this geometric sequence.

$$S_n = \frac{a_1(1 - r^n)}{1 - r}$$

38) For a given arithmetic sequence, $a_1 = 3$ and $d = 5$. Write the formula for the sequence and find the value of the 15th term. Sketch the graph, starting with $n = 1$ and without connecting the points. Find the sum of the first 15 terms.

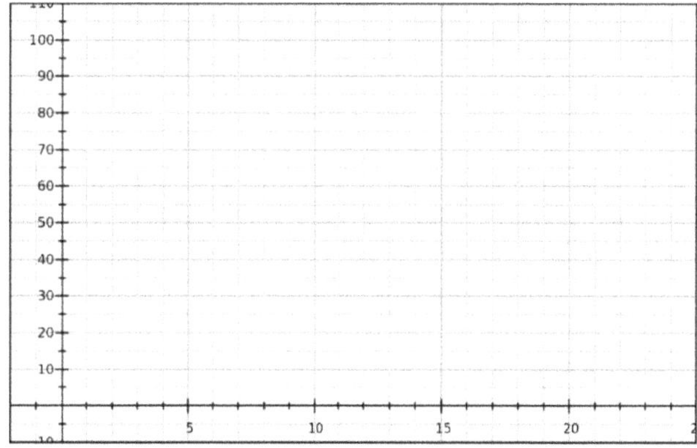

39) For a given geometric sequence, $a_1 = 6$ and $r = 1.25$. Write the formula for the sequence and use your calculator to find the value of the 18th term rounded to hundredths. Use your calculator to find the sum of the first 18 terms, rounded to tenths place.

40) For a given arithmetic sequence, $a_7 = 49$ and $a_{10} = 70$. Write the formula for the sequence and find the value of the 24th term. Find the sum of the first 24 terms.

41) For a given geometric sequence, $a_4 = 10$ and $a_5 = 15$. Write the formula for the sequence and find the value of the 15th term.

42) For a given geometric sequence, $a_1 = 3$ and $r = -2$. (You read that right.) Write down the first 5 terms of the sequence. Write the formula for the sequence and use your calculator to find the value of the 15th term. Use your calculator to find the sum of the first 10 terms. Find the sum of the first 11 terms. Compare and explain your last two answers. How can a_{15} be so large and yet these sums are so small?

43) A bacteria population grows from 1500 on day 6 to 2100 on day 10. Use your calculator to find the common ratio (rounded to thousandths). Find the formula for the sequence. Determine the population on day 20.

The Take-Home Message: Arithmetic sequences are like linear functions, and geometric sequences are like exponential functions. Arithmetic sequences feature a common difference (slope) while geometric sequences feature a common ratio (multiplier). Sequences have an initial value at $n = 1$ instead of $x = 0$, which means the formulas show a horizontal shift of 1. The domain is restricted to the natural numbers in both, and the formulas for each are written using subscripts to identify terms. There are additional formulas to find the sum of the first n terms.

Selected Answers

NS1

11)

$\frac{3}{5}$ as a fraction bar:

$\frac{1}{2}$ as a fraction bar:

Visually estimating, it seems true that $\frac{3}{5} > \frac{1}{2}$. It's also true that $\frac{3}{5} > \frac{3}{6}$ and since $\frac{3}{6} = \frac{1}{2}, \frac{3}{5} > \frac{1}{2}$. Another way to compare them is to create fraction bars made of the same number of segments and shade the fractions appropriately:

$\frac{3}{5} = \frac{6}{10}$

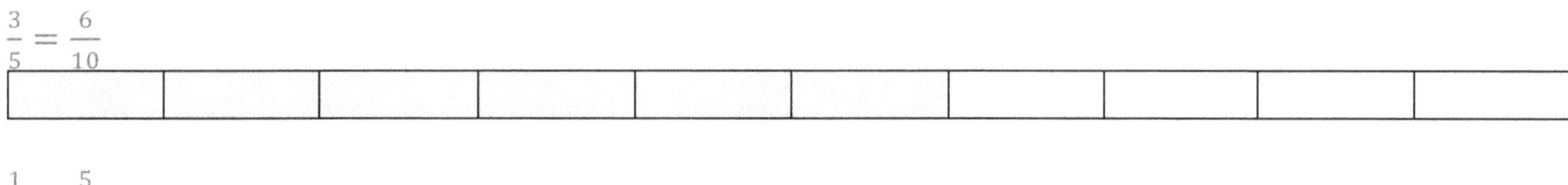

$\frac{1}{2} = \frac{5}{10}$

And $\frac{6}{10} > \frac{5}{10}$.

15)

$\frac{3}{12}$

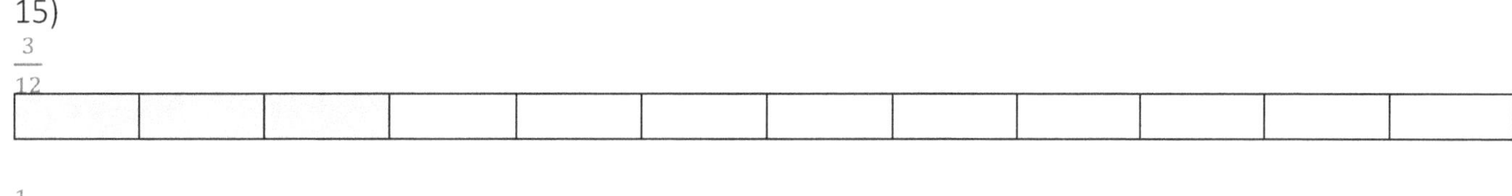

$\frac{1}{4}$

These fractions are equivalent because both the number of segments in the bar (12) and the number of shaded segments (3) get divided by 3. The result, $\frac{1}{4}$, is proportional to the original. Another way of thinking about it: We divided the numerator and denominator by 3, and the result was an equivalent fraction, proportional to the original.

19)

Decimal	Fraction	Percentage	Fraction
$0.5 = \frac{5}{10} = \frac{1}{2}$		$25\% = \frac{25}{100} = \frac{1}{4}$	
$0.\overline{3} = \frac{1}{3}$		$60\% = \frac{60}{100} = \frac{6}{10} = \frac{3}{5}$	
$1.5 = \frac{15}{10} = \frac{3}{2}$		$10\% = \frac{10}{100} = \frac{1}{10}$	
$0.\overline{6}$		$2\% = \frac{2}{100} = \cdots$	
$0.1 = \frac{1}{10}$		1%	
0.2		98%	

20)

$\frac{3}{8} = \frac{6}{16}$ $\frac{6}{16} = \frac{6}{16}$	$\frac{3}{8} < \frac{4}{8}$
$\frac{1}{2} = 0.5$	$\frac{21}{28}$ $\frac{3}{4}$ $\frac{3}{4} = \frac{3}{4}$
$\frac{150}{120}$ $\frac{45}{54}$ $\frac{15}{12}$ $\frac{5}{6}$ $\frac{5}{4}$ $\frac{5}{6}$ $1\frac{1}{4} > \frac{5}{6}$	$\frac{56}{49}$ 1.3 $\frac{8}{7}$ $\frac{13}{10}$ $\frac{80}{70} < \frac{91}{70}$
14% $\frac{28}{200}$	$\frac{24}{27} < \frac{27}{24}$ Note $\frac{27}{24}$ is an improper fraction…

21)

$\frac{5}{7}+\frac{3}{4}=\frac{20}{28}+\frac{21}{28}=\cdots$	$\frac{3}{14}+\frac{5}{7}=\frac{3}{14}+\frac{10}{14}=\cdots$
$\frac{5}{8}-\frac{3}{20}=\frac{25}{40}-\frac{6}{40}=\cdots$	$\frac{9}{28}-\frac{1}{21}=$

$0.5+\frac{3}{7}=\frac{1}{2}+\frac{3}{7}=\cdots$	$0.3+\frac{2}{11}=\frac{3}{10}+\frac{2}{11}=\cdots$
$2\frac{1}{3}+1\frac{2}{5}=3+\frac{1}{3}+\frac{2}{5}=$ $3+\frac{5}{15}+\frac{6}{15}=3\frac{11}{15}$	$\frac{5}{10}-\frac{1000}{2000}=\frac{1}{2}-\frac{1}{2}=0$

22)

$1\frac{6}{12}+3\frac{5}{15}=1\frac{1}{2}+3\frac{1}{3}=$ $1+3+\frac{1}{2}+\frac{1}{3}=\cdots$	$\frac{21}{24}-3=\frac{7}{8}-3=-2\frac{1}{8}$
$\frac{22}{11}-\frac{70}{49}=2-\frac{10}{7}=$ $2-1\frac{3}{7}=\frac{4}{7}$	$\frac{0}{13}-\frac{24}{13}+1-\frac{1}{26}=-\frac{24}{13}+\frac{25}{26}=$ $\frac{25}{26}-\frac{48}{26}=\frac{23}{26}$

NS2

2)

$\frac{6}{7}$. The numerator doubled while the denominator stayed the same. This should make sense because doubling $\frac{3}{7}$ actually means adding $\frac{3}{7}+\frac{3}{7}=\frac{6}{7}$. Sevenths are segments which we are counting: three sevenths and three more sevenths combine for a total of six sevenths. Multiplication is just repeated addition. Take the example of tripling. If we triple $\frac{3}{7}$ we get $\frac{9}{7}$, because we are just adding three sevenths plus three sevenths plus three sevenths.

3)

$\frac{2}{9}$. The numerator was halved while the denominator stayed the same. This should make sense because halving means take half of the number of ninths which we have. If we have four ninths, half of that would be two ninths.

4)

Half of $\frac{3}{5}$ is the same as half of $\frac{6}{10}$. Half of $\frac{6}{10}$ is $\frac{3}{10}$. We expanded the fraction $\frac{3}{5}$ by multiplying the numerator and denominator by 2 (doubling the number of segments in the whole fraction bar) to make it $\frac{6}{10}$. We did that so we could easily take half of 6 to arrive at the result of $\frac{3}{10}$. The quick and easy way to do this is $\frac{1}{2} \cdot \frac{3}{5} = \frac{3}{10}$, multiplying straight across, effectively doubling the number of segments in the whole fraction bar but making each one half as big. So we keep the same number of segments (3) but make them half as big. This is represented in the fraction bars below:

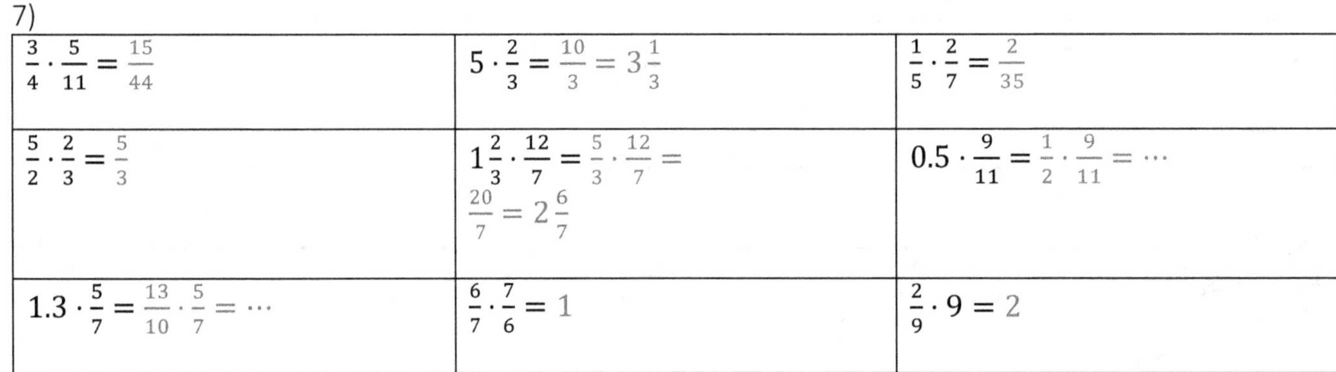

7)

$\frac{3}{4} \cdot \frac{5}{11} = \frac{15}{44}$	$5 \cdot \frac{2}{3} = \frac{10}{3} = 3\frac{1}{3}$	$\frac{1}{5} \cdot \frac{2}{7} = \frac{2}{35}$
$\frac{5}{2} \cdot \frac{2}{3} = \frac{5}{3}$	$1\frac{2}{3} \cdot \frac{12}{7} = \frac{5}{3} \cdot \frac{12}{7} =$ $\frac{20}{7} = 2\frac{6}{7}$	$0.5 \cdot \frac{9}{11} = \frac{1}{2} \cdot \frac{9}{11} = \cdots$
$1.3 \cdot \frac{5}{7} = \frac{13}{10} \cdot \frac{5}{7} = \cdots$	$\frac{6}{7} \cdot \frac{7}{6} = 1$	$\frac{2}{9} \cdot 9 = 2$

8)

$\frac{12}{3} \cdot \frac{14}{7} = \frac{4}{1} \cdot \frac{2}{1} = \cdots$	$\frac{1}{3} \cdot \frac{12}{7} \cdot 14 = \frac{1}{1} \cdot \frac{4}{1} \cdot \frac{2}{1} = \cdots$	$\frac{12}{1} \cdot \frac{1}{7} \cdot \frac{14}{1} \cdot \frac{1}{3} =$

9)

$\frac{1}{3} \cdot \frac{5}{7} \cdot \frac{12}{10} \cdot 14 =$ $\frac{1}{1} \cdot \frac{1}{7} \cdot \frac{4}{2} \cdot \frac{14}{1} =$ $2 \cdot 2 = 4$	$4 \cdot \frac{5}{6} \cdot \frac{1}{10} \cdot \frac{21}{8} =$	$\frac{12}{21} \cdot 7 \cdot \frac{1}{6} \cdot 3 =$

12)

What is half of $\frac{8}{9}$? $\frac{1}{2} \cdot \frac{8}{9} = \cdots$	What is a third of $\frac{12}{13}$?	What is half of 7 ?	What is $\frac{3}{5}$ tripled? $\frac{3}{5} \cdot \frac{3}{1} = \cdots$
What is $\frac{2}{3}$ of 42 ?	What is $\frac{3}{7}$ of $\frac{14}{27}$?	What is $\frac{5}{7}$ of $1\frac{1}{3}$? $\frac{5}{7} \cdot \frac{4}{3} = \cdots$	What is half of $\frac{6}{7}$ tripled ? $\frac{1}{2} \cdot \frac{6}{7} \cdot \frac{3}{1} = \cdots$

13)

$\frac{2}{3}\cdot\frac{3}{4}=\frac{1}{2}$	$\frac{3}{4}\cdot\frac{2}{3}=\frac{1}{2}$
$\frac{12}{20}\cdot\frac{5}{3}=1$	$\frac{32}{40}\cdot\frac{25}{20}=$
$1\frac{2}{13}\cdot 3\frac{5}{7}=\frac{15}{13}\cdot\frac{26}{7}=\frac{15}{1}\cdot\frac{2}{7}=\cdots$	$\frac{1000}{2}\left(2\frac{3}{1500}\right)=\frac{500}{1}\cdot\frac{3003}{1500}=$ $\frac{1}{1}\cdot\frac{3003}{3}=1001$
$\frac{9}{12}\cdot\frac{8}{7}=$	$\frac{56}{48}\cdot\frac{32}{12}=$
$\left(5\frac{4}{5}\right)\left(1\frac{2}{29}\right)=\frac{29}{5}\cdot\frac{31}{29}=\frac{31}{5}=6\frac{1}{5}$	$\frac{81}{52}\cdot\frac{28}{27}=$

14)

$\frac{1}{2}\cdot\frac{2}{3}\cdot\frac{3}{4}\cdot\frac{5}{6}\cdot\frac{6}{7}=\frac{5}{28}$	$\frac{100}{3}\cdot\frac{7}{17}\cdot\frac{3}{100}=$
What is $\frac{2}{3}$ of $\frac{5}{12}$?	What is $\frac{7}{8}$ of $\frac{16}{21}$? $\frac{7}{8}\cdot\frac{16}{21}=$ $\frac{1}{1}\cdot\frac{2}{3}=\frac{2}{3}$
What is $\frac{9}{24}$ of $\frac{20}{6}$?	What is $\frac{1}{2}$ of $\frac{1}{3}$ of 6 ? $\frac{1}{2}\cdot\frac{1}{3}\cdot\frac{6}{1}=\cdots$
Create two fractions which, when multiplied, return a result of 1 . $\frac{2}{3}\cdot\frac{3}{2}=1$	Create two fractions which, when multiplied, return a result of 2 . $\frac{4}{5}\cdot\frac{5}{2}=2$
Create and multiply two mixed numbers.	Create and multiply two improper fractions.

NS3

1)
$\frac{1}{16} \approx 0.06 = 6\%$

2)
10, because I would expect to win about one-sixteenth of the time. So $\frac{1}{16} \cdot 160 = 10$. However, that certainly isn't guaranteed. Probability doesn't give you a 100% certain prediction, although it does tell you the results we can expect more or less often the more times you play the game. Although 10 is the most likely single number the probability of this exact outcome is only about 13%. And although winning 20 times out of 160 is pretty improbable, it could happen. 0 out of 160 is also pretty improbable, but it could happen. The chance of losing the game all 160 times is around 0.003%.

6)
We assume that if people flip coins that the number of flips is essentially random. There are people who can flip a coin an exact number of times, so that would undermine the randomness. If the coin flip is random, then the probability of two heads is fixed.

7)
Yes. See above. There are too many variables which influence the outcome of a flipped coin—air turbulence, strength, flipping skill, flight time, catching speed, coin mass, etc. On the other hand, someone could argue that even though the outcome appears to be random, nature itself is never random, instead entirely pre-determined. Interesting idea that is beyond the scope of this mission! And since probability "works" quite well we will operate on the assumption that some experiments are random.

12)
$P(TT) = \frac{1}{4}$

$P(TT) + P(HT) + P(TH) = \frac{3}{4}$

$P(HT) + P(TH) = \frac{2}{4} = \frac{1}{2}$

$P(HH) + P(TT) = \frac{1}{2}$

13)

$P(H \text{ and } H) =$	$P(H \text{ and } T) =$	$P(T \text{ and } H) =$	$P(T \text{ and } T) =$
$\frac{3}{10} \cdot \frac{3}{10} = \frac{9}{100} =$ $0.09 = 9\%$	$\frac{3}{10} \cdot \frac{7}{10} = \frac{21}{100} =$ $0.21 = 21\%$	$\frac{7}{10} \cdot \frac{3}{10} = \frac{21}{100} =$ $0.21 = 21\%$	$\frac{7}{10} \cdot \frac{7}{10} = \frac{49}{100} =$ $0.49 = 49\%$

14)
$P(H \text{ and } H) + P(T \text{ and } T) = 58\%$

15)
$P(6 \text{ and } 6) = \frac{1}{6} \cdot \frac{1}{6} = \frac{1}{36}$

16)
5 and 6
6 and 5

$P(5 \text{ and } 6) + P(6 \text{ and } 5) = \frac{1}{36} + \frac{1}{36} = \frac{2}{36} = \frac{1}{18}$

19)
$P(5) = \frac{1}{6} \approx 0.166 = 16.6\%$

21)
$P(THH) = \frac{1}{2} \cdot \frac{1}{2} \cdot \frac{1}{2} = \frac{1}{8}$

$P(THH) + P(HTH) + P(HHT) = \frac{3}{8}$

23)
$P(R) = \frac{10}{25} = \frac{2}{5} = 0.4 = 40\%$

24)
$P(R) = \frac{15}{45} = \frac{1}{3} \approx 0.333 = 33.3\%$

25)
$P(Y \text{ and } Y) = \frac{1}{7} \cdot \frac{1}{7} = \frac{1}{49}$

NS4

2)

$\frac{4}{5} = 0.8$	$\frac{5}{6} = 0.8\overline{3}$
$\frac{6}{7} = 0.857142\overline{85714}$	$\frac{7}{12} = 0.58\overline{3}$

3)

$0.\overline{5}$ $\frac{5}{9}$	$0.\overline{52}$ $\frac{52}{99}$
$0.\overline{12}$ $\frac{4}{33}$	$0.\overline{123}$ $\frac{41}{333}$

5)

$0.5 < 0.\overline{5}$	$0.09 < 0.1$	$0.12 = 0.120$	$0.\overline{34} < 0.\overline{35}$
$\frac{1}{3} > 0.33$	$\frac{2}{3} = 0.\overline{6}$	$\frac{1}{5} < 0.\overline{2}$	$\frac{3}{10} < 0.33$
$\frac{6}{7} > 0.67$	$\frac{23}{100} = 0.23$	$\frac{456}{1000} < 0.\overline{456}$	$2 < 2.01$
$2 > 1.99$	$-2 > -2.01$	$-2 < -1.99$	$\frac{2}{3} < \frac{3}{4}$
$\frac{5}{7} > \frac{5}{8}$	$\frac{13}{14} < \frac{14}{14}$	$\frac{5}{3} < 2$	$\frac{17}{18} < \frac{18}{19}$
$\frac{48}{56} = \frac{6}{7}$	$0.\overline{18} = 2/11$	$100 = 10^2$	$\frac{1}{100} = 10^{-2}$

6)

$-\frac{99}{100} \quad -\frac{2}{3} \quad -\frac{4}{11} \quad -0.\overline{34} \quad -0.\overline{3} \quad -\frac{1}{5} \quad \frac{1}{100} \quad 0.\overline{1} \quad \frac{1}{5} \quad 0.\overline{2} \quad 0.\overline{34} \quad 0.\overline{5} \quad \frac{78}{100} \quad 0.9$

7)

$\frac{1}{2}+\frac{1}{3}=\frac{5}{6}$	$0.1-0.11=-0.01$
$1.23-1.2\overline{3}=0.00\overline{3}$	$3.2-\frac{7}{3}=\frac{13}{15}$
$3.14+3.\overline{14}=6.28\overline{14}$	$0.\overline{1}+0.\overline{2}=0.\overline{3}$
$\frac{11}{12}-\frac{12}{11}=\frac{23}{132}$	$\frac{11}{12}\cdot\frac{12}{11}=1$
$2\frac{2}{3}-2\frac{1}{7}=\frac{11}{21}$	$\frac{49}{25}\cdot\frac{100}{21}=\frac{28}{3}=9\frac{1}{3}$
$\frac{1}{10}(0.1)=\frac{1}{100}$	$\left(\frac{1}{10}\right)0.01=\frac{1}{1000}$

11)

Statement	T/F	Example
Every number which has a repeating decimal is a rational number.	T	$0.6 = \frac{3}{5}$
Every fraction which is a ratio of two non-0 integers is a rational number.	T	$\frac{2}{3}$
Every fraction which is a ratio of two non-0 integers has a repeating decimal.	F	$\frac{1}{5} = 0.2$
Some fractions which are the ratio of two integers have **terminating decimals.**	T	$\frac{3}{10}$
All integers are rational numbers.	T	$5 = \frac{5}{1}$
All rational numbers are integers.	F	$\frac{3}{4}$
All irrational numbers have non-repeating decimals.	T	π
0 divided by any non-0 integer is a rational number.	T	$\frac{0}{5} = 0$
Any non-0 integer divided by 0 is 0.	F	$\frac{4}{0}$ undefined
There are real numbers which aren't either rational or irrational.	F	

16)

$\frac{3}{10}$ rat	0.1 rat	$0.\overline{78}$ rat	5 int	4.3 rat	$2\frac{1}{3}$ rat
$\frac{5}{1}$ nat	$\frac{0}{6}$ depends on definition nat	$\sqrt{2}$ irrat	$\frac{\sqrt{2}}{3}$ irrat	$\sqrt{-3}$ imag	$\sqrt{25}$ nat
$\frac{27}{27}$ nat	5^3 nat	$(\sqrt{6})^2$ nat	$\frac{5}{9}$ rat	$\frac{45}{9}$ rat	$\frac{\sqrt{36}}{2}$ nat

NS5

2)

Fraction	Multiply by	What this looks like	Equivalent Result
$\frac{2}{3}$	5	$\frac{2}{3} \cdot \frac{5}{5}$	$\frac{10}{15}$ (expanded)
$\frac{3}{4}$	7	$\frac{3}{4} \cdot \frac{7}{7}$	$\frac{21}{28}$ (expanded)
$\frac{8}{10}$	$\frac{1}{2}$	$\frac{8}{10} \cdot \frac{\left(\frac{1}{2}\right)}{\left(\frac{1}{2}\right)}$	$\frac{4}{5}$ (reduced)
$\frac{9}{12}$	$\frac{1}{3}$	$\frac{9}{12} \cdot \frac{\left(\frac{1}{3}\right)}{\left(\frac{1}{3}\right)}$	$\frac{3}{4}$ (reduced)

$\frac{\left(\frac{2}{3}\right)}{\left(\frac{4}{5}\right)}$	$\frac{5}{4}$	$\frac{\left(\frac{2}{3}\right)}{\left(\frac{4}{5}\right)} \cdot \frac{\left(\frac{5}{4}\right)}{\left(\frac{5}{4}\right)}$	$\frac{5}{6}$
$\frac{2}{\left(\frac{6}{7}\right)}$	7	$\frac{2}{\left(\frac{6}{7}\right)} \cdot \frac{7}{7} = \frac{14}{6}$	$\frac{7}{2}$
$\frac{\left(\frac{10}{9}\right)}{5}$	$\frac{1}{5}$	$\frac{\left(\frac{10}{9}\right)}{5} \cdot \frac{\left(\frac{1}{5}\right)}{\left(\frac{1}{5}\right)}$	$\frac{2}{9}$

5)

$\frac{\left(\frac{2}{3}\right)}{2} \cdot \frac{\left(\frac{1}{2}\right)}{\left(\frac{1}{2}\right)} = \frac{1}{3}$	$\frac{\left(\frac{3}{4}\right)}{5} \cdot \frac{\left(\frac{1}{5}\right)}{\left(\frac{1}{5}\right)} = \frac{3}{20}$
$\frac{5}{\left(\frac{6}{7}\right)} \cdot \frac{\left(\frac{7}{6}\right)}{\left(\frac{7}{6}\right)} = \frac{35}{6}$	$\frac{\left(\frac{2}{3}\right)}{\left(\frac{\left(\frac{3}{4}\right)}{5}\right)}$ First: $\frac{\left(\frac{3}{4}\right)}{5} \cdot \frac{\left(\frac{1}{5}\right)}{\left(\frac{1}{5}\right)} = \frac{3}{20}$ $\frac{\left(\frac{2}{3}\right)}{\left(\frac{3}{20}\right)} \cdot \frac{\left(\frac{20}{3}\right)}{\left(\frac{20}{3}\right)} = \frac{40}{9}$

7)

$\frac{1}{\frac{2}{\left(\frac{3}{\left(\frac{4}{5}\right)}\right)}}$ First: $\frac{3}{\left(\frac{4}{5}\right)} \cdot \frac{\left(\frac{5}{4}\right)}{\left(\frac{5}{4}\right)} = \frac{15}{4}$ $\frac{1}{\left(\frac{2}{\left(\frac{15}{4}\right)}\right)}$ Next: $\frac{2}{\left(\frac{15}{4}\right)} \cdot \frac{\left(\frac{4}{15}\right)}{\left(\frac{4}{15}\right)} = \frac{8}{15}$ $\frac{1}{\left(\frac{8}{15}\right)} \cdot \frac{\left(\frac{15}{8}\right)}{\left(\frac{15}{8}\right)} = \frac{15}{8}$	$\frac{5}{\left(\frac{4}{\left(\frac{3}{\left(\frac{2}{1}\right)}\right)}\right)}$

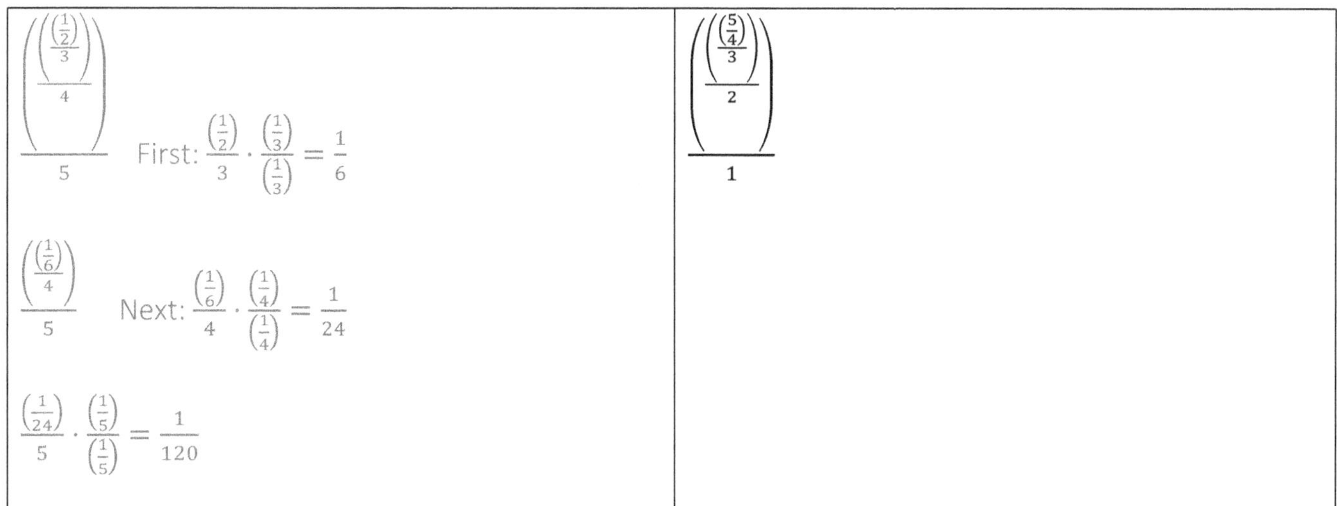

NS6

3)

Note: A little Googling and some rounding gave me these results. I encourage you to research more!

ant height—human height	65, 650, 6500
human mass—largest dinosaur mass	5.66, 56.6, 566
average 2 story house height—tallest skyscraper height	138, 1380, 13800
Usain Bolt top speed—cheetah top speed	2.13, 21.3, 213
housecat daily calories—human daily calories	1, 10, 100
fastest bicycle speed—passenger jet cruising speed	6.72, 67.2, 672
tallest tree in the world—tallest building in Madrid	2.15, 21.5, 215
population of Spain—population of US	0.68, 6.8, 68
total road length Spain—total road length US	0.96, 9.6, 96
American egg consumption per person—French egg consumption	0.97, 9.7, 97
walking speed in Madrid—walking speed in Dubai	0.729, 7.29, 72.9

4)

9500 mm

5)

9492.5 mm

6)

divide: $\frac{9500}{7.5} \approx 1267$

7)

$4 \cdot 4 \cdot 4 = 4^3$	$5^3 \cdot 5^4 = 5^7$	$\frac{6^5}{6^2} = 6^3$	$\frac{6^2}{6^5} = \frac{1}{6^3} = 6^{-3}$	$250 \cdot 5^{-3} = 2$	$250 \cdot \frac{1}{5^3} = 2$

10)
How many times as big is…

…fish D compared to fish A? __4^3______
…fish B compared to fish B? _1_______
…fish G compared to fish D? __4^3______
…fish E compared to fish C? _4^2________
…fish C compared to fish F? __$\frac{1}{4^3} = 4^{-3}$______

12)
$1 \cdot 6^4$

16)
10^3

17)
$10^0 = 1$

18)
$\frac{1}{10^3} = 10^{-3}$

26)

$2 \cdot 7^2 < 2 \cdot 7^5$ The 2nd number is 7^3 times as big.	$4 \cdot 10^7 > 4 \cdot 10^4$ The 2nd number is $\frac{1}{10^3}$ times as big.	$0.0007 = 7 \cdot 10^{-4}$
$3 \cdot 10^4 < 3 \cdot 10^5$ 2nd 10^1 times as big as 1st.	$7 \cdot 12^{-7} > 7 \cdot 12^{-9}$ 2nd 10^{-2} times as big as 1st.	$6 \cdot 2^{-3} > 6 \cdot 2^{-7}$ 2nd 2^{-4} times as big as 1st.
$\frac{1}{1000}$ $\frac{1}{10000}$ $10^{-3} > 10^{-4}$ 2nd is 10^{-1} times as big as 1st.	$\frac{1}{100}$ 10^{-2}	0.1 $\frac{1}{10}$ $10^{-1} = 10^{-1}$
$\frac{1}{7^5}$ 7^{-6} $7^{-5} > 7^{-6}$ 2nd is 7^{-1} times as big as 1st.	$3 \cdot 10^1$ 30 $3 \cdot 10^1 = 3 \cdot 10^1$	$3 \cdot 10^{-1}$ 0.3
$17 \cdot 6^{-4}$ $17 \cdot 6^{10}$	$6^{-2} < 6^2$ 2nd is 6^4 times as big as 1st.	0.003 $3 \cdot 10^{-3}$
0.03 $3 \cdot 10^{-2}$	0.3 $3 \cdot 10^{-1}$	$3 = 3 \cdot 10^0$

NS7

1)

$3^2 =$	$\sqrt{9} =$
$4^2 =$	$\sqrt{16} =$
$1^2 =$	$\sqrt{1} =$
$\left(\frac{1}{2}\right)^2 =$	$\sqrt{\frac{1}{4}} = \frac{1}{2}$
$\left(\frac{1}{3}\right)^2 =$	$\sqrt{\frac{1}{3}} = \frac{1}{\sqrt{3}}$
$\left(\frac{2}{3}\right)^2 =$	$\sqrt{\frac{4}{9}} = \frac{2}{3}$
$(2+3)^2 = 5^2 = 25$	$\sqrt{5} = \sqrt{5}$
$\left(\frac{1}{2+3}\right)^2 =$	$\sqrt{\frac{1}{25}} = \frac{1}{5}$
$\left(\frac{1}{2}+\frac{1}{3}\right)^2 = \left(\frac{5}{6}\right)^2 = \frac{25}{36}$	$\sqrt{\frac{25}{36}} =$
$\left(\frac{1}{2}-\frac{1}{3}\right)^2 =$	$\sqrt{\frac{1}{36}} =$
$\left(\frac{1}{4}\cdot\frac{1}{3}\right)^2 = \frac{1}{144}$	$\sqrt{\frac{1}{144}} = \frac{1}{12}$
$(3\cdot 10^7)^2 = 9\cdot 10^{14}$	$\sqrt{9\cdot 10^{14}} =$

7)

Note that the top speed for a sailboat is highway speed for a car: 121 km/hr. That is a sailboat! And, the fasted motorized boat achieved a ridiculous top speed in the year 1978!

	length of boat (feet)	Speed (km/hr)	year record set
fastest sailboat	**40**, 50, 60	71, **121**, 171	1992, 2002, **2012**
fastest motorized boat	7, 17 **27**	311, 411, **511**	**1978**, 1993, 2008

11)

Multiplying Roots	Expanding A Fraction	Rationalizing a Denominator
$\sqrt{5}\cdot\sqrt{5} = 5$	$\frac{3}{5}\cdot\frac{2}{2} = \frac{6}{10}$	$\frac{3}{\sqrt{2}}\cdot\frac{\sqrt{2}}{\sqrt{2}} = \frac{3\sqrt{2}}{2}$
$\sqrt{7}\cdot\sqrt{7} = 7$	$\frac{6}{11}\cdot\frac{5}{5} = \frac{30}{33}$	$\frac{7}{\sqrt{5}}\cdot\frac{\sqrt{5}}{\sqrt{5}} = \frac{7\sqrt{5}}{5}$
$\sqrt{3}\cdot\sqrt{15} = 9\sqrt{5}$	$\frac{2}{5}\cdot\frac{35}{35} = \frac{70}{175}$	$\frac{6}{\sqrt{2}}\cdot\frac{\sqrt{2}}{\sqrt{2}} = \frac{6\sqrt{2}}{2} = 3\sqrt{2}$

13)

$\sqrt{\frac{1}{2}} = \frac{\sqrt{1}}{\sqrt{2}} = \frac{1}{\sqrt{2}} =$ $\frac{1}{\sqrt{2}}\left(\frac{\sqrt{2}}{\sqrt{2}}\right) =$ $\frac{\sqrt{2}}{2}$	$\sqrt{\frac{1}{3}} = \frac{1}{\sqrt{3}} \cdot \frac{\sqrt{3}}{\sqrt{3}} = \frac{\sqrt{3}}{3}$
$\sqrt{\frac{2}{3}} = \frac{\sqrt{2}}{\sqrt{3}} =$ $\frac{\sqrt{2}}{\sqrt{3}}\left(\frac{\sqrt{3}}{\sqrt{3}}\right) =$ $\frac{\sqrt{6}}{3}$	$\sqrt{\frac{5}{7}} = \frac{\sqrt{5}}{\sqrt{7}} \cdot \frac{\sqrt{7}}{\sqrt{7}} = \frac{\sqrt{35}}{7}$
$\sqrt{\frac{9}{10}} = \frac{3}{\sqrt{10}} \cdot \frac{\sqrt{10}}{\sqrt{10}} = \frac{3\sqrt{10}}{10}$	$\sqrt{\frac{25}{7}} = \frac{5}{\sqrt{7}} \cdot \frac{\sqrt{7}}{\sqrt{7}} = \frac{5\sqrt{7}}{7}$
$\frac{\sqrt{6}}{\sqrt{10}} = \frac{\sqrt{3}}{\sqrt{5}} \cdot \frac{\sqrt{5}}{\sqrt{5}} = \frac{\sqrt{15}}{5}$	$\frac{\sqrt{1}}{\sqrt{12}} = \frac{1}{2\sqrt{3}} \cdot \frac{\sqrt{3}}{\sqrt{3}} = \frac{\sqrt{3}}{6}$

NS8

1)

$2b \cap \sim 2b$ (Hint: ∩ means "or")	$\sqrt{-1}\; 2^3\; \sum \pi$ (Hint: $\sqrt{-1}$ is represented with *i*, and the $\sum$ symbol means "sum"
$\left(\frac{5}{2} - \frac{8}{3}\right)^2 = \left(\frac{15}{6} - \frac{16}{6}\right)^2 =$ $\left(-\frac{1}{6}\right)^2 = \frac{1}{36}$	$2(3+4)^1 = 2(7)^1 =$ $2(7) = 14$
$\left(\frac{2-7}{1+4}\right)^2 + \frac{1}{50} = \left(-\frac{5}{5}\right)^2 + \frac{1}{50} =$ $1 + \frac{1}{50} =$ $1\frac{1}{50}$	$\left(\sqrt{4} + \sqrt{9}\right)^2 = (2+3)^2 =$ $(5)^2 = 25$
$\left(-1 + \frac{2}{4} - 3\right)^2 = \left(-1 + \frac{1}{2} - 3\right)^2 =$	$\frac{3}{5-3} \cdot \frac{5}{3-2} - \frac{2}{3} = \left(\frac{3}{2} \cdot \frac{5}{1}\right) - \frac{2}{3} =$

$\left(\frac{1}{2}-4\right)^2=\left(-3\frac{1}{2}\right)^2=$ $\left(-\frac{7}{2}\right)^2=\frac{49}{4}=12\frac{1}{4}$	$\frac{15}{2}-\frac{2}{3}=\frac{45}{6}-\frac{4}{6}=$ $\frac{41}{6}=6\frac{5}{6}$

3)

$\left(\sqrt{2+3}\right)^2=5$	$\frac{10}{2+3}+\frac{2}{10}\cdot\frac{3}{10}=\frac{10}{5}+\frac{5}{100}=$ $2+\frac{1}{20}=2\frac{1}{20}$
$\frac{1}{\sqrt{2}}+\frac{2}{\sqrt{2}}=\frac{\sqrt{2}}{2}+\frac{2\sqrt{2}}{2}=$ $\frac{3\sqrt{2}}{2}$	$\left(\frac{1}{\sqrt{3}}-\frac{2}{\sqrt{3}}\right)^2=\left(\frac{-1}{\sqrt{3}}\right)^2=$ $=\frac{1}{3}$
$\left(-\frac{3}{2}-\frac{5}{3}\right)^3=$	$\frac{2(7-6)}{4(2-5)}+\frac{1}{2}=$

4)

1st expression	2nd expression	compare and contrast
$(2+3)^2=5^2=25$	$2^2+3^2=4+9=13$	The sum must be completed first in the first expression.
$\sqrt{4+5}=\sqrt{9}=3$	$\sqrt{4}+\sqrt{5}=2+\sqrt{5}$	The sum must be completed first in the first expression.
$7\left(3\sqrt{2}+5\right)=21\sqrt{2}+35$	$7\left(3\sqrt{2}\right)+5=21\sqrt{2}$	Distribute the 7 in the first expression. Remember that $7\cdot 3\sqrt{2}=21\sqrt{2}$. Multiplication isn't distributive over multiplication.
$\frac{5\sqrt{3}+10}{5}=\sqrt{3}+2$	$\frac{5\sqrt{3}}{5}+10=\sqrt{3}+10$	Divide both by 5 in the first expression. Division (multiplying by a fraction) is distributive over addition.
$\frac{\sqrt{24}}{3}=$	$\frac{\sqrt{24}}{\sqrt{3}}=$	

$\left(\frac{1}{3}+\frac{1}{5}\right)^2 =$	$\left(\frac{1}{3}\right)^2 + \left(\frac{1}{5}\right)^2 =$	Add the fractions first in the first expression. Square the fractions first in the second.
$(3\sqrt{5})^2 + 7 = 9 \cdot 5 + 7 =$ $45 + 7 =$ 52	$(3\sqrt{5} + 7)^2 =$	Square $3\sqrt{5}$ first. That means squaring 3 and $\sqrt{5}$. Because $(3\sqrt{5})^2 = (3\sqrt{5}) \cdot (3\sqrt{5}) =$ $3 \cdot 3 \cdot \sqrt{5} \cdot \sqrt{5} =$ $9 \cdot 5 = 45$ Nothing to be done in the second expression until you learn how to square a binomial.

5)

expression	pitfall
$(5 + 10)^2 =$	It can be very tempting to square 5 and 10 idependently first. This is wrong, because powers or roots are not distributive over addition.
$5(4\sqrt{7} + 2) = 20\sqrt{7} + 10$	Distribute the 5. Multiply the 5 times the 4 but not times the $\sqrt{7}$. Multiplication is distributive over addition. Multiplication is not distributive over multiplication.
$\sqrt{9 + 16} =$	Addition first.
$\frac{\sqrt{15}}{5} =$	Nothing to be done here.
$\frac{5+12\sqrt{3}}{5} = 1 + \frac{12\sqrt{3}}{5}$	Division (multiplication by a fraction) is distributive over addition. In this case there is no great benefit but you can still do it as shown.
$\frac{\sqrt{8+10}}{2} =$	Addition first.

$\frac{1}{2}(12+7\sqrt{3})=$	If you distribute only multiply $\frac{1}{2}$ times the 12 and the 7.
$3(4+5)^2=$	Add first, then square, then multiply by 3.
$(3\sqrt{7})^2=9\cdot 7=63$	Square both.

6)

$\pi+\pi=2\pi$	$\sqrt{3}+\sqrt{3}=2\sqrt{3}$
$(\sqrt{3}+\sqrt{3})^2=(2\sqrt{3})^2=4\cdot 3=12$	$\left(\frac{1}{\sqrt{5}}+\frac{1}{\sqrt{5}}\right)^2=$
$(2\cdot 10^3)(2\cdot 10^4)=4\cdot 10^7$	$(2+3)(2+4)=$
$3(5+1)^2+1=$	$3\cdot 2^3=$
$\left(\frac{3}{5}\right)^2\left(\frac{1}{3}\right)+\left(\frac{4}{50}\right)^1=$	$\left(\frac{1}{\sqrt{5}}+2\right)^2=\left(\frac{\sqrt{5}}{5}+\frac{10}{5}\right)=\frac{10+\sqrt{5}}{5}$

NS9

6)

$5y+6y=y(5+6)=y11=11y$	$3\sqrt{2}+5\sqrt{2}=(3+5)\sqrt{2}=8\sqrt{2}$
$7\pi+8\pi=(8+7)\pi=15\pi$	$\frac{1}{2}e+\frac{2}{3}e=e\left(\frac{1}{2}+\frac{2}{3}\right)=\frac{7}{6}e$

7)

Expression	GCF	Factored result, simplified
$2x^2+3x^2$	x^2	$(2+3)x^2=5x^2$
$4y^3+20y^4$	$4y^3$	$4y^3(1+5y)$
$30\sqrt{2}+35\sqrt{2}$	$5\sqrt{2}$	$(6+7)5\sqrt{2}=13\cdot 5\sqrt{2}=65\sqrt{2}$
$12\sqrt{5}+18\sqrt{10}$	$6\sqrt{5}$	$6\sqrt{5}(2+3\sqrt{2})$

$\sqrt{2}+\sqrt{6}$	$\sqrt{2}$	$\sqrt{2}(1+\sqrt{3})$
$\frac{1}{2}+\frac{1}{6}$	$\frac{1}{2}$	$\frac{1}{2}\left(1+\frac{1}{3}\right)=\frac{1}{2}\left(\frac{4}{3}\right)=\frac{2}{3}$

8)

Distribute, then simplify	Factor, then simplify
$7(1+3)=7+21=28$	$14+21=7(2+3)=7(5)=35$
$3(2\cdot 10^2+4\cdot 10^2)=6\cdot 10^2+12\cdot 10^2=$ $18\cdot 10^2$	$42\cdot 10^8+11\cdot 10^8=10^8(42+11)=$ $53\cdot 10^8$
$2(\sqrt{2}+\sqrt{3})=2\sqrt{2}+2\sqrt{3}$	$3\sqrt{5}+6\sqrt{7}=3(\sqrt{5}+2\sqrt{7})$
$2(4\sqrt{6}-3)=8\sqrt{6}-6$	$10\sqrt{2}-45=5(2\sqrt{2}-9)$
$2\sqrt{3}(\sqrt{3}+\sqrt{5})=6+2\sqrt{15}$	$2\sqrt{6}+\sqrt{12}=\sqrt{6}(2+\sqrt{2})$
$2\sqrt{5}(1-3\sqrt{2})=2\sqrt{5}-6\sqrt{10}$	$2\sqrt{7}+3\sqrt{14}=\sqrt{7}(2+3\sqrt{2})$
$5(6+2\sqrt{2})=30+10\sqrt{2}$	$21+14\sqrt{3}=7(3+2\sqrt{3})$

$\frac{1}{5}\left(1+\frac{1}{2}\right)=\frac{1}{5}+\frac{1}{10}=\frac{3}{10}$	$\frac{1}{4}-\frac{1}{8}=\frac{1}{4}\left(1-\frac{1}{2}\right)=$ $\frac{1}{4}\cdot\frac{1}{2}=\frac{1}{8}$
$\frac{2}{3}\left(3+\frac{2}{3}\right)=2+\frac{4}{9}=2\frac{4}{9}$	$\frac{7}{9}-\frac{2}{3}=\frac{1}{3}\left(\frac{7}{3}-2\right)=\frac{1}{3}\cdot\frac{1}{3}=\frac{1}{9}$
$\sqrt{3}(4\sqrt{3}+5\sqrt{6})=12+5\sqrt{18}=$ $12+15\sqrt{2}$	$3\sqrt{7}+6\sqrt{35}=3\sqrt{7}(1+2\sqrt{5})$

11)

Distribute, then simplify	Factor, then simplify
$2(4+5x)+6x+4=$ $8+10x+6x+4=$ $16x+12$	$\frac{(3\sqrt{2}+5\sqrt{2})}{\sqrt{2}}=\frac{\sqrt{2}(3+5)}{\sqrt{2}}=$ 8
$4(2+3\sqrt{2})+5+\sqrt{2}=$ $8+12\sqrt{2}+5+\sqrt{2}=$ $13+13\sqrt{2}$	$\frac{2\pi^2+3\pi}{\pi}=\frac{\pi(2\pi+3)}{\pi}=$ $2\pi+3$
$4\cdot 10^2(3+2\cdot 10^2)+5\cdot 10^2=$	$2x+3x+4x=$

$12 \cdot 10^2 + 8 \cdot 10^4 + 5 \cdot 10^2 =$ $17 \cdot 10^2 + 8 \cdot 10^4$	$(2+3+4)x = 9x$
$3(4e+3) + e + 1 = 12e + 9 + e + 1 =$ $13e + 10$	$\frac{e^2+3e}{4e} = \frac{e(e+3)}{4e} = \frac{e+3}{4}$

NS10

1)

The product of powers rule works like this: $2^2 \cdot 2^3 = 2^5$. You add exponents because:

$2^2 \cdot 2^3 = 2 \cdot 2 \cdot 2 \cdot 2 \cdot 2 = 2^5$

You need to remember that an exponent is simply counting the number of times something is multiplied.

The quotient of powers rule works like this $\frac{2^7}{2^4} = 2^3$. You subtract exponents because:

$$\frac{2^7}{2^4} = \frac{2 \cdot 2 \cdot 2 \cdot 2 \cdot 2 \cdot 2 \cdot 2}{2 \cdot 2 \cdot 2 \cdot 2} = 2 \cdot 2 \cdot 2 \cdot 2 = 2^4$$

You are cancelling 2's.

The power to a power rule works like this: $(2^3)^2 = 2^6$ You multiply exponents because:

$(2^3)^2 = 2^3 \cdot 2^3 \cdot 2^3 = 2^6$

The rule for negative exponents works like this: $4^{-7} = \frac{1}{4^7}$ This is because:

$4^7 \cdot 4^{-7} = 4^0 = 1$ and $4^7 \cdot \frac{1}{4^7} = 1$, which demonstrates that 4^{-7} and $\frac{1}{4^7}$ are equivalent.

6)

	this means:	example
Exponents can be natural numbers.	Exponents can be $0,1,2, \ldots$ The exponent just counts the number of times you are multiplying the number by itself (and 1).	$2^0 = 1$ $2^1 = 2$ $2^2 = 4$ $2^3 = 8$
Exponents can be negative integers.	Negative exponents express how many times 1 is being divided by the base.	$2^{-1} = \frac{1}{2}$ $2^{-2} = \frac{1}{4}$ $2^{-3} = \frac{1}{8}$
Exponents can be rational non-integers.	Non-integer exponents can be interpreted as multiplying a number by	$2^{2.3} \approx 4.92$ $2^{\frac{1}{2}} = \sqrt{2}$

	itself a fractional number of times, or as rooting, or as rooting combined with raising to a power.	$2^{0.5} = \sqrt{2}$ $2^{\frac{3}{2}} = (2^3)^{\frac{1}{2}} = \sqrt{2^3}$

7)

March is earthquake month, because of a relationship between unstable weather and seismic activity.	F
The 2004 earthquake in Indonesia made the earth more round.	T
In the San Francisco earthquake of 1906, most buildings were lost to fire.	T
The 2010 earthquake in Chile shortened the day.	T
Oil extraction can cause earthquakes.	T
The cumulative impact of falling leaves in autumn can cause earthquakes.	F
There are $5 \cdot 10^5$ earthquakes every year.	T
The German philosopher Immanuel Kant thought that the 1755 earthquake which destroyed Lisbon was caused by earth farts.	T
Just as the sun and the moon pull on the water in the oceans to cause the tides. they can pull on the earth's plates to cause earthquakes.	T
All the earthquakes described above were magnitude 10 or above.	F
The 2004 Indian Ocean earthquake (subject of the movie, The Impossible) caused a 30 meter high tsunami.	T
In 2011 a magnitude 5.1 earthquake centered in Murcia was felt in Madrid.	T
In 2013, Valencia was hit with a magnitude 8.7 earthquake which caused many forest fires.	F
A 6.0 magnitude earthquake release the energy equivalent of the Hiroshima atomic bomb.	T
A magnitude 9.0 earthquake releases 10^5 times the energy of a magnitude 6.0 earthquake.	F
The longer the geologic fault (where two plates meet), the more powerful the earthquake.	T
A magnitude 10.5 earthquake is probable in your lifetime.	F

9)

$9^7 9^{-7} = 9^0 = 1$	$(3 \cdot 10^5) \cdot (4 \cdot 10^6) = 12 \cdot 10^{11} =$ $1.2 \cdot 10^{12}$
$\left(10^5 \cdot \frac{1}{10^3}\right) = 10^2$	$\frac{3^7}{3^5} \cdot \frac{3^4}{3^8} = 3^{-2}$
$4^{-9} 4^8 = 4^{-1} = \frac{1}{4}$	$\frac{4^{-4}}{4^{-6}} = \frac{4^6}{4^4} = 4^2$
$\frac{\left(\frac{1}{2^5}\right)}{\left(\frac{1}{2^7}\right)} = \frac{2^7}{2^5} = 2^2 = 4$	$4^6 \cdot \frac{1}{4^9} =$
$\frac{10^{-7}}{10^4} \cdot \frac{10^6}{10^4} =$	$6^0 6^{-1} 6^{-2} \cdot \frac{1}{6^2} \cdot \frac{1}{6^1} \cdot \frac{1}{6^0} =$
$\frac{2^3}{\left(\frac{2^4}{2^5}\right)} = 2^3 \cdot \frac{2^5}{2^4} = 2^3 2^1 = 2^4$	$\frac{\left(\frac{8^3}{8^7}\right)}{8^2} =$
$\frac{5^7}{4^7} \cdot \frac{4^6}{5^6} =$	$\frac{10^7}{10^5} \cdot \frac{9^6}{9^8} =$

12)

1st expression	2nd expression	compare and contrast
$(4+9)^2 =$	$\sqrt{4+9} =$	Add first in both cases.
$(5^2)^3 = 5^6$	$\left(5^{\frac{1}{2}}\right)^{\left(\frac{1}{3}\right)} = 5^{\frac{1}{6}}$	Power to a power means multiply exponents.
$\left(\sqrt{5}\right)^2 = 5$	$(5^{\frac{1}{2}})$^2 $= 5$	Power to a power means multiply exponents. A square root squared returns the base.
$\left(\sqrt{6}\right)^4 = \sqrt{6}\sqrt{6}\sqrt{6}\sqrt{6} =$ $6 \cdot 6 = 36$	$\left(6^{\frac{1}{2}}\right)^4 = 6^2 = 36$	Power to a power multiply exponents.
$\sqrt{7}\sqrt{7}\sqrt{7} = 7\sqrt{7}$	$\left(7^{\frac{1}{2}}\right)^3 = 7^{\frac{1}{2}} 7^{\frac{1}{2}} 7^{\frac{1}{2}} = 7 \cdot 7^{\frac{1}{2}} =$ $7\sqrt{7}$	Multiplying powers with same base, add exponents.
$\left(\sqrt[3]{10}\right)^6 =$	$\left(10^{\frac{1}{3}}\right)^6 =$	
$5\sqrt{3} \cdot 2\sqrt{3} = 30$	$5\left(7^{\frac{1}{3}}\right)(2)\left(7^{\frac{2}{3}}\right) = 10 \cdot 7$ $= 70$	Multiplying powers with same base, add exponents.
$\sqrt{(11)^6} =$	$(7^6)^{\frac{1}{2}} =$	
$5^2 5^3 =$	$5^{\frac{1}{2}} 5^{\frac{1}{3}} =$	
$\sqrt[5]{7^{10}} =$	$(7^{10})^{\frac{1}{5}} =$	

EE1

2)

x	y	z	expression	result
2	5	7	$3x$	6
2	1	5	$3x + 4y = 3(2) + 4(1) = 10$	10
0	-2	3	xy	0
-4	3	0	$xy + z$	-12
-5	3	4	$2x + 4y + z$	
1	2	3	$\frac{xy}{2z}$	$\frac{1}{3}$
-7	2	4	$x^2 + y - z = 49 + 2 - 4 = \cdots$	
-1	1	5	$3x^2 + 5y^3 - z = 3(-1)^2 + 5(1)^3 - 5 = 3$	3

5)

x	y	z	expression	result
$\frac{1}{2}$	$\frac{2}{3}$	3	$4x + 9y = 4\left(\frac{1}{2}\right) + 9\left(\frac{2}{3}\right) = 2 + 6 = 8$	8
$\frac{3}{5}$	1	0	$\frac{1}{2}x + \frac{2}{3}y = \frac{1}{2}\left(\frac{3}{5}\right) + \frac{2}{3}(1) = \frac{3}{10} + \frac{2}{3} = \cdots$	
$\frac{4}{5}$	$\frac{6}{7}$	$\frac{0}{3}$	$xyz = \left(\frac{4}{5}\right)\left(\frac{6}{7}\right)\left(\frac{0}{3}\right) = \left(\frac{4}{5}\right)\left(\frac{6}{7}\right)(0) = 0$	
0.75	2	0.5	$xy + z = \left(\frac{3}{4}\right)(2) + \frac{1}{2} = \frac{3}{2} + \frac{1}{2} = \cdots$	
5	3	4	$\frac{1}{x} + \frac{1}{y} = \frac{1}{5} + \frac{1}{3} = \cdots$	
$\frac{8}{27}$	$\frac{9}{4}$	5	$\frac{xy}{z} = \frac{\left(\frac{8}{27}\right)\left(\frac{9}{4}\right)}{5} = \frac{\left(\frac{2}{3}\right)}{5} = \frac{\left(\frac{2}{3}\right)}{\left(\frac{5}{1}\right)} = \cdots$	

13)

operation	arithmetic with numbers	simplifying an expression
multiplication and addition	$4(5) + 3(5) =$ $(4 + 3)(5) =$ $(7)(5) = 35$	$2x + 3x =$ $(2 + 3)x =$ $5x$
multiplication and subtraction	$6(7) - 2(7) =$ $(6 - 2)7 =$ $4(7) = \cdots$	$6x - 2x =$ $(6 - 2)x = 4x$
multiplication	$5 \cdot 5 \cdot 5 =$ 5^3	$y \cdot y \cdot y = y^3$

division and addition	$\frac{3}{7}+\frac{5}{7}=(3+5)\left(\frac{1}{7}\right)=8\left(\frac{1}{7}\right)=\frac{8}{7}$	$\frac{3}{z}+\frac{5}{z}=(3+5)\left(\frac{1}{z}\right)=8\left(\frac{1}{z}\right)=\frac{8}{z}$
multiplication and addition	$2(3)+5(3)+7(10)=$ $(2+5)3+70$ $=7(3)+70$ $=21+70=91$	$2x+3x+7y=$ $=(2+3)x+7y$ $=5x+7y$
multiplication and division	$\frac{10(3)}{14}=$ $\left(\frac{2}{2}\right)\left(\frac{(5)(3)}{7}\right)=\frac{15}{7}$	$\frac{xyz}{7x}=\frac{x}{x}\cdot\frac{yz}{7}=\frac{yz}{7}$
multiplication, exponentiation and addition	$3\cdot 2^3+5\cdot 2^3=$ $(3+5)\cdot 2^3=$ $7\cdot 2^3=56$	$3x^3+5x^3=(3+5)x^3=8x^3$
division and division	$\frac{\frac{8}{3}}{\left(\frac{2}{7}\right)}=\frac{\frac{8}{3}}{\left(\frac{2}{7}\right)}\frac{\left(\frac{7}{2}\right)}{\left(\frac{7}{2}\right)}=\frac{28}{3}=9\frac{1}{3}$	$\frac{\frac{x}{y}}{\left(\frac{y}{z}\right)}=\frac{\frac{x}{y}}{\left(\frac{y}{z}\right)}\cdot\frac{\left(\frac{z}{y}\right)}{\left(\frac{z}{y}\right)}=\frac{xz}{y^2}$

16)

$7x+4x-3x=8x$	$\frac{5x+3x}{16xy}=\frac{8x}{16xy}=\frac{1}{2y}$
$2x(x+5x)=2x(6x)=12x^2$	$3\cdot 2xyzyxy=6x^2y^3z$
$\frac{xx}{xxxx}=\frac{1}{x^2}=x^{-2}$	$\frac{x^2+x}{x}=\frac{x(x+1)}{x}=x+1$
$2x+5y-7x+3z+8y=-5x+13y+3z$	$2x^2+5x^2-3y+4y-7=7x^2+y-7$
$\frac{14x^2}{y}\cdot\frac{y^3}{7}=\frac{2x^2}{1}\cdot\frac{y^2}{1}=2x^2y^2$	$\frac{3}{2x}+\frac{5y}{2x}=\frac{3+5y}{2x}$
$6(x^2)(x^3)=6x^5$	$3(x^2)^3=3(x^2)(x^2)(x^2)=3x^6$
$xy^2+2xy^2=(1+2)(xy^2)=3xy^2$	$x^2-2x^2+3x+5x-7+4$ $=-x^2+8x-3$

17)

x	y	z	expression	result
1	5	-2	$2x+3z$	-4
$\frac{1}{2}$	-1	5	$\frac{(x-y)}{z}=\frac{\left(\frac{1}{2}+1\right)}{5}=\frac{\left(\frac{3}{2}\right)}{\left(\frac{5}{1}\right)}=\left(\frac{3}{2}\right)\left(\frac{1}{5}\right)=\frac{3}{10}$	$\frac{3}{10}$
12	20	-1	$\frac{xy}{3z}=\frac{(12)(20)}{3(-1)}=\frac{(4)(20)}{-1}=-80$	
$-\frac{4}{3}$	3	0	$\frac{(3x-z)}{-y}=\frac{3\left(-\frac{4}{3}\right)-0}{-3}=-\frac{4}{-3}=\frac{4}{3}=1\frac{1}{3}$	
$-5\frac{1}{4}$	$3\frac{1}{2}$	2	$2x+4y+z=2\left(-5\frac{1}{4}\right)+4\left(3\frac{1}{2}\right)+2$	$5\frac{1}{2}$

			$= 2\left(-5 - \frac{1}{4}\right) + 4\left(3 + \frac{1}{2}\right) + 2$ $= -10 - \frac{1}{2} + 12 + 2 + 2$ $= 5\frac{1}{2}$	
1	2	$-\frac{2}{3}$	$\frac{x}{y} + \frac{z}{2} = \frac{1}{2} + \frac{\left(-\frac{2}{3}\right)}{2}$ $= \frac{1}{2} + \left(-\frac{2}{3}\right) \cdot \frac{1}{2}$ $= \frac{1}{2} - \frac{1}{3} = \cdots$	
−2	1	$\frac{2}{3}$	$x^2 + y - z$ $= 4 + 1 - \frac{2}{3} = \cdots$	

18)

$x + 2x - 3x = 0x = 0$	$\frac{4x + 5x}{16xy} = \frac{9x}{16xy} = \frac{9}{16y}$
$2(4x + x) = 10x$	$3x \cdot 2x \cdot 5x = 30x^3$
$\frac{x^5}{x^2} = x^3$	$\frac{2x^2 + 5x^2}{x^3} = \frac{7x^2}{x^3} = \frac{7}{x}$
$5x + y - x - z + 5y = 4x + 6y - z$	$\frac{1}{2}x^2 + 5x^2 - 3y + \frac{1}{2}y = \frac{11}{2}x^2 - \frac{5}{2}y$
$\frac{12x^2}{y^3} \cdot \frac{y^3}{6x^3} = \frac{12}{6} \cdot \frac{y^3}{y^3} \cdot \frac{x^2}{x^3} = \frac{2}{x}$	$\frac{y}{2y} + \frac{5y}{5y} = \frac{1}{2} + 1 = 1\frac{1}{2}$
$5 \cdot x \cdot y + 3 \cdot x \cdot y = (5 + 3)xy = 8xy$	$3(x^4)^5 = 3x^{20}$
$x^2y^3 + 2x^2y^3 = (1 + 2)x^2y^3 = 3x^2y^3$	$x^2 - 2x^2 + 3x + 5x - 7 + 4$ $= -x^2 + 8x - 3$

EE2

12)

$4x + 5 = 6$ $x = \frac{1}{4}$	$3x - 2 = 8$ $x = \frac{10}{3} = 1\frac{1}{3}$
$x -$ $\frac{4}{1} \cdot \frac{(x-5)}{4} = 7 \cdot \frac{4}{1}$ $x - 5 = 18$ $x = 23$	$\frac{x-3}{5} = -6$
$\frac{2x-4}{3} = 8$ $2x - 4 = 24$	$\frac{(-3x+8)}{7} = 4$ $-3x + 8 = 28$

$2x = 28$ $x = 14$	$-3x = 20$ $x = \frac{20}{-3} = -6\frac{2}{3}$
$-2(x+5) = -8$ $x + 5 = \frac{-8}{-2}$ $x + 5 = 4$ $x = -1$	$5(3x+4) - 7 = 9$ $5(3x+4) = 16$ $3x + 4 = \frac{16}{5}$ $3x = 3\frac{1}{5} - 4$ $3x = -\frac{4}{5}$ $x = -\frac{4}{5} \cdot \frac{1}{3}$ $x = -\frac{4}{15}$

13)

$2x + \frac{1}{2} = 7$ $2x = 6\frac{1}{2}$ $x = 3\frac{1}{4}$	$3x - \frac{2}{3} = 1$ $3x = 1\frac{2}{3}$ $3x = \frac{5}{3}$ $x = \frac{5}{3} \cdot \frac{1}{3} = \frac{5}{9}$
$\frac{1}{3}x + 4 = \frac{1}{5}$ $\frac{1}{3}x = -3\frac{4}{5}$ $x = -\frac{19}{5} \cdot 3$ $x = -\frac{57}{5} = -11\frac{2}{5}$	$\frac{3}{7}x - \frac{3}{4} = -\frac{5}{4}$ $\frac{3}{7}x = -\frac{5}{4} + \frac{3}{4}$ $\frac{3}{7}x = -\frac{2}{4}$ $\frac{3}{7}x = -\frac{1}{2}$ ...
$\frac{3}{4}(x-5) = -2$ $x - 5 = -2 \cdot \frac{4}{3}$ ...	$\frac{6}{7}(2x-3) = 12$

14)

Input	add 8	multiply by 4	subtract 3	= 10
x	$x + 8$	$4(x+8)$	$4(x+8) - 3$	$4(x+8) - 3 = 10$
Input	subtract 4	divide by 3	add 4	= 11
y	$4y$	$\frac{4y}{3}$	$\frac{4y}{3} + 4$	$\frac{4y}{3} + 4 = 11$
Input	multiply by 3	add 7	divide by 4	= 12
z				

15)

$5x - 8 = 10$	$-2x + 12 = -3$ $-2x = -15$ $x = -15 \cdot \left(-\frac{1}{2}\right)$ $x = \frac{15}{2} = 7\frac{1}{2}$
$3x - \frac{1}{2} = 4$	$4x + \frac{2}{3} = 2$

16)

$2(x - 5) = 6$	$3(5 + x) = 4$
$4\left(\frac{1}{2} + x\right) = 8$	$-2\left(x - \frac{1}{3}\right) = -\frac{16}{3}$ $x - \frac{1}{3} = -\frac{16}{3} \cdot \left(-\frac{1}{2}\right)$ $x - \frac{1}{3} = \frac{8}{3}$ $x = \frac{9}{3}$ $x = 3$
$-5(-2x + 4) = 7$	$3(2x - 5) + 7 = 8$
$\frac{3}{5}(x - 2) - 5 = 10$ $\frac{3}{5}(x - 2) = 15$ $(x - 2) = 15 \cdot \frac{5}{3}$ ...	$5(x)^2 = 80$ $x^2 = 16$ $x = 4 \text{ or } -4$

17)

$\frac{2}{3}x - \frac{7}{9} = 8$ $\frac{2}{3}x = 8\frac{7}{9}$ $x = \frac{79}{9} \cdot \frac{3}{2} = \frac{79}{6} = 13\frac{1}{6}$	$\frac{2}{3}x - 8 = \frac{7}{9}$
$3(x - 5) = 9$	$-9 = -3(x - 5)$
$4x - 3x = \frac{5}{6} - \frac{4}{7}$ $1x = \frac{35}{42} - \frac{24}{42}$ $x = \frac{11}{42}$	$1\frac{1}{7} + 2\frac{1}{3} = x - 2x$
$\frac{1}{2}x - \frac{1}{3}x = 5(2 + 1)$	$\frac{7}{14} - \frac{3}{21} = \frac{10}{20}x + \frac{40}{80}x$

$\frac{3}{6}x - \frac{2}{6}x = 15$ $\frac{1}{6}x = 15$ $x = 90$	$\frac{1}{2} - \frac{1}{7} = \frac{1}{2}x + \frac{1}{2}x$ ...

EE3

1) Fill in the blanks using the words in the word bank below. Many expressions are **polynomials**. Some of the simplest polynomials are linear. Most equations you have solved involved linear expressions. which are just a particular type of polynomial. When you solved those equations you used a predictable algorithm. You did some of the same steps from one problem to the next because all the equations were linear. So recognizing the sort of equation or expression you are dealing with is important, so that you can use the correct algorithm to manipulate the expression or solve an equation involving that type of expression.

3)
In biology we learn there are very simple animals like sponges, and more complex ones like jellyfish, and super-complex ones like fish and so on. But animals form a kingdom and every animal shares certain defining characteristics with every other animal. For example, all animals are multi-cellular, their cells have nuclei and they have to eat to survive. Other characteristics were added or subtracted on top of this basic structure. For example, sometimes animals have added features (like feathers). But no matter how simple or complex an animal is, it will still be an animal.

4)
In algebra we learn there are simple polynomials expressions which we call linear, more complex ones we call **quadratic**, and even more complex ones we call cubic and so on. But polynomials form a special category of algebraic expression, and every polynomial shares certain defining characteristics with every other polynomial. For example, all polynomials are composed of **terms** that are a number called a **coefficient** times a variable to an exponent. These are connected by addition or subtraction, and the exponents can't be negative. In the course of creating a polynomial, we can add or eliminate as many terms as we want to make the polynomial simple or complex, but it will still be a polynomial.

5)

$5x^3 - 3x$	cubic, binomial	$4x^2 - 3x + 1$	
$x^2 - 4x$	quadratic, binomial	$5x^3 + 1x - 7$	
$3x^2$	quadratic, monomial	$4x - 3$	
$-2x^3$	cubic, monomial	$4x$	
11	constant, monomial	x^2	
$x^3 + 3x^2 - x + 4$		$6x^4 + 8x$	**degree** 4, binomial

6)

The expression $6x + 7$ has 2 terms, which means it's a binomial. The exponent of the variable in the first term is 1, which means the expression is linear. There is no variable in the second term, which means that term is constant. It's called constant because no matter what number we substitute for x the second term is always constant.
The expression $5x^2 + 6x + 7$ has 3 terms, thus it's a trinomial. The exponent of the variable in the first term is 2, which means the expression is quadratic. The exponent of the variable in the second term is 1, which means that term is linear. There is no variable in the third term, which means that term is constant.

8)

$x^3 + x^4 + x - 2x^2$ $x^4 + x^3 - 2x^2 + x$	$5(x+2) - 3$ $5x + 10 - 3$ $5x + 7$
$3x^2 + 2(3x - 4) + 3x - 4$	$4(x-2) + 3(x^2 - 7)$ $4x - 8 + 3x^2 - 21$ $3x^2 + 4x - 29$
$4x(x+2) - 3x(x+5)$ $4x^2 + 8x - 3x^2 - 15x$ $x^2 - 7x$	$5(x-2) + 4x^2 - 3x$

9)

$3x^2(x+2) - 3x(4)$ $3x^3 + 6x^2 - 12x$	$(3x-5)4$ $4(3x-5)$ $12x - 20$
$2x(x^2 \cdot 3) + 7(x)$ $2x \cdot 3x^2 + 7x$ $6x^3 + 7x$	$5x(x^3 \cdot 4 \cdot x) + x^2(8)$ $5x(4x^4) + 8x^2$ $20x^5 + 8x^2$
$5 \cdot x(x-8) + 3 \cdot 4$ $5x(x-8) + 12$ $5x^2 - 40x + 12$	$\frac{2}{3}x\left(\frac{x}{3} + 7\right)$ $\frac{2}{3}x\left(\frac{x}{3} + 7\right)$ $\frac{2x^2}{9} + \frac{14x}{3}$ $\frac{2}{9}x^2 + \frac{14}{3}x$
$\frac{3}{5}x + \frac{2}{3}x$ $\frac{9}{15}x + \frac{10}{15}x = \frac{19}{15}x$	$\frac{x}{5} + \frac{x}{7} + 8x^2$ $\frac{7x}{35} + \frac{5x}{35} + 8x^2$ $\frac{12x}{35} + 8x^2$ $8x^2 + \frac{12}{35}x$

16)

$5(x-3)$	$3x(2x+5)$ $6x^2+15x$
$4x(2x^2+3x)+8$ $8x^3+12x^2+8$	$2(x-4)+3(x-5)$ $2x-8+3x-15$ $5x-23$
$-4x(2x-1)+3$	$(x^2)^3+5-3x$ x^6+5-3x x^6-3x+5
$xxxxx+xxxx+xxx-xx-x-x^0$	$5x^3x^2-3x^1x+4x^0x^1x^2$ $5x^5-3x^2+4x^3$ ...
$5(x-3)+(4-x)x$ $5x-15+4x-x^2$ $-x^2+9x-15$	$5\big(-(-x)\big)+3$ $5x+3$
$x(-x)(-x)(-3)+6x^2$ $-3x^3+6x^2$	$3(4x)+5x(4x)-1(4x)^0$ $12x+20x^2-1$ ...

17)

$\frac{1}{2}(x-2)+\frac{3}{5}(x-5)$ $\frac{1}{2}x-1+\frac{3}{5}x-3$ $\frac{5}{10}x+\frac{6}{10}x-4$ $\frac{11}{10}x-4$	$\frac{2}{3}x-\frac{5}{6}x$
$3x\left(\frac{1}{2}x+4\right)-x\left(x^2-\frac{2}{3}\right)$	$\frac{5}{6}\left(\frac{3}{10}\right)\left(\frac{x}{1}\right)+x$ $\frac{1}{4}x+x$ $\frac{5}{4}x$
$\frac{8}{9}x+\frac{7x}{3}-x$ $\frac{24}{27}x+\frac{63}{27}x-\frac{27}{27}x$ $\frac{60}{27}x$ $\frac{20}{9}x$	$\frac{5}{3}(4x)\left(\frac{6}{25}\right)+3x$ $\left(\frac{4x}{1}\right)\left(\frac{2}{5}\right)+3x$ $\frac{8}{5}x+\frac{15}{5}x$ $\frac{23}{5}x$

$6\left(5x-\frac{4}{12}\right)-2x^2+x^2$	$-\left(-\frac{5}{7}\right)(14x-21)+x^0$ $\frac{5}{7}(14x-21)+1$ $10x-\left(\frac{5}{7}\right)\left(\frac{21}{1}\right)+1$ $10x-15+1$ $10x-14$

EE4

1)

$5x-3=7$ $5x=10$ $x=2$	$9x+2=8$ $9x=6$ $x=\frac{6}{9}$ $x=\frac{2}{3}$
$\frac{3}{10}x-7=8$	$\frac{2}{3}x-\frac{3}{4}=-1$
$0.5x+0.7=1.2$ $0.5x=0.5$ $x=1$	$x+2\cdot10^4=5\cdot10^4$ $x=5\cdot10^4-2\cdot10^4$ $x=(5-2)\cdot10^4$ $x=3\cdot10^4$

2)

$x+x+2=4+5$ $2x+2=9$ ...	$2(x-3)=5(2+4)$ $2(x-3)=30$ $x-3=15$ $x=18$
$3(x-5)=4\left(\frac{2}{3}+\frac{1}{2}\right)$ $3x-15=\frac{8}{3}+2$ $3x=15+2\frac{2}{3}+2$ $3x=19\frac{2}{3}$ $3x=\frac{59}{3}$ $3x=\frac{59}{9}=6\frac{5}{9}$	$-2x+3x^2-3x^2+5=-10$
$\frac{3}{2}(x-12)=90x-90x$ $\frac{3}{2}x-18=0$	$4x^3+3x^2-2x-3x^2-4x^3=-1$

$\frac{3}{2}x = 18$ ...	
$-3(12x^2 - 5) + 36x^2 = 5x$	$(2x)^2 - 4x^2 + 5 = 4$ $(2x)(2x) - 4x^2 = -1$ $4x^2 - 4x^2 = -1$ $0 = -1$ That is a false statement. So this equation is always false, no matter what you possibly plug in for x. Therefore we say this equation has no solution.

5)

$3x + 4 = 7x$ $4 = 4x$ (subtract $3x$ both sides) $x = 1$ (divide both sides by 4)	$5x = 3x - 2$
$6x - 4 = 2x + 8$	$x + 2x = 5x - 4$ $3x = 5x - 4$ $-2x = -4$ ...
$3(x - 5) = -4(2x + 7)$ $3x - 15 = -8x - 28$ $11x - 15 = -28$ $11x = -13$ $x = -\frac{13}{11}$	$4x^2 + 3x = 4x^2 + 5x - 9$ $3x = 5x - 9$ $-2x = -9$ $x = \frac{-9}{-2}$ $x = \frac{9}{2} = 4\frac{1}{2}$
$\frac{1}{2}x - \frac{8}{9} = \frac{5}{2}x$ $-\frac{8}{9} = \frac{4}{2}x$ $-\frac{8}{9} = 2x$ ...	$6(x - 8) + 2x = 8x$ $6x - 48 + 2x = 8x$ $8x - 48 = 8x$ $-48 = 0$ This is a false statement. This means the equation is always false, no matter what you plug in for x. We say this equation has no solution.

8)

$3(y + 2) - 7 = 9$ $3y + 6 - 7 = 9$ $3y - 1 = 9$ $3y = 10$ $y = \frac{10}{3}$ $y = 3\frac{1}{3}$	$2y + 3y = 10$

$\frac{12}{14}y+\frac{1}{7}y=3$ $\frac{6}{7}y+\frac{1}{7}y=3$ $\frac{7}{7}y=3$ $y=3$	$\frac{3y}{5}+\frac{8y}{5}=22$ $\frac{11y}{5}=22$ $11y=22(5)$ $y=\frac{22(5)}{11}$ $y=10$
$-\frac{4}{7}(14y+2)=\frac{1}{7}$ $14y+2=\frac{1}{7}\left(-\frac{7}{4}\right)$ $14y+2=-\frac{1}{4}$ $14y=-\frac{9}{4}$ $y=-\frac{9}{4}\left(\frac{1}{14}\right)$ $y=-\frac{9}{56}$	$(3y-10)2=3(4-5)$ $6y-20=3(-1)$ $6y-20=-3$ $6y=17$ $y=\frac{17}{6}$

EE5

5)

$\frac{2}{3}x-\frac{1}{6}=\frac{4}{12}x-2$ $\frac{12}{1}\left(\frac{2}{3}x-\frac{1}{6}\right)=\frac{12}{1}\left(\frac{4}{12}x-2\right)$ $\frac{12}{1}\cdot\frac{2}{3}x-\frac{12}{1}\cdot\frac{1}{6}=\frac{12}{1}\cdot\frac{4}{12}-\frac{12}{1}\cdot\frac{2}{1}$ $8x-2=4x-24$ $4x-2=-24$ $4x=-22$ $x=-\frac{22}{4}$ $x=-\frac{11}{2}=-5\frac{1}{2}$	$-\frac{2}{5}x+\frac{1}{10}=\frac{3}{10}x-\frac{3}{20}$

7)

$-8x-9=-7x-14$ $8x+9=7x+14$ …	$-\frac{1}{3}x-2=-\frac{2}{3}x-5$ $\frac{1}{3}x+2=\frac{2}{3}x+5$ $x+6=2x+15$ …

$-\frac{3}{5}x - 7 = -2$	$-x = -5$

8)

$\frac{2}{11}x - \frac{3}{22} = \frac{1}{11}$ $\frac{22}{1}\left(\frac{2}{11}x - \frac{3}{22}\right) = \frac{1}{11}\left(\frac{22}{1}\right)$ $4x - 3 = 2$ ...	$-\frac{5}{7}x - \frac{3}{14} = -\frac{2}{21}$ $\frac{5}{7}x + \frac{3}{14} = \frac{2}{21}$ $\frac{42}{1}\left(\frac{5}{7}x + \frac{3}{14}\right) = \frac{2}{21}\left(\frac{42}{1}\right)$ $6x + 3 = 4$ ...
$\frac{1}{10^3}x + \frac{1}{10^2} = \frac{1}{10}$ $\frac{10^3}{1}\left(\frac{1}{10^3}x + \frac{1}{10^2}\right) = \frac{1}{10}\left(\frac{10^3}{1}\right)$ $x + 10 = 10^2$ $x = 90$	$\frac{1}{6^4}x - \frac{1}{6^3} = \frac{1}{36}$ $\frac{6^4}{1}\left(\frac{1}{6^4}x - \frac{1}{6^3}\right) = \frac{1}{6^2}\left(\frac{6^4}{1}\right)$ $x - 6 = 36$ ...
$\frac{1}{\sqrt{2}}x + \frac{5}{\sqrt{2}} = \frac{7}{\sqrt{2}}$ $\sqrt{2}\left(\frac{1}{\sqrt{2}}x + \frac{5}{\sqrt{2}}\right) = \frac{7}{\sqrt{2}}(\sqrt{2})$ $x + 5 = 7$ ...	$-\frac{3x}{5} - \frac{1}{25} = \frac{3}{5^2}$

9)

$5(x-4) = 12\left(\frac{1}{2}x - 2\right)$	$\frac{2}{3}x - \frac{5}{3} = \frac{1}{3}$ $2x - 5 = 1$ ...
$-3x = -5 - 2x$	$\frac{4x}{7} - \frac{3x}{7} = \frac{5}{14}$ $\frac{14}{1}\left(\frac{x}{7}\right) = \frac{5}{14}\left(\frac{14}{1}\right)$ ...
$\frac{x-5}{10^3} = 10^{-2}$ $\frac{10^3}{1} \cdot \frac{(x-5)}{10^3} = 10^{-2}(10^3)$ $x - 5 = 10$ ...	$\frac{x}{\sqrt{3}} + \frac{2}{\sqrt{3}} = \frac{5}{\sqrt{3}}$ $\frac{\sqrt{3}}{1}\left(\frac{x}{\sqrt{3}} + \frac{2}{\sqrt{3}}\right) = \left(\frac{5}{\sqrt{3}}\right)\frac{\sqrt{3}}{1}$ $x + 2 = 5$ ...

10)

$\frac{1}{10^4}(x-2)=\frac{3}{10^3}$ $\frac{10^4}{1}\cdot\frac{1}{10^4}(x-2)=\frac{3}{10^3}\cdot\frac{10^4}{1}$ $x-2=30$ …	$x3+x4=28$ $7x=28$
$x\sqrt{3}=5$ $x=\frac{5}{\sqrt{3}}$ $x=\frac{5\sqrt{3}}{3}$	$(-\sqrt{5})x=-7$
$(2\cdot 10^6)x=4\cdot 10^7$ $x=\frac{4\cdot 10^7}{2\cdot 10^6}$ $x=2\cdot 10$ $x=20$	$-\frac{7}{12}x-\frac{8}{12}x=-3$ $\frac{7}{12}x+\frac{8}{12}x=3$ $\frac{12}{1}(\frac{7}{12}x+\frac{8}{12}x=3\left(\frac{12}{1}\right)$ $7x+8x=36$ …

EE6

3)

Equation
$D=\frac{m}{V}$ 7
$E=mc^2$ 5
$F=ma$ 1
$C=\pi d$ 3
$s=\frac{1}{2}at^2+v_it$ 10
$v_f=v_i+at$ 12
$v_f^2=v_i^2+2as$ 4
$a^2+b^2=c^2$ 11
$p=mv$ 6
$W=Fs$

8
$P = \frac{W}{t}$ 2
$F = \frac{Gm_1m_2}{r^2}$ 9

6)

Equation	$v_f = v_i + at$	$s = \frac{1}{2}at^2 + v_it$	$P = \frac{W}{t}$
Given values	$v_f = 42\ m/s$ $a = 4\ m/s^2$ $t = 5\ s$	$s = 42\ m$ $a = 3\ m/s^2$ $t = 4\ s$	$P = 6\ W$ $t = 5\ s$
Equation with values substituted	$42 = v_i + 4(5)$ $v_i = 20\frac{m}{s}$	$42 = \frac{1}{2}(3)(4)^2 + v_i(4)$ $42 = 24 + 4v_i$ $18 = 4v_i$ $v_i = 4.5\ m/s$	
Solution	$v_i = 20\ m/s$		

7)

Quantity to solve for	Density	Energy	Acceleration
Given values	$m = 5.00\ kg$ $V = 3.00\ m^3$	$m = 1.00 \cdot 10^{-9}\ kg$ $c = 3.00 \cdot 10^8\ m/s$	$F = 12\ N$ $m = 88.0\ kg$
Equation with values substituted	$D = \frac{m}{V}$ $D = \frac{5}{3}$ $D = 1\frac{2}{3}\ kg/m^3$	$E = mc^2$ $E = 10^{-9} \cdot (3 \cdot 10^8)^2$ $E = 10^{-9} \cdot 9 \cdot 10^{16}$ $E = 9 \cdot 10^7\ J$	
Solution			

Quantity to solve for	pi	Distance	Mass
Given values	measure C and d for a circle and compute	$a = 10.0\ m/s^2$ $t = 4.00\ s$ $v_i = 5.00\ m/s$	$G = 7.00 \cdot 10^{-11}$ $m_1 = 1.00 \cdot 10^{11}\ kg$ $r = 2\ m$ $F = 1.00\ N$
Equation with values substituted		$s = \frac{1}{2}at^2 + v_it$ $s = \frac{1}{2}(10)(4)^2 + 5(4)$ $s = 80 + 20$ $s = 100\ m$	$F = \frac{Gm_1m_2}{r^2}$ $1 = \frac{(7 \cdot 10^{-11})(10^{11})m_2}{2^2}$ $1 = \frac{7}{4}m_2$ $m_2 = \frac{4}{7}\ kg$
Solution			

10)

$a = \frac{F}{m}$

11)

$F = m\boldsymbol{a}$	$D = \frac{\boldsymbol{m}}{V}$ $\left(\frac{V}{1}\right)(D) = \frac{m}{V}\left(\frac{V}{1}\right)$ $VD = m$ $m = VD$
$W = F\boldsymbol{s}$	$v_f = \boldsymbol{v_i} + at$ $v_i = v_f - at$
$v_f = v_i + \boldsymbol{at}$ $v_f - v_i = at$ $\frac{v_f - v_i}{a} = t$ $t = \frac{v_f - v_i}{a}$	$F = \frac{G\boldsymbol{m_1}m_2}{r^2}$ $Fr^2 = Gm_1m_2$ $m_1 = \frac{Fr^2}{Gm_2}$
$\boldsymbol{a}bc = d + 3$	$a\boldsymbol{b} + c = d - 4$ $ab = d - 4 - c$ $b = \frac{d-c-4}{a}$

12)

$W = \boldsymbol{F}s$ $F = \frac{W}{s}$	$p = m\boldsymbol{v}$ $v = \frac{p}{m}$
$v_f^2 = v_i^2 + 2\boldsymbol{a}s$ $a = \frac{(v_f^2 - v_i^2)}{2s}$	$A = \boldsymbol{\pi}r^2$ $\pi = \frac{A}{r^2}$
$E = \boldsymbol{m}c^2$ $m = \frac{E}{c^2}$	$P = \frac{W}{\boldsymbol{t}}$ $Pt = W$ $t = \frac{W}{P}$
$\boldsymbol{a^2} + b^2 = c^2$ (Solve for a^2.) $a^2 = c^2 - b^2$	$y = mx + \boldsymbol{b}$ $b = y - mx$

$m = \frac{y_2 - y_1}{x_2 - x_1}$ $m(x_2 - x_1) = y_2 - y_1$ $y_2 = m(x_2 - x_1) + y_1$	$abc = 123$ $c = \frac{123}{ab}$

LF1

2)

Fill in the blank using the words from below. As the **input** increases, the **output** increases. For every increase of 1 in input, the output increases by 2. This is because the first thing the genie does is double the input. Interestingly, even if you give the genie 0 candies, you still get 9 candies as output. $2x + 9$ is a polynomial, specifically, it is linear and a binomial. The doubling shows up in the expression $2x + 9$ as the coefficient of the linear term. The 9 which gets added on shows up as the constant term.

5)

The output (y) increases as the input (x) increases. Specifically, as x increases by 1 y increases by 2. This is seen as the steepness of the graph. When you give the expression an x of 0 it returns a y of 9. This is evident on the graph as the point $(0,9)$, where the graph crosses the y-axis.

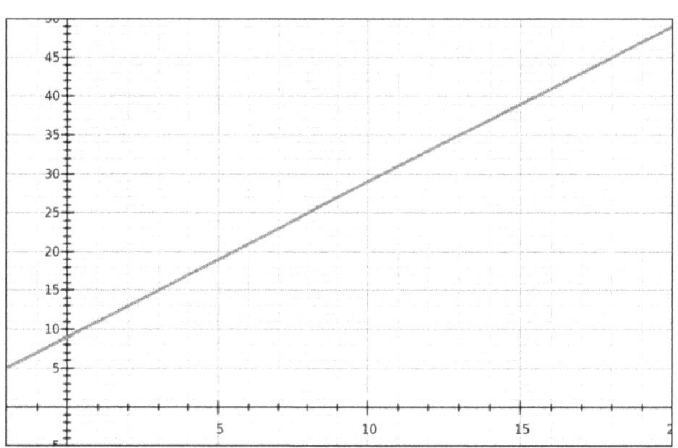

7)

Genies like $40x + 1$ and $100x + 7$ both reward even small numbers of input, and they quickly return large values for output when input grows. Genies like $2x + 3$ are ok, but not as rewarding as the previous ones. Genies like $-20x + 7$ return negative quantities of candies for many inputs. If you show up with 0 you get 7, but if you show up with 1 you get -13. We can interpret this as meaning you have a candy deficit of 13, meaning after he does hi magic you actually owe the genie 13 candies!

10) The steeper graph is g.

11) $5\frac{5}{8}$ is the input which produces the same output in both. The output produced is $61\frac{1}{4}$. If you look at the graph you see something close to this. Later you'll learn to get an exact answer algebraically.

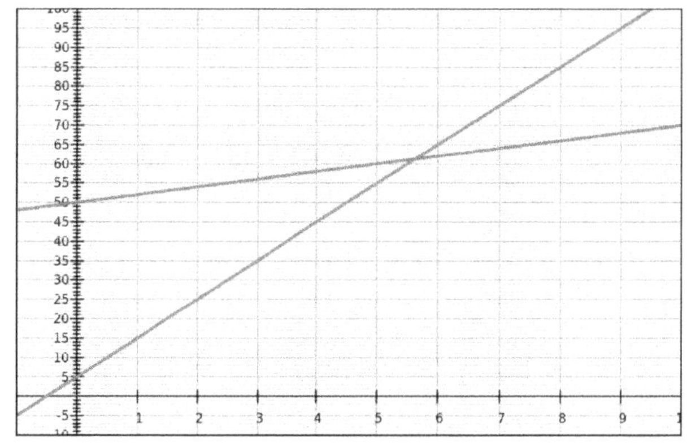

18)

3

19)

1

20)
$(15, 65)$

21)
Fill in the blanks. Functions accept values as input and return values as output. Functions are named using letters like f and g and each function can be written as an equation like $y = 2x + 3$. The relationship between the input and the output of a function can be shown in a table or in a graph.

23)
$y = 3x + 7$

24)
$y = 2x + 5$

26)
1.5 $\quad 1\frac{2}{3}$
16 $\quad 14\frac{2}{3}$
(Your answers should be approximations somewhat close to these. You'll soon learn algebra to find these answers exactly.)

28)
9.8 5.5
27 7.7
(These values are just my approximations based on the graph.)

31)

For function f:	Given two functions:
Approximately what input returns an output of 15? (Decimals good.) $15 = 2x + 7$ (Substitute desired output for y.) $2x = 8$ $x = 4$ (Now you solve the equation.)	Approximately what single input will produce the same output for two functions? (Decimals good.) $2x + 7 = x - 3$ (Set the expressions equal.) $x = -10$ (Now you solve the equation.)
What output is returned for an input of 18? $y_1 = 2(18) + 7$ (Substitute input for x.) (Now you simplify.)	What output does the input above produce? $y = (-10) - 3$ (Substitute the above input for x in either equation.) (Now you simplify.)

33)

Equation	Question this equation is trying to answer
$3x + 20 = 100$	What input produces an output of 100 for f ?
$x + 50 = 100$	What input produces an output of 100 for g ?
$3x + 20 = x + 50$	What single input produces the same output in both?

LF2

2)

Genie	Function	How much output increases when input increases by 1
f	$y = 3x + 20$	3
g	$y = 7x + 100$	7
h	$y = x + 50$	1

3)
The coefficient of the linear term is the same as the increase in output when input increases by 1. The given example shows that this is the case when you try it with consecutive inputs like 5 and 6, and thanks to distribution you can show that this is the case when you input x then $x + 1$.

4)
$\frac{1}{3}, \frac{2}{3}, 15, \frac{7}{3}, \frac{11}{3}$

For f, g *and* i it was easy to find how much the output increased by when input increased by 3 because the denominator of the coefficient was 3 in each case. h was easy too because it was a whole number, but it would have been better to ask how much the output increases by when the input increases by 1. For j it would have been better to ask how much the output increases by when the input increases by 6. The answer is 11. Finally, the output of the first function increases by $\frac{1}{3}$ when the input increases by 3.

9)
For f the output increases by 5 when input increases by 2. That means the rate of change is $\frac{5}{2}$.

10)

x	y
2	35
12	60

11)
Increase of output of 25 for an increase of input of 10.
Therefore a rate of change of $\frac{25}{10} = 2.5$.

12)
$50, 2.5, 2.5, 2.5$

18)
600
5

19)
$\frac{600}{5} = 120$

21)
$600, \Delta y, 5, \Delta x$

25)
$y = 60x$

27)
$\frac{478-178}{4-1} = \frac{300}{3} = 100$

29)
$(0,78)$

30)
$y = 100x + 78$

35)
17.75 km/hr
$y = 17.75x + 39.75$

40)
$y = 45x + 65$

42)
$y = 11x + 7$

44)
$y = \frac{7}{5}x + \frac{1}{5}$

LF3

1)

Genie f $y_1 = 5x + 15$

Genie g $y = 15x + 50$

2)

Genie f $y = 20x$

Genie g $y = 20x + 10$

3)

Genie f $y = 10x + 12$

Genie g $y = 15x + 16$

6)

$y = 3x + 30$

8)

$y = 2x + 40$

15)

	x	y
A	3	7
B	8	22
C	10	28

Show work here	Algorithm
$m = \frac{15}{5} = 3$	1) Find slope
$y = mx + b$	2) Write general slope-intercept form for a line.
$7 = 3(3) + b$	3) Substitute known information.
$b = -2$	4) Solve for b.
$y = 3x - 2$	5) Write specific function.
$y = 3(10) - 2 = 28$ ☺	6) Confirm function is correct using other point.

17)

Genie f $y = 7x - 21$

Name	x	y_1
Alice	3	0
Bob	5	14
Chris	7	28

Genie g $y = 51x + 4$

Name	x	y_2
Adam	0	4
Billy	1	55
Cathy	2	106

LF4

1)

Genie f $y_1 = 5x + 15$

Genie g $y = 15x + 50$

2)

Genie f $y = 20x$

Genie g $y = 20x + 10$

3)

Genie f $y = 10x + 12$

Genie g $y = 15x + 16$

6)
$y = 3x + 30$

8)
$y = 2x + 40$

15)

Show work here	Algorithm
$m = \frac{15}{5} = 3$	1) Find slope
$y = mx + b$	2) Write general slope-intercept form for a line.
$7 = 3(3) + b$	3) Substitute known information.
$b = -2$	4) Solve for b.
$y = 3x - 2$	5) Write specific function.
$y = 3(10) - 2 = 28$ ☺	6) Confirm function is correct using other point.

	x	y
A	3	7
B	8	22
C	10	28

17)

Genie f $y = 7x - 21$

Name	x	y_1
Alice	3	0
Bob	5	14
Chris	7	28

Genie g $y = 51x + 4$

Name	x	y_2
Adam	0	4
Billy	1	55
Cathy	2	106

LF5

4)

$30x + 50 = 25x + 100$

$5x = 50$

$x = 10$

$y = 350$

$(10,350)$

6)

$y = -1x + 1200$

9)

$0 = -1x + 1200$

$x = 1200$

12)

$0 = -\frac{1}{2}x + 650$

$\frac{1}{2}x = 650$

$x = 1300$

13)

$-\frac{1}{2}x + 650 = -\frac{1}{3}x + 500$

multiplying both sides by -6 gives:

$3x - 3900 = 2x - 3000$

$x = 900$

Substitute this into one of the functions:

$y_1 = -\frac{1}{2}(900) + 650$

$y_1 = 200$

You can also substitute this x value into the other function to confirm the same result is returned for both.

14)

∠1 = _26.6° _____ ∠2 = __18.4°

16)
$m_1 =$ ___ -2.1_________
$m_2 =$ _____ -0.5_________

17)
$A(536,0)$
$B(1500,0)$

19)
$m \approx -0.27$

22)
$(0,1050)$
$(500,700)$
$m = -\frac{350}{500} = -\frac{7}{10}$
$y = -\frac{7}{10}x + 1050$

LF6

1)
$y = 100$

2)
$6x + 4 = 100$
…

3)
$y = 124$

4)
$m = \frac{12-12}{5-0} = \frac{0}{5} = 0$
$y = 0x + 12$
$y = 12$

9)
$0 = 2x + 3$
...

$0 = \frac{1}{2}x + 3$
...

$0 = 3$
no solution
That means no x-intercept for the third function.

LF7

1)

function	type
$y = -\frac{2}{3}x + 7$	linear
$y = 5$	constant
$y = x^2 + 5x - 9$	quadratic
$y = \sqrt{x}$	square root
$y = \frac{10}{x}$	rational
$y = \sin(x)$	trigonometric
$y = 7x^3 - 8x$	cubic
$y = 2^x$	exponential

5)
$y = 3\sqrt{x}$ doesn't accept negative numbers as input because we cannot square root a negative number without producing an imaginary result. If you're not sure about this, try to determine a result for $\sqrt{-9}$.

6)
$y = \frac{10}{x}$ doesn't accept 0 as input because $\frac{10}{0}$ is undefined. If you're not sure about this, try to determine how many times 0 goes into 10.

7)

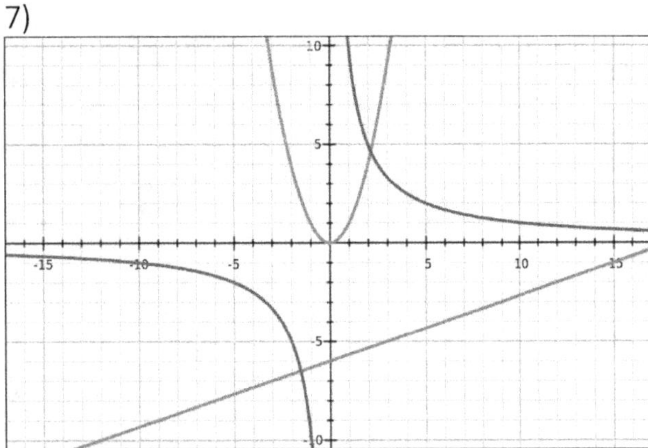

12)

function	rule
$y = \frac{4}{x}$	cannot accept 0 as input.
$y = \sqrt{x}$	cannot accept negative numbers as input.
$y = 6\sqrt{x}$	cannot accept negative numbers as input.
$y = \frac{12}{x}$	cannot accept 0 as input.

13)

function	This function never produces a y-value below:
$y = x^2$	0. because squaring returns positive numbers. The lowest we can go is 0 by substituting 0 as input, which returns 0 for y.
$y = x^2 + 5$	5. because squaring returns positive numbers. The lowest we can go is 5 by substituting 0 as input.
$y = \sqrt{x}$	0. We can't substitute negative numbers and the square root of a positive number is a positive number. So the lowest we can go is 0, which we get by substituting 0 for x.
$y = x^4$	0. We get this by substituting 0 for x. When we substitute negative numbers the function returns positive values.

16)

function	$y = x^2 + 3$	$y = \frac{10}{x}$
graph		

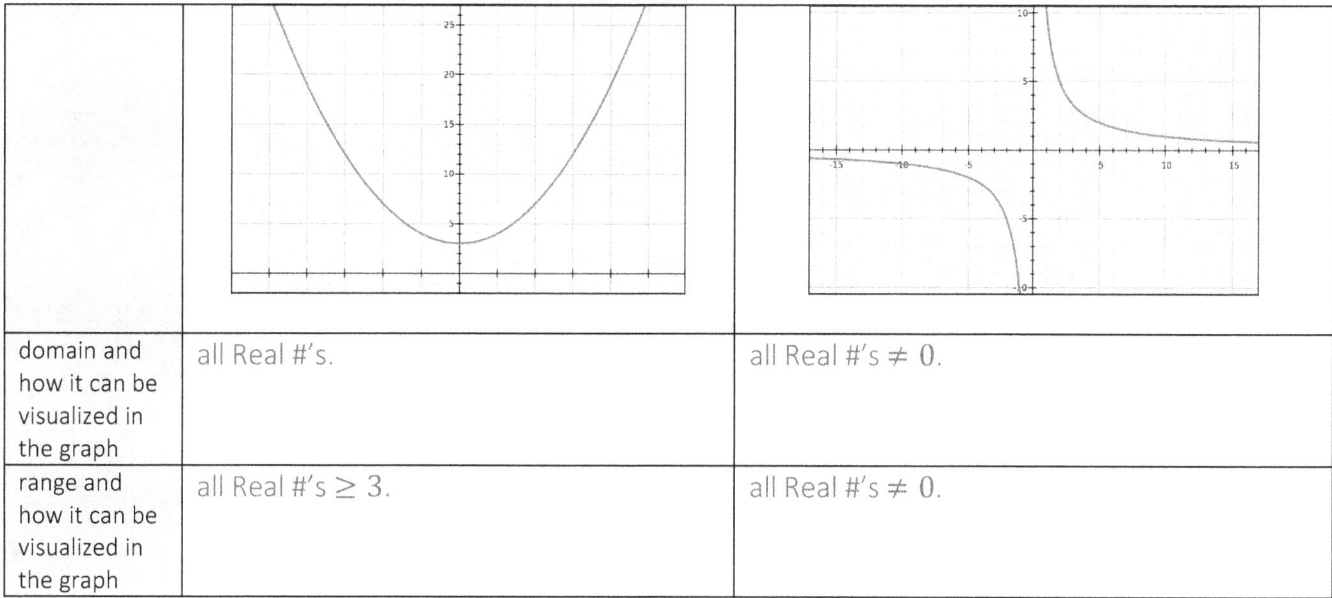

domain and how it can be visualized in the graph	all Real #'s.	all Real #'s $\neq 0$.
range and how it can be visualized in the graph	all Real #'s ≥ 3.	all Real #'s $\neq 0$.

17)

function	domain	range
$y = 2x$	all Real numbers	all Real numbers
$y = x^2$	all Real numbers	all Real numbers ≥ 0
$y = \sqrt{x}$	all Real numbers ≥ 0	all Real numbers ≥ 0
$y = 7$	all Real numbers	7
$y = -\frac{3}{7}x + 8$	all Real numbers	all Real numbers
$y = x^3$	all Real numbers	all Real numbers
$y = \frac{10}{x}$	all Reals $\neq 0$	all Reals $\neq 0$
$y = x^4$	all Real numbers	all Real numbers ≥ 0
$x = 5$	5	all Real numbers

LF8

6)

$\left(16\frac{2}{3}, 1333\frac{1}{3}\right)$

10)

Example
$y = 2x - 5$ $y = 4x + 17$
$2x - 5 = 4x + 17$ $2x = -22$ $x = -11$
$y = 2(-11) - 5$ $y = -27$
$y = 4(-11) + 17$ $y = -27$ ☺
$(-11, -27)$

11)

$-2x - 5 = 4x + 7$
$6x = -12$
$x = -2$
$y = -1$
$(-2, -1)$

13)
$1.5x - 5 = 3.5x + 7$
$2x = -12$
$x = -6$
$y = -14$
$(-6, -14)$

LF9

1)

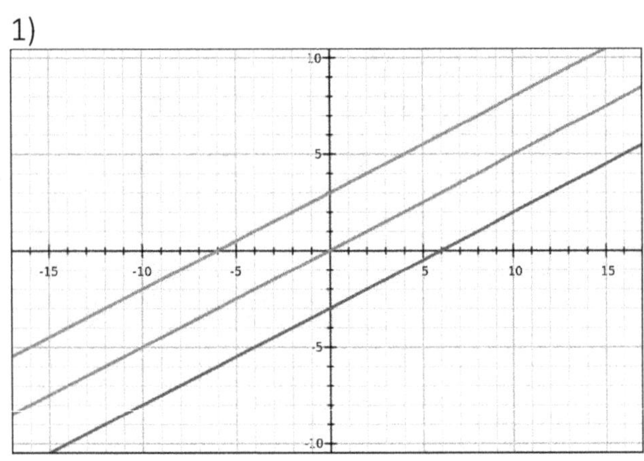

4)

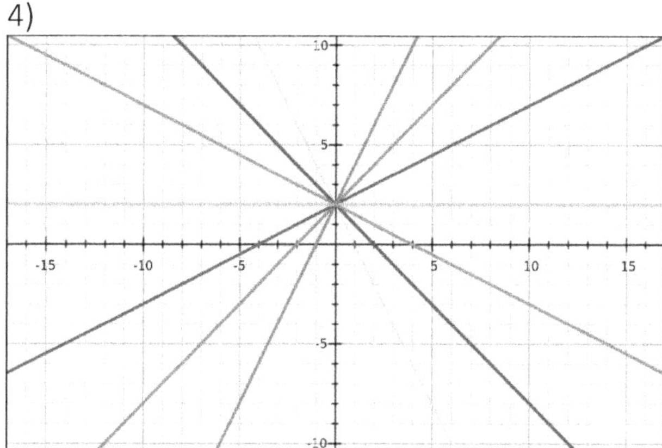

7)

equivalent equations	description of algebra
$\boldsymbol{y = 2x + 3}$	original equation
$\boldsymbol{2y = 4x + 6}$	multiplied both sides by $\mathbf{2}$
$\boldsymbol{2y - 4x = 6}$	subtracted $4x$ both sides
$\boldsymbol{-4x + 2y = 6}$	rearranged first two terms using commutative property of addition
$4x - 2y = -6$	multiplied both sides by $\mathbf{-1}$ (be sure to distribute where appropriate!)
$\boldsymbol{40x - 20y = -60}$	multiplied both sides by 10
$2x - y = -3$	multiplied both sides by $\boldsymbol{\frac{1}{20}}$

9)

$3x + 4y = 42$

11)

$\boldsymbol{y = \frac{1}{3}x + 7}$

$3y = x + 21$

$3y - x = 21$

$-x + 3y = 21$

$\boldsymbol{2x + 3y = 6}$

$\frac{2}{3}x + y = 2$

$y = -\frac{2}{3}x + 2$

12)

$\boldsymbol{x}$	$\boldsymbol{y}$
$\mathbf{0}$	2
3	$\mathbf{0}$

14)

$\boldsymbol{2x + 5y = 7}$	$\boldsymbol{5y = 4x + 8}$	$\boldsymbol{y = 5x - 9}$

standard	neither	slope-intercept
$\frac{1}{2}x + 7 = 8y$ neither	$\frac{2}{3}x - 7y = \frac{1}{2}$ standard	$4x - 7 = y$ slope-intercept
$x = 2y - 4$ neither	$y = 6$ slope-intercept	$3x = y$ slope-intercept

17)

Fill in the blanks using the word list below: The substitution method is based on the idea that at the intersection point of two lines, the y-values of both functions are the same . So at that point "y" in one function is the same as in the second function. We solve for x to find the x-value that returns the same y-value in both functions.

18)

Example	Steps
$-2x + y = 5$ $x + 3y = 1$	Make sure you've got two equations in two variables. Make sure you are being asked to solve the system, or find the intersection point, or determine when the functions have the same y-value.
$-2x + y = 5$ $y = 5 + 2x$	Use algebra to re-arrange one of the equations so that either x or y isolated. (In this example, isolate y.)
$x + 3(5 + 2x) = 1$	Substitute the expression for y in the above line in place of y in the other equation.
$x + 15 + 6x = 1$ $7x = -14$ $x = -2$	Solve the resulting equation for the single remaining variable. (In this example, solve for x.)
$y = 5 + 2(-2)$ $y = 1$	Substitute your value for x into either function to find the corresponding y value. Write your solution as an ordered pair.
$-2 + 3(1) = 1$ $-2 + 3 = 1$ $1 = 1$ ☺	Optionally, you can confirm your solution is correct by substituting the ordered pair into the other equation to see that it makes that equation true.
	Optionally, you can confirm your solution is correct by graphing the two functions and finding the intersection point visually.

19)

$-2x - 5y = -7$ $-2x + y = 4$ $2x + 5y = 7$ $-2x + y = 4$ $y = 2x + 4$ $2x + 5(2x + 4) = 7$ $2x + 10x + 20 = 7$ $12x = -13$ $x = -\frac{13}{12}$ $y = 2\left(-\frac{13}{12}\right) + 4$	$y = 2x - 5$ $4x - 3y = 1$

$y = -\frac{13}{6} + 4$ $y = \frac{11}{6}$ $\left(-\frac{13}{12}, \frac{11}{6}\right)$	
$x = 3y - 1$ $2x - y = 2$ $2(3y - 1) - y = 2$ …	$y = 3$ $y = 5x - 7$ $3 = 5x - 7$ $x = 2$ $y = 3$ $(2,3)$

22)

Example	Steps
$x = 11 - 2y$ $3x - 2y = 1$	Make sure you've got two equations in two variables. Make sure you are being asked to solve the system, or find the intersection point, or determine when the functions have the same y-value.
$x + 2y = 11$ $3x - 2y = 1$	Re-arrange the equations so that x terms, y terms and constants are vertically aligned. (Standard form works.)
$4x + 0y = 12$ $4x = 12$	In the "easy elimination" method, you'll quickly notice that adding (or subtracting) the two equations will eliminate x's or y's. Add or subtract straight down to accomplish this.
$x = 3$	Now you can solve for the single variable that remained after elimination.
$3(3) - 2y = 1$ $9 - 2y = 1$ $2y = 8$ $y = 4$	Substitute this into either equation to find $\mathbf{y}$.
$(3,4)$	Write solution as an ordered pair.
$3 = 11 - 2(4)$ $3 = 11 - 8$ $3 = 3$ ☺	Optionally, confirm your solution by substitution in the other equation.
	Optionally, confirm your solution by graphing.

24)

$4x-5y=8$ $-4x+2y=1$	$3x+4y=1$ $2x+3y=-1$
$x+y=2$ $-3x+2y=5$	$\left(\frac{8}{1}\right)\left(\frac{1}{2}x+\frac{1}{4}y\right)=\left(\frac{1}{8}\right)\left(\frac{8}{1}\right)$ $-4x+y=7$ $4x+2y=1$ $-4x+y=7$ …

LF10

1)

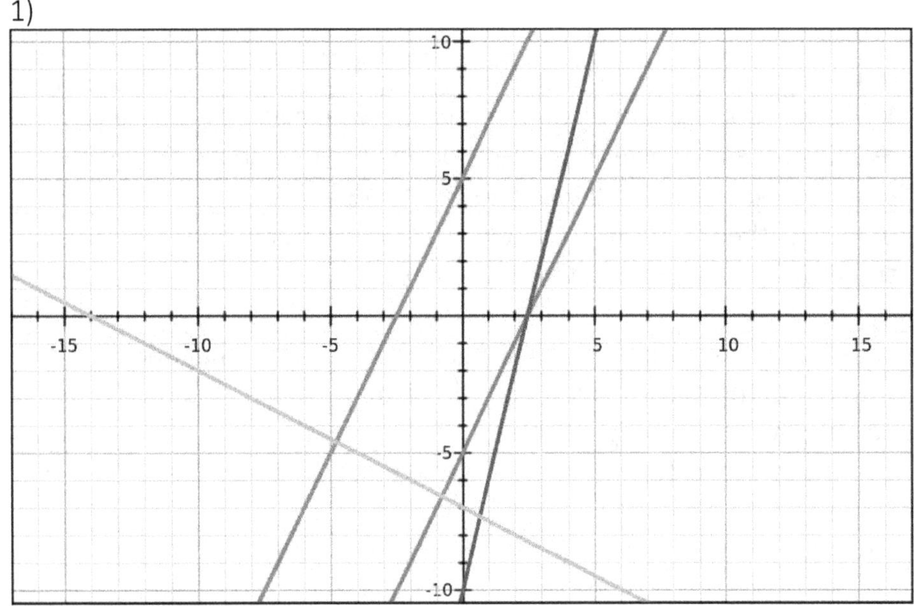

10)

Slope y_1 is $\frac{2}{5}$.

Slope y_2 is -3.

Slope y_3 is $-\frac{5}{2}$.

y_3 is perpendicular to y_1.

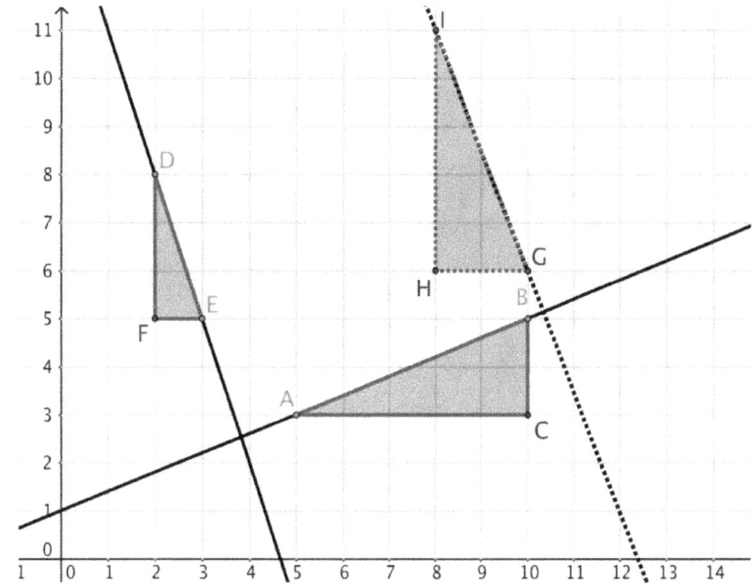

12)

Slope of y_1 is $-\frac{2}{3}$.
Slope of y_2 is $\frac{3}{2}$.
The lines are perpendicular.

13)

$y_1 = -\frac{2}{3}x + 8$

$y_2 = \frac{3}{2}x - \frac{23}{2}$

19)

$y = -\frac{7}{3}x + 3$

21)

$y = \frac{3}{7}x - 1$

22)

$y = -\frac{7}{3}x + 8$

24)

$y = -5x + 54$

LF11

3)

label	point online (x, y)	label	point offline (but with same x-value) (x, y')	$y' > y$ or $y' < y$ or $y' = y$?	Is the offline point above or below the line?
A	$(-6,2)$	A'	$(-6,8)$	$y' > y$	above
B	$(-4,3)$	B'	$(-4,6)$	>	above
C	$(-2,4)$	C'	$(-2,4)$	=	on
D		D'		>	above
E		E'		<	below
F		F'		<	below
G		G'		=	on
H		H'		>	above
I		I'		<	below
J		J'		<	below

4)
Point C, $(-2,4)$ is on the line.
$4 = \frac{1}{2}(-2) + 5$
$4 = -1 + 5$
$4 = 4$ ☺

5)
Point A, $(-6,8)$ is above the line.
$8 > \frac{1}{2}(-6) + 5$
$8 > -3 + 5$
$8 > 2$ ☺

13)

equation or inequality	shading	line dotted or solid	Why?
$y \geq \frac{4}{7}x + 5$	above	solid	both points which are on the line and points above the line make the inequality true
$y > 8x$	below	dotted	only points above the line make the inequality true
$y < -10x + 5$	below	dotted	only points below...
$y = \frac{1}{3}x + 11$	none	solid	only points on...
$y \leq 0.6x$	below	solid	points below and points on...
$y = -2x - 54$	none	solid	only points on

17)

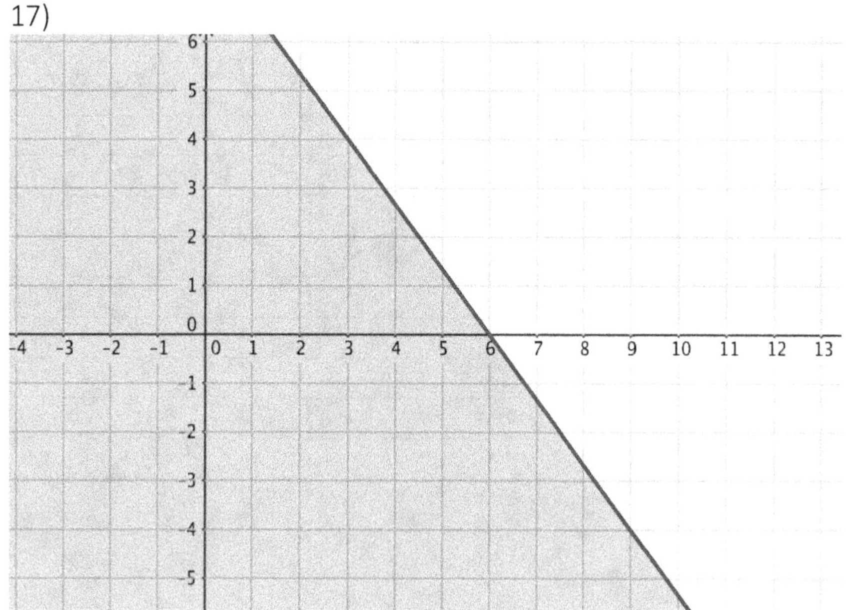

18)

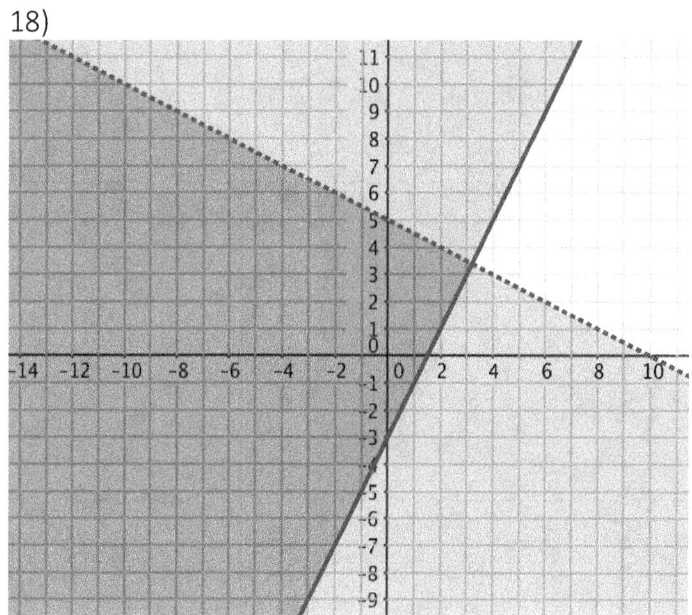

DS1

1)

your height
the temperature outside
the batting average your favorite player
the current time
the number of students in the room

2)

what we are trying to assess	what we might measure to assess this
a football player's performance	% passes completed
the healthiness of a meal	calories of the meal
the weather	the current humidity
musical skill	speed with which student correctly plays a scale
air quality	particulate %
performance in school	grades
quality of a book	number of positive reviews on amazon
talent of a hip hop artist	number of tracks sold
reliability of an airline	% crashes
quality of a chair	% returned

3)
number of people visiting google
number of coffees sold at a particular Starbucks
amount of electricity being used by the school
sulfur dioxide concentration in the air
number of people entering a national park

5)

Statistic		Additional Information Desired
Number of hours of sleep for a teenager to feel well rested.	8-10 hrs	How do they measure how well-rested a person is?
Increase in difficulty for a teen to fall asleep if you add 2 hours of evening screen time.	20%	Does that include television? How is difficulty falling asleep measured?
Average hours a day spent by teenagers online (including mobile) or watching TV.	7.5 hrs	Does that include when the phone is connected but the teen isn't looking at the phone?
Average pulse rate of teenager.	60-90 bpm	Why is the range so big?
Age at which brain reaches full maturity.	25 years	How is that measured? What is the range of values?
Reduction in life span due to smoking.	10 years	What is the relationship between quantity of smoking and life span reduction?
Reduction in life span from sitting 3 hours a day.	2 years	How do they measure sitting? Doesn't everyone sit 3 hours a day?
Grade improvement for teens from increasing exercise to recommended 1 hour per day.	One letter grade	So was this a study of students who did little or no exercise? What about students who already exercise—can they increase their grades by increasing exercise?

9)

Probably you noticed that mean was higher for the boys than the girls. This difference might vary based on age. Perhaps when you were all younger the girls achieved a greater mean height. Or perhaps as you all get older the average heights get closer, or further from each other. Where was the overall mean relative to the mean for the boys and the mean for the girls? Were there more or less boys than girls? If there had been equal numbers of boys and girls, how would the mean overall have shifted?

13)

You can find the middle 50% by dividing the data into quarters. So if there were 20 students in the class, you would say that the middle 50% is roughly from the 5th student to the 15th student. This is a rough method at the moment, but the formal way of doing this is pretty similar.

14)

This is a pretty important idea. A smaller group is a smaller sample from our classes. So the mean could be pretty far away from the actual mean for the class, or pretty close. A bigger sample (the double group) is more likely to give you a mean which is closer to the actual mean.

DS2

1)

Caffeine consumption, how recently you've eaten, how recently you've exercised, whether or not you have a cold, whether or not you are excited or tired, etc...

2)

Mass, age, gender, habits, etc...

6)

In the general population, it is thought that the mean for adolescent girls is about 5 bpm more than the mean for adolescent boys.

7)

That larger sample size probably resulted in a sample mean closer to the actual mean for the class. Probably!

10)

mean	68.3
median	69
mode	
Quartile1 (the median of the lower half)	61
Q3 (the median of the upper half)	
Interquartile Range (Q3-Q1)	
Outliers (any values that are more than 1.5 times the IQR "beyond" Q1 or Q3)	26

11)
I don't know that I would expect a person to be in the middle 50% since they just as likely to be in the box as out (hence the middle 50%). But we might expect to be in that middle range and be comforted by being close to the median. On the other hand, we also know that a resting heart rate of 60 is probably preferable to a resting heart rate of 120! So we might prefer to be more on the left of the median than on the right. I think that the only one that is concerning is the 26! Maybe 95, but even that isn't necessarily outside of acceptable. It's hard to be concerned when you only have the heart rate and not additional data!

12)
25%

14)

Highest mid-day temperature in July, 2007	21
Was that temperature an outlier?	yes
Median temperature in July, 2007	15
Median temperature in July 2008	
Range of values above the median in July 2008	16 to 26
Range of values below the median in July 2008	
Number of outliers in December 2008	
Range of values in July 2009	13 to 26
Month, year with greatest range of values	
Month, year with smallest range of values	Sept 2008
Month, year with lowest temperature recorded	
Month, year with highest temp recorded	
Month, year with greatest IQR	May 2008 ?
Month, year with smallest IQR	June 2008 ?
Two adjacent months with almost the same median	

16)

Temps seem more predictable and tightly packed in August September October. December Jan Feb temps seem more spread out. Obviously the temps go up in summer, down in winter.

17)

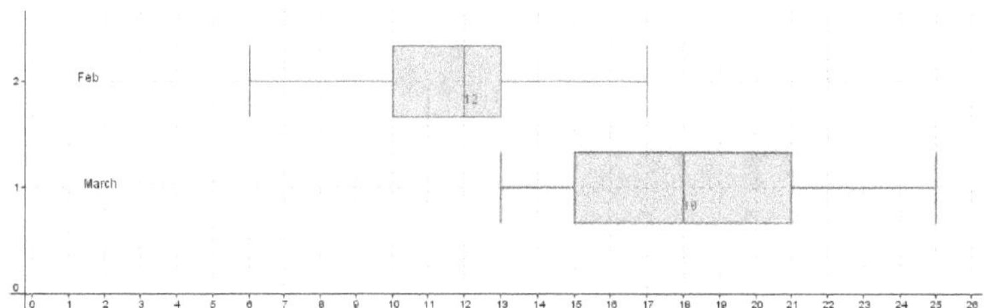

18)
Obviously warming up in March, and more variation in March.

19)

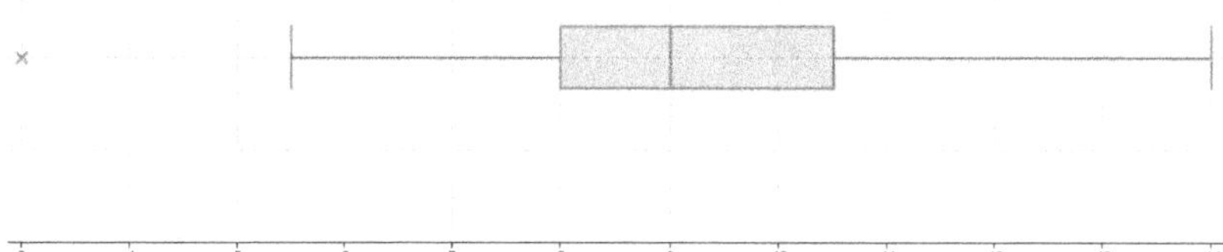

DS3

2)

$$Mean = \frac{0.5 + 0.7(3) + 0.8(4) + \cdots}{40}$$

3)
Pretty evenly distributed. Nobody wants to be selling coffee too high (nobody will buy) or too low (they won't make any money).

5)

price	freq.	prob.
0.5	1	$\frac{1}{40}$
0.7	3	$\frac{3}{40}$
0.8	4	
0.9	7	
1.0	10	$\frac{1}{4} = 0.25$
1.1	8	
1.2	4	
1.3	2	
1.4	1	

9)
Outcome of 3 heads occurred 6 times.
$\frac{6}{40} = \frac{3}{20}$ was the relative frequency.
The probability is $\frac{6}{40} = \frac{3}{20} = \frac{15}{100} = 0.15 = 15\%$

10)

Note how the graph of the probability distribution has the same shape (just squished vertically) as the original frequency plot.

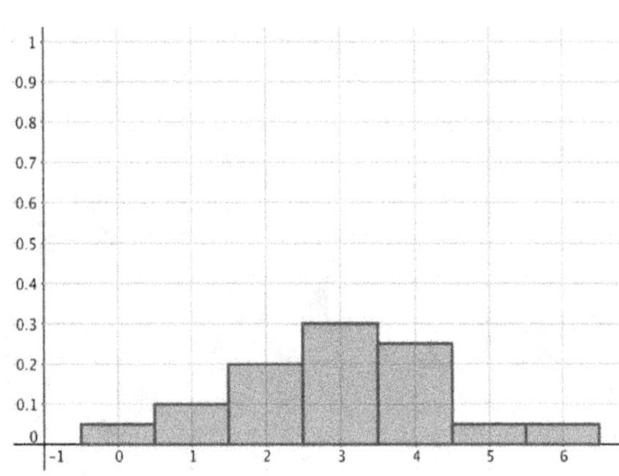

12)
Doesn't look normal. Not at all symmetrical around the mean and not consistently decreasing as you go further from the mean. Mean is 1.6 but frequency increases as you go to 1.

13)

# heads	Prob.
0	$\frac{5}{20} = \frac{1}{4} = 0.25$
1	
2	
3	
4	
5	
6	

16)
Two heads 27 times. $\frac{27}{100} = 0.27$. Looks somewhat normally distributed. Tails could be more symmetrical.

17)
Two heads 340 times. Probability: $\frac{340}{1500} = \frac{34}{150} = 0.2\overline{26} \approx 0.227 = 22.7\%$
This looks normally distributed. Mean in the middle, symmetrical tails either side.

18)
Probability of 4 heads $\approx 0.2344 = 23.44\%$

24)
$0.81 \leq price \leq 1.17$

25 espressos had prices between these two values. You may have gotten a different answer if you rounded or if you took some fraction of the espressos that cost 0.80 or 1.20. 25 espressos gives a probability of $\frac{25}{40} = \frac{5}{8} = 0.625 = 62.5\%$.

27)
The mean is 41.6. With a standard deviation of 2.2 That gives us a ± 1 SD range of:

$39.4 \leq size \leq 43.6$

30 pairs of shoes had sizes of 40, 41, 42 or 43. That gives a probability of $\frac{30}{50} = \frac{3}{5} = 0.6 = 60\%$.

28)
$37.2 \leq size \leq 45.8$
$\frac{47}{50} = 0.94 = 94\%$

31)
$\frac{10}{50} = \frac{1}{5} = 0.2 = 20\%$

32)
$\frac{8}{50} = \frac{4}{25} = 0.16 = 16\%$

33)
$\frac{41}{50} = 0.82 = 82\%$

DS4

19)
The standard deviation should be smaller when the sample size increases. If you have a small sample size, random variability can give you a wider range of results. For example, if you have a sample size of 3 you could pretty easily get no green candies, but that isn't reflecting reality. But if you have a sample size of 100, it's less likely that you'll get no green candies. In general, outcomes that are further from the truth are less likely as sample size increases.

20)
The percentage of ratios within ± 1 SD of the mean should be close to 68%. This number doesn't change as long as the data is distributed roughly normally.

22)
<u>20 students</u>
$\frac{12}{20} = \frac{3}{5} = 0.6 = 60\%$
The mean is pretty close to the mode, though they are different.
Very roughly normal. (Doesn't decrease consistently and symmetrically on either side of mean...)
<u>50 students</u>
$\frac{40}{50} = \frac{4}{5} = 0.8 = 80\%$
Mean is much closer to mode.
Roughly normal. (Not quite symmetrical...)

25)
They all look roughly normal. As the sample size increases, the distributions cluster more tightly around the mean, and the standard deviation gets smaller. However, the percentage of points between ± 1 SD of the mean stays roughly the same.

27)
$\frac{62}{100} = 62\%$
$\frac{61}{100} = 61\%$
$\frac{68}{100} = 68\%$
Pretty close to each other. The last one has the smallest SD therefore provides a better estimate.

28)
About the same mean but a much smaller standard deviation.
29)
Because ± 1 SD encloses about 68% of the sample ratios, the chance that any single ratio from any single sample is between those values is 68%.

30)

	What happens when you increase sample size?	What happens when you increase the number of samples?
standard deviation	The standard deviation gets smaller. Our 68% confidence in our estimate for the parameter for the population is within a smaller range of values.	It has no impact on the standard deviation, but it does make the plot of the data from the samples look more obviously bell shaped.
sketch (before)	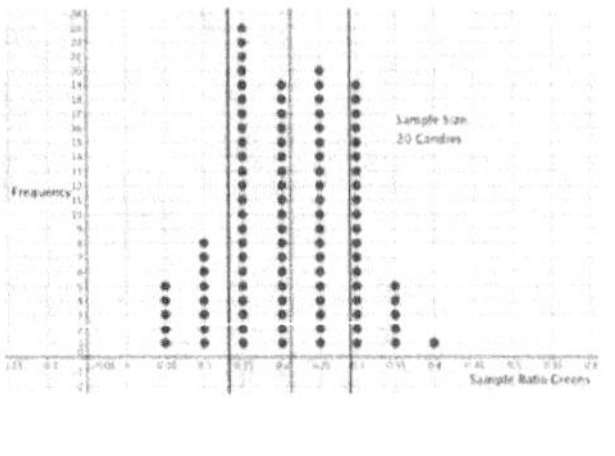	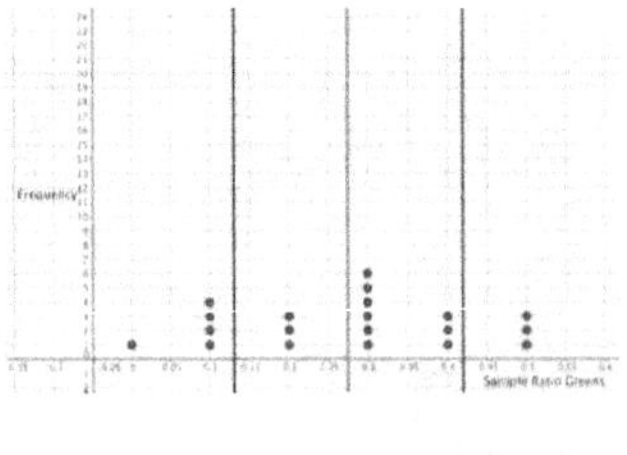

sketch (after)	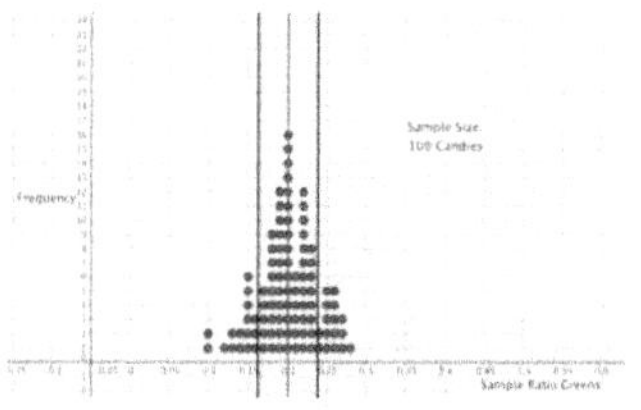	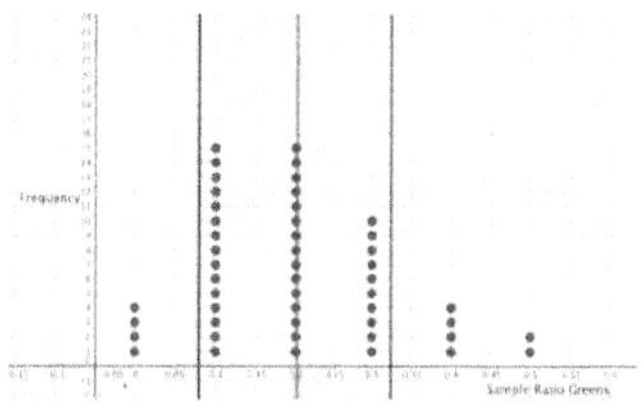
description	As the sample size increases, the sample ratios tend to be closer to the mean.	As the number of samples increases the standard deviation does not change.

31)
2.06 meters is the most common. 2 is more common than 1.9, which you can see because the curve which shows the frequency of this height is higher at 2 than at 1.9.

33)
-1 SD $= 1.98$
$+1$ SD $= 2.14$
about 68% of the heights are between these two values.
Area under the whole curve is 100%.

34)

mean	average.
standard deviation	A measure of variation from the mean.
± 1 SD	The values which are one standard deviation from the mean in both directions. Values between ± 1 SD comprise about 68% of the sample.
sample	A selection of values from the population. For example, the masses of each of the fries in your sample.
statistic	A summary number about a sample. For example, the average mass of 20 french fries you bought.
population	All the data you are selecting from. For example, the masses of all the fries in the world.
parameter	A summary number about the entire population. For example, the average mass of all the fries in the world.
increasing sample size	The standard deviation decreases if you are drawing from a roughly normal population. For example, if we take a bigger sample of fries, they will be distributed more tightly around the mean of the sample.
increasing number of samples	The standard deviation doesn't change.
estimating parameter of a population with 68% confidence	We can say there is a 68% probability that the parameter is between ± 1 SD of the statistic from the sample.

35)
Mean = 36.6 grams
SD= 1.16 grams

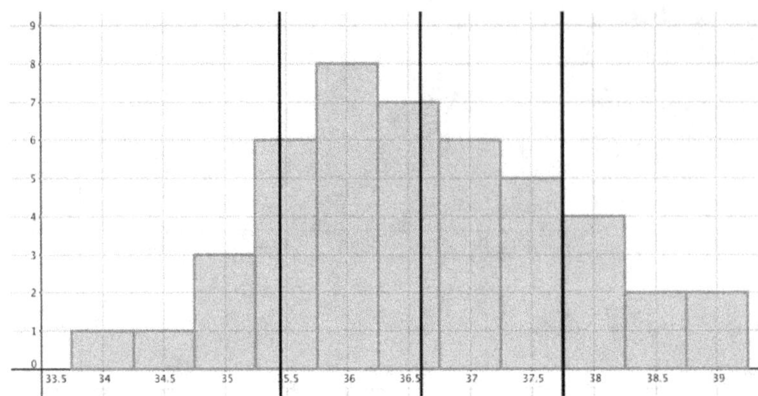

36)
35.45 and 37.75

DS5

4)

Correlation	A relationship evident in the graph. As one variable changes the other changes correspondingly.
Strong Positive	As age increases in youth, vocabulary increases.
Strong Negative	As marathon training increases, resting heart rate decreases.
Weak Positive	As maximum education level increases, salaries increase.
Weak Negative	As time passes the cost of computers decrease.
No correlation	The more vegetables I eat, the taller my students get.

5)
Doesn't look like much of a correlation. As height increases the points don't increase. We do notice that there are a few players at the extremes of height, but the scores of players that are short and that are tall vary similarly. Maybe you could say that there is a a bit more variation in the middle heights, but that doesn't mean there is a correlation between height and scoring. We would need to see evidence that as one increases, the other clearly increases or decreases

6)
There definitely seems to be a strong positive correlation between rebounds and height. It seems like as the height increases, the rebounds increase. In fact, the rebounds seem to accelerate, that is, increase at a faster rate, as height increases. that makes it seem quadratic rather than linear.

7)
Roughly seems like an increase of 8 rebounds for an increase in height of 10 inches. Which reduces to 0.8 rebounds per inch.

8)
There are lots of other factors! That's one reason there is still variation even though rebounds and height are strongly correlated. Other factors would be just player skill, playing time, age of players, strength, mass, etc. All of these contribute to variation. Some of them might actually correlate with height though, like mass definitely would correlate with height. So then the question is, are the rebounds caused by the height, or by the mass, or both, or neither? I think increasing height does cause an increase in rebounds, because being taller does make it easier to pull down a rebound.

9)

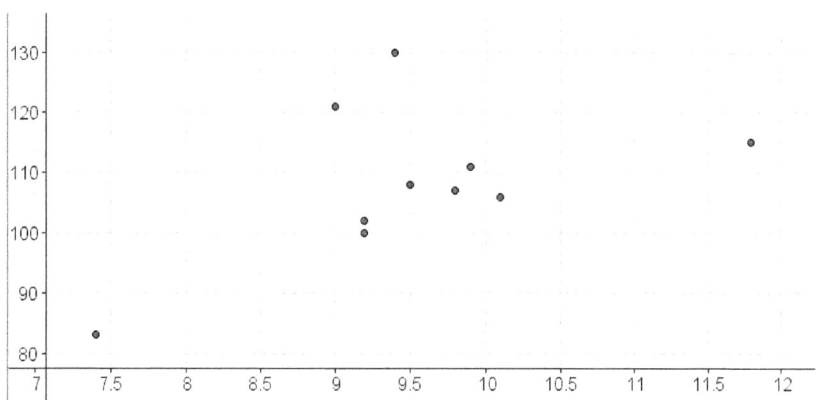

10)
I doubt that one causes the other. Seems like both generally have increased over time, so time has caused them both to increase. More specifically increases in both are probably caused by increased demand (more people with more money wanting more food) and perhaps by improved farming techniques. It also might be true that a bad year for one is matched by a bad year for the other, due to weather variation (even though coffee isn't grown much in the US).

11)
Strong negative correlation.

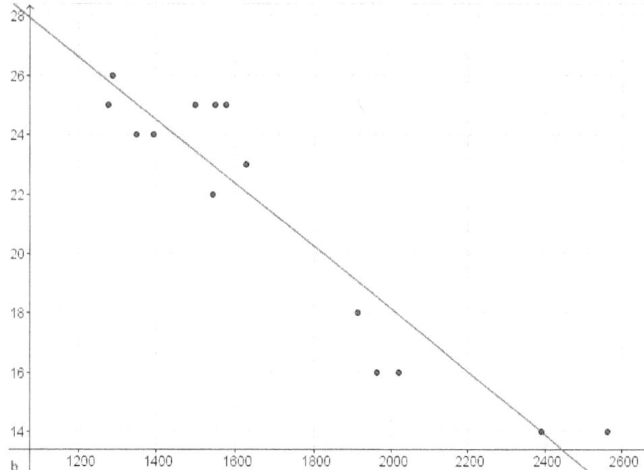

12)
Probably. Seems like as the mass increases in our sample, the fuel efficiency decreases. The increased mass requires more fuel to accelerate, and probably presents more surface area and thus suffers more air resistance.

16)
$y = -0.01x + 39$

17)
$r = -0.94$

20)
Pretty strong positive correlation.

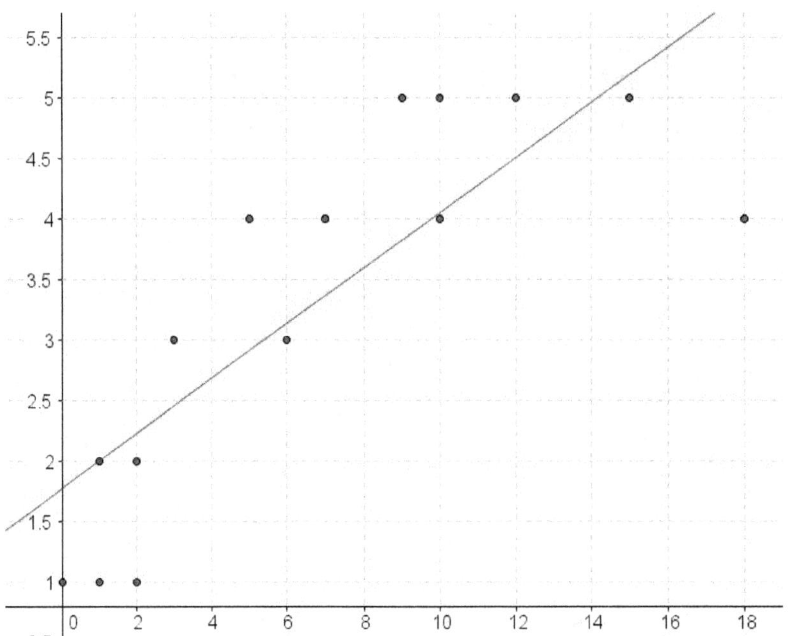

21)
$y = 0.23x + 1.78$
$r = 0.81$

23)
Potential problems: People are reporting on how many novels they read and they might not be giving accurate answers. There is no controlling for the quality of the novels. There is no accounting for other reading they might have done. Or writing. The vocab test could be flawed. Other factors influence vocab development, like quality of the novels, or amount of writing, or even listening, or learning a foreign language, or being on the debate team. Suspicious data points as indicated above.

24)

1-variable vs. 2 variable statistics	Sometimes we are interested not just in one random variable, but in two random variables and how one affects the other.	1 var: Number of windows in restaurants. 2 vars: Number of windows in restaurants and cost of a Coca-Cola in that restaurant.
correlation	A relationship between two variables.	As the amount of spray paint sold increases, the amount of graffiti increases. This doesn't necessarily imply causality.
causality	A relationship between two variables where the value of one variable has a direct influence on the value of the other.	As the number of donuts I eat increases, my times in 10K runs increase.
linear regression	A statistical process of finding the line of best fit for 2-variable data.	See the chapter for examples.
line of best fit	The line which minimizes the distance between the data and the line.	
trend line	Line of best fit.	
correlation coefficient	A measure of how well the line fits the data. Bigger than 0.8 is strong positive correlation, between 0.5 and 0.8 is weak positive.	

EQ1

1)

A genie takes the number of m and m's you give him, multiplies them by two and adds 3.	$y = 2x + 3$ x is input, the number of m and m's you give him. y is output, the number he returns.
A gym membership costs 20 euros to start and then 15 euros a month after that.	$y = 15x + 20$ x is the number of months, y is the total cost you've paid to be in the gym for that many months.
A wingsuit flier descends 2 meters vertically for every 5 meters of horizontal movement. He jumps from a height of 500 meters.	$y = -\frac{2}{5}x + 500$ x is the horizontal distance travelled by the flier, y is the vertical distance.
A car is racing at 150 km per hour with a 100 km head start over its competitors.	$y = 150x + 100$ x is the number of hours of travel, y is the distance travelled including the headstart.
The cost of a special variety of apple in 2000 (year 0) was 10 cents. Now, in the year 2015, the same apple costs 50 cents.	$y = \frac{40}{15}x + 10$ $y = \frac{8}{3}x + 10$ x is the year (assuming 2000 is year 0) and y is the cost.
A scatterplot shows that as morning rainfall in cm increases, more students are late to school.	$y = 10x + 15$ x is the amount of rainfall. y is the number of students late. You may have made different assumptions, but I am assuming a slope of 10, which is 10 more students late for every cm of increase in rainfall. I am also assuming that even if rainfall is 0 there are 15 students late.

2)

Equation	Meaning	Example Equation	Type of function
$D = \frac{m}{V}$	Density is mass over Volume	D as a function of V: $D = \frac{6}{V}$	rational
$E = mc^2$	Energy is mass times the speed of light squared	E as a function of m: $E = (9 \cdot 10^{16})m$	linear
$F = ma$	Force is mass times acceleration	F as a function of m: $F = 5m$	linear
$C = \pi d$	Circumference of a circle is pi times the diameter	C as a function of d: $C = \pi d$	linear
$s = \frac{1}{2}at^2 + v_i t$	Displacement is half the acceleration times time squared, plus initial velocity times time	s as a function of t: $s = 5t^2 + 12t$	quadratic

$y = \sqrt{x} + d$	this is an abstract function, no particular meaning in this case	y as a function of x $y = \sqrt{x} + 17$	square root
$y = 2^x + e$	this is an abstract function, no particular meaning in this case	y as a function of x: $y = 2^x + e$	exponential
$y = a(3)^x + 5$	this is an abstract function, no particular meaning in this case	y as a function of x: $y = 15(3)^x + 5$	exponential

4)

candies	Genie	Magic Pebbles	Expression	Value
5	doubler (times 2)	3	$5(2)^3$	40
10	tripler (times 3)	2		
7	times 4	1		
100	times $\frac{1}{2}$	4	$100\left(\frac{1}{2}\right)^4$	$\frac{25}{4} = 6\frac{1}{4}$
12	times 1.5	3	$12(1.5)^3$	40.5
8	times 10	4		$8 \cdot 10^4$
1	times 6	3		
1900	times 1	8		1900
900	times $\frac{1}{3}$	5		
114	times 2.5	x	$114(2.5)^x$	
130	times 15	x		

5)

a is the initial amount, the amount you have when x is 0. b is the multiplier. if the multiplier is above 1 the function is increasing, if it is below 1 (and above 0) the function is decreasing. x is the input, for example, the number of magic pebbles you bring, or the number of years you keep the money in the bank to earn interest.

6)

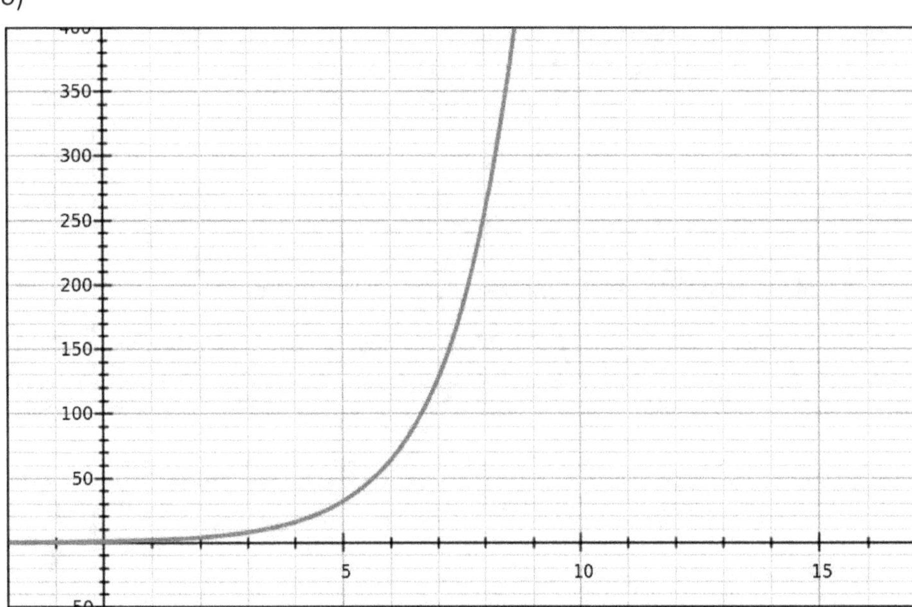

7)

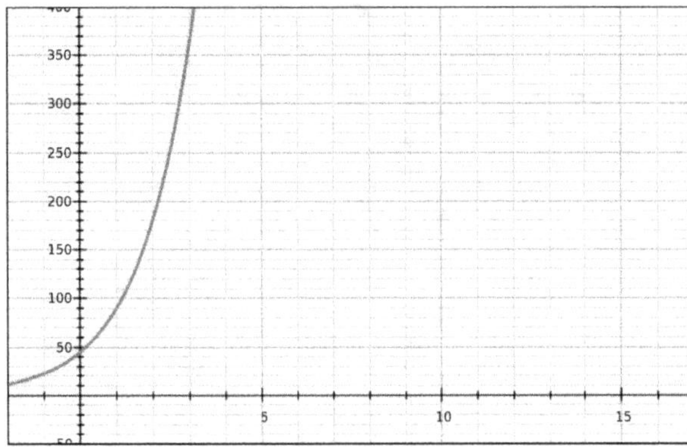

8)

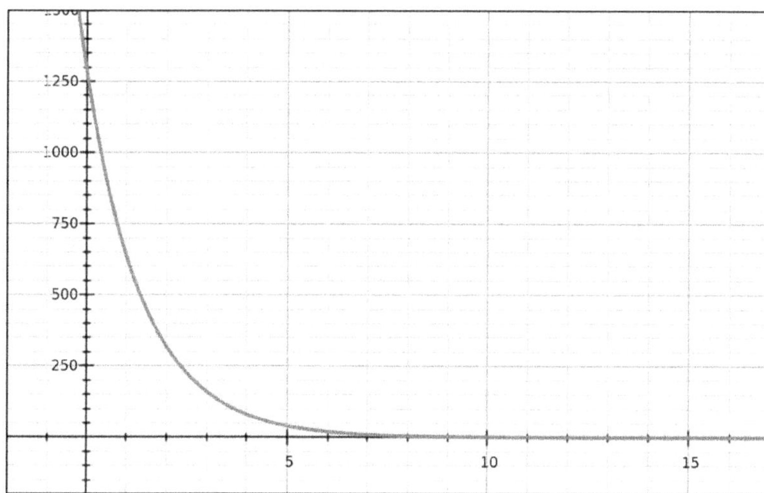

9)

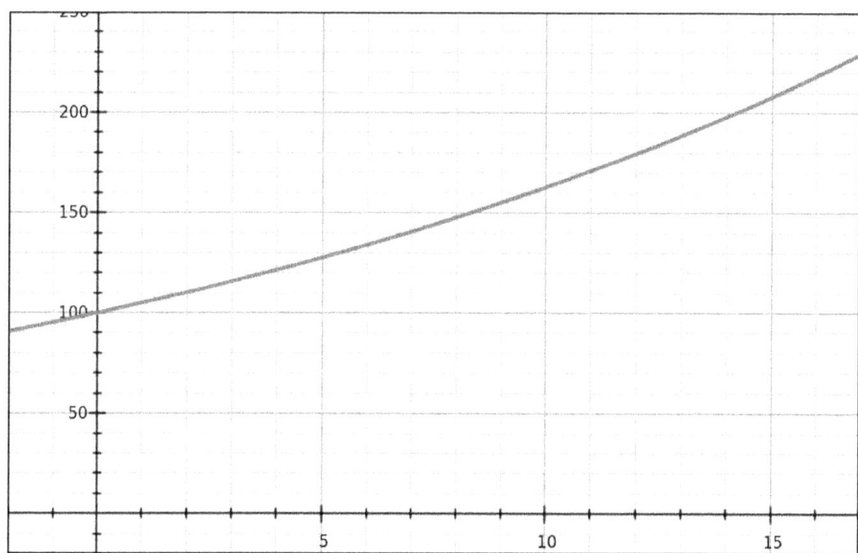

10)

Scenario	Equation
A genie takes your candies, multiplies them by 3 and adds 20.	y=3x+20
A doubling genie has 5 candies and doubles them based on the number of magic pebbles you give him.	$y = 5(2)^x$
A car is racing at 140 km/hr with a 200 km head start.	$y = 140x + 200$
A gym costs 45 euros per month with an initial cost of 25 euros.	$y = 45x + 25$
60 bacteria double every day.	$y = 60(2)^x$
100 bacteria triple every day.	
3000 bacteria increase by 14% every day.	$y = 3000(1.14)^x$
2450 bacteria die off by $\frac{1}{2}$ every day.	
1200 bacteria decrease by 70% every day.	$y = 1200(.30)^x$
1500 dollars are put in a bank account and the value increases by 8% every year.	$y = 1500(1.08)^x$
A genie takes the square root of the number of candies you give him then adds 500.	$y = \sqrt{x} + 500$
A genie squares the number of candiess you give him then subtracts 12.	

11)

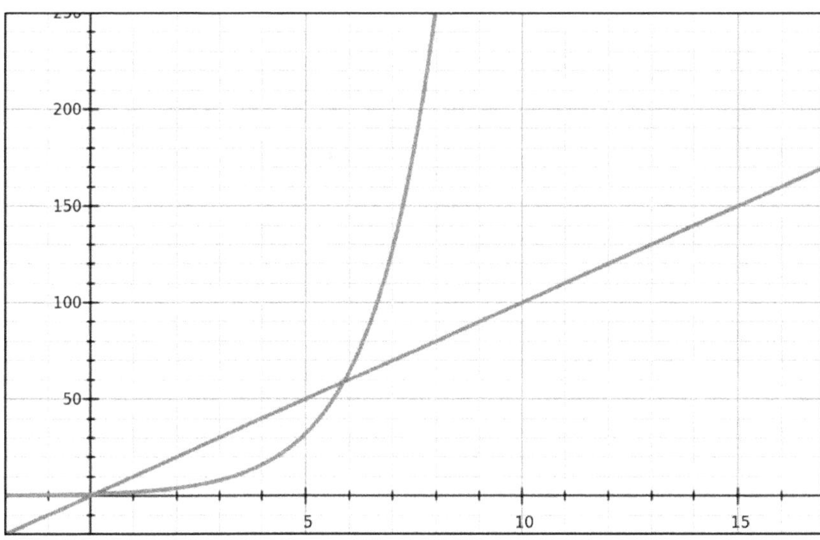

15)

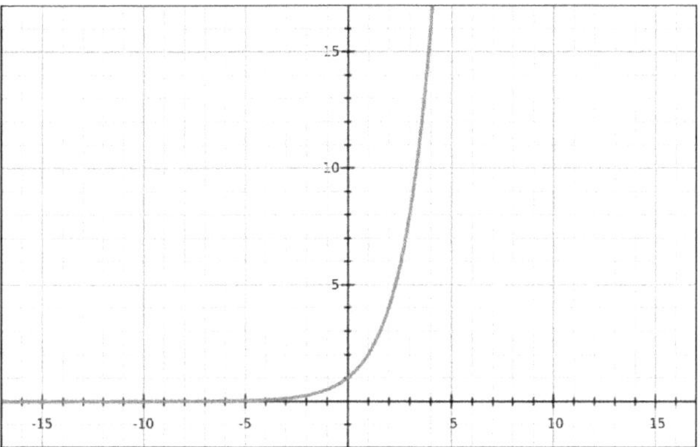

17)

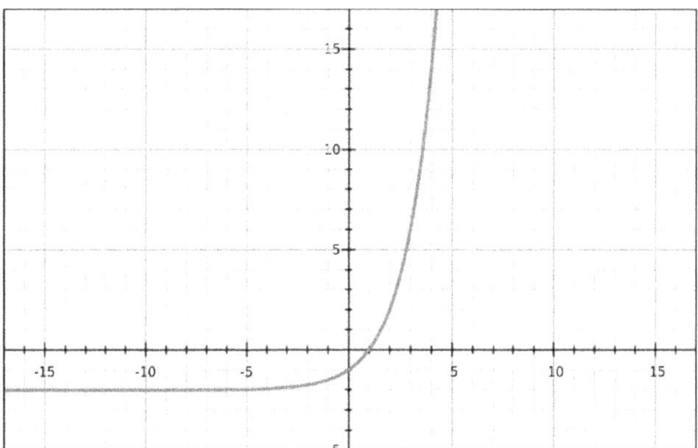

EQ2

1)

Year	Larry's Plan (Euros)	Year	Elizabeth's Plan (Euros)
0	300	0	300
	400		330
	500		363
	600		399.3
x		x	
30	3300	30	5234.8

2)

$y_L = 100x + 300$
$y_E = 300(1.1)^x$

4)

The linear function increases at a constant rate, and that rate of increase is faster than for the exponential function when the x values are small. The exponential function increases at a faster and faster rate, so that when x is bigger the exponential function eventually overtakes the linear function.

6)

The quadratic function is still winning after 5 years, because the exponential function is increasing very slowly and losing to the linear function. The linear function is just plodding along with its constant slope.

7)

The quadratic function is still ahead after 30 years, but the exponential function is now increasing faster than it was before and has passed the linear function.

8)

The exponential function overtakes the quadratic function! All exponential functions with a multiplier greater than 1 will eventually overtake any quadratic function.

9)

Name	Function	Type	Meaning of numbers and variables
Larry	$y = 100x + 300$	linear	x is the number of years that have passed and y is the amount of money at that time. The initial amount (for $x = 0$) is 300. The slope is 100—that's the amount of increase in money every year.
Elizabeth	$y = 300(1.1)^x$	exponential	x and y mean the same as above. The initial amount (for $x = 0$) is 300. The increase is 10% every year, which means you multiply by 1.1. That's so that every year you have 110% of what you had the previous year.
Q-Bert	$y = 300x^2$	quadratic	x and y mean the same as above. The initial value is 300 but that's actually only the case at $x = 1$. At $x = 0$ you still have \$0.

12)

$y = 100 + 2x$ linear Now. Actually this is the worst one at all stages, but if you have to go to this one at some point, get it out of the way early.	$y = 100(2)^x$ exponential Long term. In 30 years you'll have more money than anyone on earth.	$y = 100x^2$ quadratic Medium Term. It's already behind the exponential function at this point, but not by a ton.
$y = 20(5)^x$ exponential Long term, probably. Once again, you'll be the richest person in the world in 30 years. But the other ones return trivial amounts of money...	$y = 20 + 5x$ linear Medium Term, I suppose.	$y = 20(5)^2$ Now. This is a constant function, the amount returned never changes.
$y = 600\left(\frac{1}{2}\right)^2$ Constant. Medium term by process of elimination.	$y = 600 + \frac{1}{2}x$ Linear. Long term, but that doesn't really help you that much.	$y = 600\left(\frac{1}{2}\right)^x$ Exponential. Now. This one loses money as time passes.

17)

The first one is linear because it's increasing at a constant rate. For every increase of 1 for x, y increases by 2. The function is $2x + 15$ because the initial value is 15.

The second one is exponential. Each value for y is the previous value multiplied by 2. The function is $g(x) = 15(2)^x$ because the initial value is 15.

The third one is quadratic, $h(x) = 15x^2$.

19)
$y = x^2$ because you can see $(0,0)$; $(1,1)$; $(2,4)$; $(-2,4)$ and so on on the graph.

$y = x + 5$ for the linear function because it has a y-intercept of 5 and a slope of 1.

23)
$y = x^2 - 7$
$y = 3x$

24)
$y = 10(1.5)^x$
$y = 5x + 20$

EQ3

1)

x	y_1	y_2
0	600	1600
1	672	1672
2	752.64	1752.64
3		
4		
6		

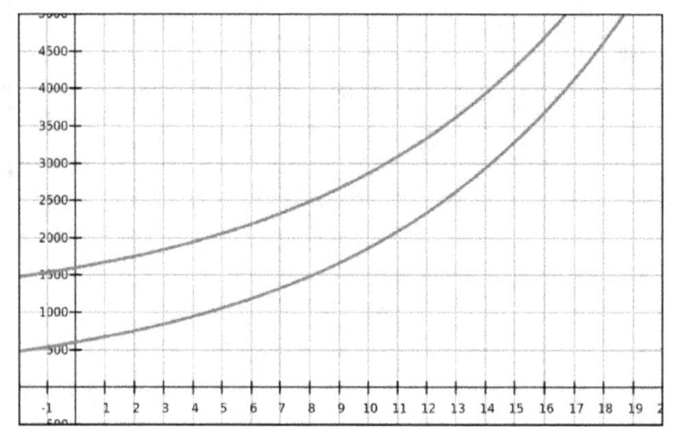

3)

11)
$f(x) = x^2$
$i(x) = (x + 4)^2$

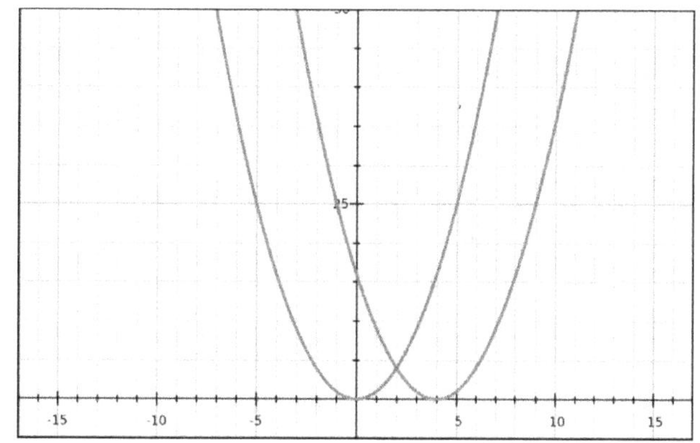

12)

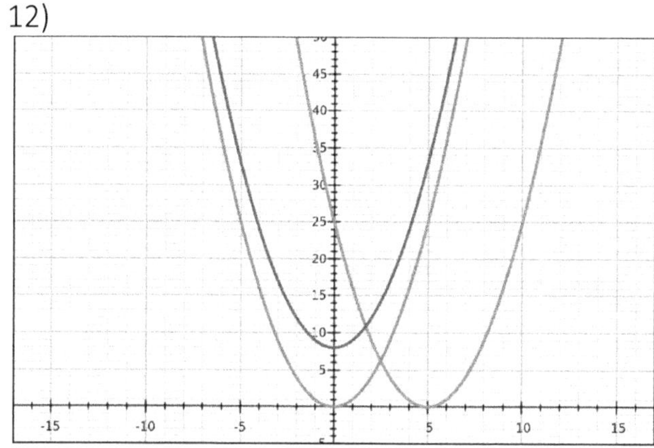

The pre-image is $f(x) = x^2$. The image $i(x) = x^2 + 8$ is a vertical shift of 8 units. The function $f(x)$ produces y-values which are then increased by 8 units, thus shifting the graph upwards. $j(x) = (x - 5)^2$ is a horizontal shift of 5 units. One way to understand this counter intuitive result is that the vertex of $f(x) = x^2$ is (0,0). For $j(x) = (x - 5)^2$ we need to input an x-value of 5 in order to produce the same minimum y-value of 0. So if the vertex for $j(x)$ is (5,0), that represents a horizontal shift of 5 units.

15)

<table>
<tr><th></th><th>f</th><th>g</th><th>h</th></tr>
<tr><td>Pre-image</td><td>$f(x) = x^3$</td><td><table><tr><td>x</td><td>y</td></tr><tr><td>2</td><td>7</td></tr><tr><td>5</td><td>1</td></tr></table></td><td></td></tr>
<tr><td>Image</td><td>$i(x) = (x + 8)^3$</td><td><table><tr><td>x</td><td>y</td></tr><tr><td>-6</td><td>7</td></tr><tr><td>-3</td><td>1</td></tr></table></td><td></td></tr>
</table>

16)

<table>
<tr><th></th><th>f</th><th>g</th><th>h</th></tr>
<tr><td>Pre-image</td><td>$f(x) = \sqrt{x}$</td><td><table><tr><th>x</th><th>y</th></tr><tr><td>4</td><td>1</td></tr><tr><td>−7</td><td>4</td></tr><tr><td>12</td><td>−2</td></tr></table></td><td>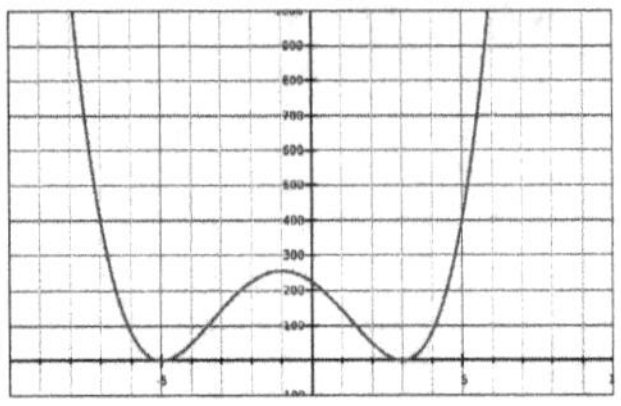</td></tr>
<tr><td>Image</td><td>$i(x) = \sqrt{x-7}$</td><td><table><tr><th>x</th><th>y</th></tr><tr><td>2</td><td>1</td></tr><tr><td>−9</td><td>4</td></tr><tr><td>10</td><td>−2</td></tr></table></td><td>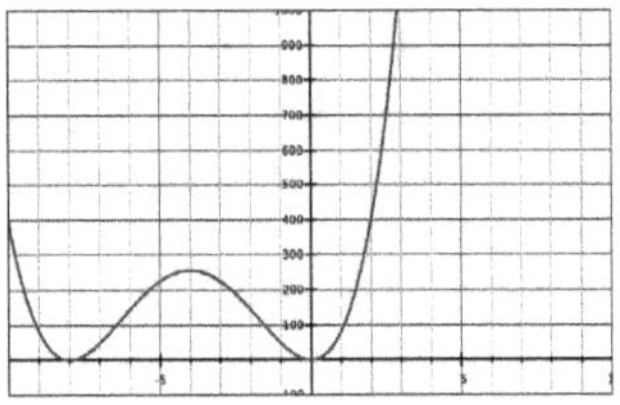</td></tr>
<tr><td>Description</td><td>horizontal shift 7 units</td><td>horizontal shift −2 units</td><td>horizontal shift −3 units</td></tr>
</table>

17)

$g(x) = x^2$	vertical shift 4 units and horizontal shift −17 units	$i(x) = (x-4)^2 - 17$

20)

For $j(x) = (x-3)^2 + 6$, the vertex is (3,6). This is because the lowest y-value this function can produce is 6, and that y-value is produced by "neutralizing" the $(x-3)^2$. This is accomplished by inputting an x-value of 3. Any other x-value won't neutralize the $(x-3)^2$. (Try other values to confirm this.) So the method of finding the vertex here is, determine the x which neutralizes the squared component of the function, then take the value "outside" as the y-value.

21)

function	after horizontal shift 12 only	after vertical shift −4 only	after both
$y = h(x)$	$y = h(x-12)$	$y = h(x) - 4$	$y = h(x-12) + 4$
$y = p(x)$			

23)

function	$f(x+5)$	$f(x)+9$	$f(x+5)+9$
$y=f(x)=x^3$	$y=(x+5)^3$	$y=x^3+9$	$y=(x+5)^3+9$
$y=f(x)=2x+7$	$y=2(x+5)+7$	$y=2x+7+9$	$y=2(x+5)+7+9$
description of transformation	horizontal shift -5 units	vertical shift 9 units	horizontal shift -5 units and vertical shift 9 units

24)

function	$f(x+5)$	$f(x)+9$	$f(x+5)+9$
$y=f(x)=x^2+3x$	$y=(x+5)^2+3(x+5)$	$y=x^2+3x+9$	$y=(x+5)^2+3(x+5)+9$
$y=f(x)=7x^3$	$y=7(x+5)^3$	$7x^3+9$	$7(x+5)^3+9$
$y=f(x)=(x+4)^2$	$(x+5+4)^2$	$(x+4)^2+9$	$(x+5+4)^2+9$
$y=f(x)=8\sqrt{x}$	$8\sqrt{x+5}$	$9+8\sqrt{x}$	$9+8\sqrt{x+5}$
$y=f(x)=-2x+1$	$-2(x+5)+1$	$-2x+1+9$	$-2(x+5)+1+9$
$y=f(x)=2(3)^x$	$2(3)^{x+5}$	$2(3)^x+9$	$2(3)^{x+5}+9$

EQ4

1)

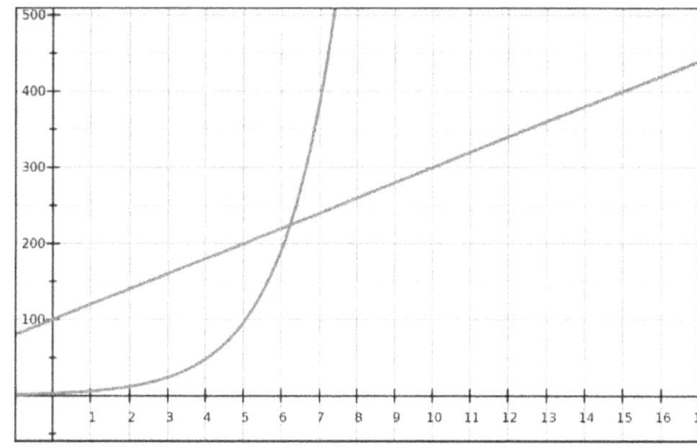

$y=3(2)^x$
After 16 days, 196608 bacteria.
This is an exponential function

2)
$y = 20x + 100$
This is a linear function.
Exponential function surpasses linear function on day 7. Exponential has 384 vs 240 bacteria on day 7.

4)

Scenario	meaning of x	meaning of y	Equation
A magic genie takes the candies you have in your pocket, multiplies them by 4 and adds 15, then returns them to you.	the number of candies in your pocket	the number of candies the genie returns	$y = 4x + 15$
57 bacteria in a petri dish double every day.	days	# bacteria	$y = 57(2)^x$
A car is racing at 180 km/hr with an 80 km head start.	hours	km traveled	$y = 180x + 80$
A gym costs 35 euros per month with an initial cost of 17 euros.	months	cumulative cost	$y = 35x + 17$
60 bacteria triple every day.	days	# bacteria	$y = 60(3)^x$
$3000 in a bank account increases by 14% every year.	years	money in account	$y = 3000(1.14)^x$
2450 bacteria die off by half every day.	days	# bacteria	$y = 2450\left(\frac{1}{2}\right)^x$
Your friend is much smaller than you, and eats half the calories that you eat for lunch, plus a 200 calorie soda.	calories I eat for lunch	calories my friend eats for lunch	$y = \frac{1}{2}x + 200$
1800 bacteria decrease by 60% every day.	days	#bacteria	$y = 1800(.4)^x$
1500 dollars are put in a bank account and the value increases by $400 every year.	years	money in account	$y = 1500 + 400x$

5)
For a linear function, $y = mx + b$, b is the y-intercept. It is the value which is returned for an input of 0. For an exponential function, $y = a(b)^x$, a is the y-intercept, the value which is returned for an input of 0. they both represent the initial value for the function. In the linear function, b is added to a linear term. In the exponential function, a is multiplied times b (a multipier) raised to a power.

6)

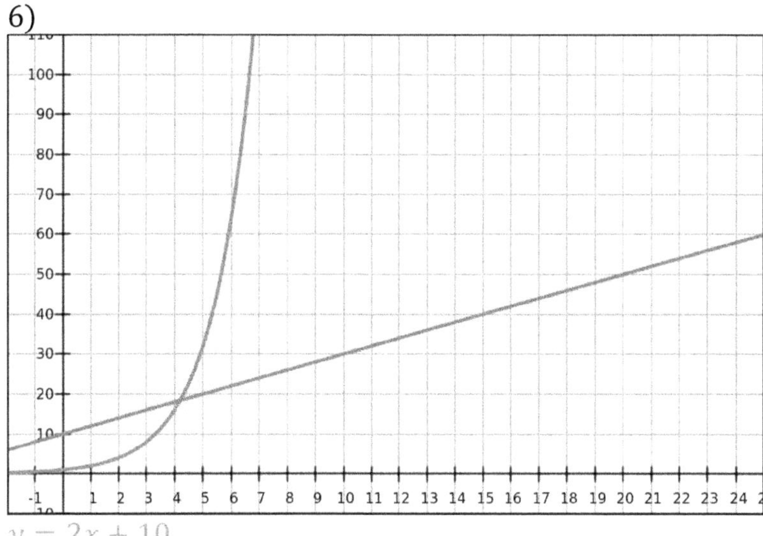

$y = 2x + 10$

$y = 1(2)^x$

game 5.

8)

For the linear function, when the game number increases by 1, the points scored increase by the addition of 2 points. For the exponential function, when the game number increases by 1, the points scored increase by multiplication by 2.

9)

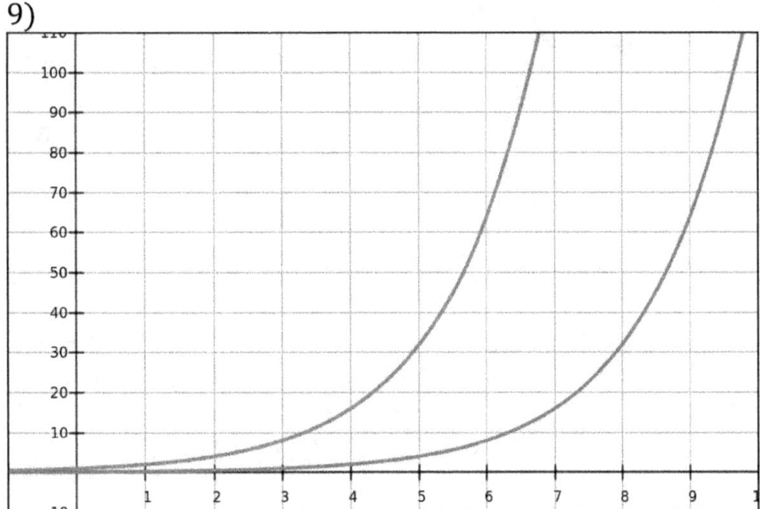

10)
point A: (0,1)
point B: (3,1)
Point B is 3 units to the right of point A. This is because, in order to generate the same y-value as in $f(x)$, we have to plug in 3 to $i(x)$ instead of 1. For any point you choose from $f(x)$, if we plug in an x value 3 units greater into the $i(x)$, we get the same y-value. In fact, we can generalize this:

$f(x) = 2^x$

For example, suppose $x = c$. That means, $f(c) = 2^c$.
That also means:
$i(c + 3) = 2^{c+3-3}$
$i(c + 3) = 2^c$

So when we put c into f, we get 2^c
And when we put $c + 3$ into i, we get 2^c, the same y for a different x.

Visually this means a horizontal shift, which you can totally see in the tables. the first table starts with 0,1,2,3 while the second starts with 3,4,5,6 returning the same y-values.

Of course you can also see it in the graph. Pick any y-value on the first graph, walk three units right and you'll be on the second graph.

11)

pre-image	transformation	image
$f(x) = 4(3)^x$	horizontal shift 7 units	$i(x) = 4(3)^{x-7}$
$g(x) = 12x - 8$	horizontal shift -9 units	$j(x) = 12(x + 9) - 8$
$h(x) = 1.5^x$	horizontal shift -8 units	$y = 1.5^{x+8}$
$p(x) = 3x + 5$	horizontal shift 12 units	$y = 3(x - 12) + 5$
$q(x) = 5(0.9)^x$	horizontal shift -19 units	$y = 5(0.9)^{x+19}$
$g(x) = \frac{1}{5}x$	horizontal shift 4 units	$y = \frac{1}{5}(x - 4)$

12)
$i(x) = \frac{1}{2}(x - 4) + 3$
See #10 for description of why it works this way.

15)
$i(x) = 1(2)^{x-14}$

17)
$T(x) = 24 + 3(x - 1)$
$K(x) = 1(2)^{x-1}$

18)
Which function is linear or exponential? What is the initial value? In what game does the player start his season? If it's a linear function, what's the slope? If it's an exponential function, what is the multiplier?

20)

type	arithmetic sequence	geometric sequence
example scenario	Steph scores 3 points in his first game, and his points increase by 7 each game after that.	James scores 6 points in his first game, and his points double each game after that.
table for this example	<table><tr><th>x</th><th>y</th></tr><tr><td>1</td><td>3</td></tr><tr><td>2</td><td>7</td></tr></table>	<table><tr><th>x</th><th>y</th></tr><tr><td>1</td><td>6</td></tr><tr><td>2</td><td>12</td></tr></table>
function for this example	*If he had scored 3 points in game 0:* $S(x) = mx + b$ $S(x) = 7x + 3$ *Now horizontally shift that 1 unit, so that he scores 6 points in game 1, not game 0:* $S(x) = 7(x - 1) + 3$	*If he had scored 6 points in game 0:* $J(x) = a(b)^x$ $J(x) = 6(2)^x$ *Now horizontally shift that 1 unit, so that he scores 6 points in game 1, not game 0:* $J(x) = 6(2)^{x-1}$
sequence notation	$a_n = a_1 + (n - 1)d$	$a_n = a_1(r)^{n-1}$
meaning of variables in sequence notation	Notice that the **formula** for a_n is the same as for $S(x)$ above, just slightly re-arranged with d in place of m and a_1 in place of b. a_n is read "a sub n" and represents the nth **term** (or output) in the sequence. a_1 is "a sub 1," the first term in the sequence. d is the **common difference** (or slope). It's the amount that is added to a term to get the next one. n is the **term number** (or input). In our example it's the game number, starting with game 1.	Notice how this is the same as $J(x)$ above, with a_1 in place of a and r in place of b. a_n is the nth term in the sequence. $a_1$is the first term in the sequence. r is the **common ratio** (or multiplier). It's the amount a term is multiplied by to get the next one. n is the term number.
example in sequence notation	$a_n = 3 + 7(n - 1)$	$a_n = 6(2)^{n-1}$

22)

description	type	formula	9th term
Tim scores 5 points in his first game and double each game thereafter.	geometric	$a_n = 5(2)^{n-1}$	$a_9 = 80$
Kevin scores 27 points in his first game and his score increases by 2 each game thereafter.	arithmetic	$a_n = 27 + (n-1)(2)$	$a_9 = 43$
A vending machine sells 4 candy bars the first day it's installed, and its sales increase by 5 candy bars each day thereafter.	arith		
A coffee shop sells 5 coffees on the first day its open, and its sales triple each day thereafter.	geo	$a_n = 5(3)^{n-1}$	Use calculator.
Serena makes $7000 in sales of her biography the first day its on sale. Her sales increase 30% each day after that.	geo		Use calculator.
A gamer scores 1200 points the first day she plays a game. Her score improves 300 points every day thereafter.	arith		
An arithmetic sequence has an initial value of $a_1 = 14$ with a common difference $d = 3$.	arith	$a_n = a_1 + (n-1)(3)$	
A geometric sequence has an initial value of $a_1 = 15$ and common ratio of $r = 1.6$.			Use calculator.
$\{2,5,8,11,\ldots\}$	arith	$a_n = 2 + (n-1)(3)$	
$\{2,6,18,54,162,\ldots\}$	geo		Use calculator.

23)

Finding the common difference.	Finding the initial value.	Writing the formula for the sequence.
2	$a_n = a_1 + (n-1)d$ $a_9 = a_1 + (9-1)2$ $35 = a_1 + (8)2$	$a_n = 19 + (n-1)2$

24)

Finding the common difference.	Finding the initial value.	Writing the formula for the sequence.
$m = \frac{29500 - 25000}{7-5}$ $m = 2250$ $d = 2250$	$a_n = a_1 + (n-1)d$ $a_5 = a_1 + (5-1)2250$ $25000 = a_1 + (4)(2250)$ $a_1 = 11000$	

25)
$r = 1.1$
$a_1 \approx 157.72$

26)
$d = 4$
$a_1 = 26$

27)
$r = 1.5$
$a_1 = 18$

28)
$d = -8$
$a_1 = 105$

29)
$r = 1.2$
$a_1 \approx 7.23$

35)
$S_{20} = 440$

36)
$S_{20} = 670$
37)
$S_{20} = 1048575$

38)
$S_{15} = 570$

39)
$S_{18} \approx 1308.3$

40)
$S_{24} = 2100$

41)
$a_{15} \approx 864.98$

42)
$S_{11} = 2049$

43)
$a_{20} \approx 5064$

www.ingramcontent.com/pod-product-compliance
Lightning Source LLC
Chambersburg PA
CBHW080756180526
45168CB00006B/2227

* 9 7 8 1 5 1 6 9 8 1 0 6 9 *